COMPUTATIONAL METHODS FOR TURBULENT, TRANSONIC, AND VISCOUS FLOWS

Edited by
J. A. Essers
von Karman Institute for Fluid Dynamics

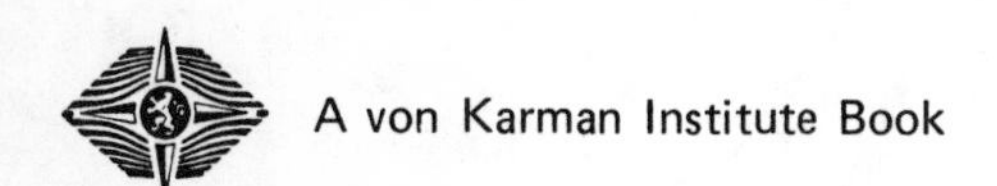

HEMISPHERE PUBLISHING CORPORATION
Washington New York London

DISTRIBUTION OUTSIDE NORTH AMERICA
SPRINGER-VERLAG
Berlin Heidelberg New York Tokyo

COMPUTATIONAL METHODS FOR TURBULENT, TRANSONIC, AND VISCOUS FLOWS

Copyright © 1983 by Hemisphere Publishing Corporation. All rights reserved.
Printed in the United States of America. Except as permitted under the United States Copyright Act of 1976, no part of this publication may be reproduced or distributed in any form or by any means, or stored in a data base or retrieval system, without the prior written permission of the publisher.

1 2 3 4 5 6 7 8 9 0 B C B C 8 9 8 7 6 5 4 3

Library of Congress Cataloging in Publication Data
Main entry under title:

Computational methods for turbulent, transonic, and viscous flows.

"A von Karman Institute Book."
Bibliography: p.
Includes index.
1. Fluid dynamics—Addresses, essays, lectures.
2. Numerical analysis—Addresses, essays, lectures.
I. Essers, Jean-André, date.
QA911.C624 1983 532'.051'015194 83-187
ISBN 0-89116-273-9 Hemisphere Publishing Corporation

DISTRIBUTION OUTSIDE NORTH AMERICA:
ISBN 3-540-12549-3 Springer-Verlag Berlin

Contents

Contributors

TIMOTHY J. BAKER
Aircraft Research Association Ltd.
Bedford, England

R. THOMAS DAVIS
Department of Aerospace Engineering and Applied Mechanics
University of Cincinnati
Cincinnati, Ohio 45221 USA

ERIK DICK
Department of Machinery
State University at Gent
Gent, Belgium

J. H. FERZIGER
Thermosciences Division
Department of Mechanical Engineering
Stanford University
Stanford, California 94305 USA

WOLFGANG HACKBUSCH
Institut für Informatik und Praktische Mathematik
Christian-Albrechts-Universität
Olshausenstrasse 40-60
D-2300 Kiel 1,
Federal Republic of Germany

R. I. ISSA
Department of Mineral Resources Engineering
Imperial College of Science and Technology
London, England

Preface

This book comprises the notes for a course on computational fluid dynamics held March 30–April 3, 1981, at the von Karman Institute. Lectures were presented by specialists from the United States and Europe active in this field. The following topics are treated in detail:

- body-fitted grid-generation techniques,
- recent developments in fast multi-grid solvers for the solution of linear and nonlinear elliptic equations,
- advances in higher-level simulations of turbulent flows,
- numerical methods for the calculation of incompressible viscous flows solving the Navier–Stokes equations,
- iterative techniques for the solution of the full potential equations at transonic regime,
- development of new relaxation methods for Euler equations and their application to steady transonic flows.

We gratefully acknowledge the support of the Advisory Group for Aerospace Research and Development (AGARD) of NATO.

Jean-André Essers

Numerical Methods for Coordinate Generation Based on a Mapping Technique

R. THOMAS DAVIS

ABSTRACT

A coordinate generation method for use in computational fluid dynamics problems is developed which is much simpler, more accurate, and more flexible than currently existing methods. This approach is based on numerical integration of Schwarz-Christoffel transformations for general curved surfaces. It is shown to be second-order accurate in mesh size with extensions to higher-order accuracy levels identified. In addition, this method directly provides the two dimensional incompressible potential flow solution for flow past complex body shapes including flows with free streamlines. Example symmetric cases are given for flow past a hexagon, a six point cross, a NACA airfoil, and free streamline flow past a circular cylinder. Finally, the method is extended to more general cases and, in particular, to flows with circulation and channel flows.

1. INTRODUCTION

In order to numerically solve viscous flow problems at high Reynolds number efficiently and accurately, it is necessary that one find a suitable coordinate system in which to do the computations. The problem of generating a proper coordinate system is difficult since the coordinate system should be determined such that it minimizes errors in the viscous flow solver and at the same time does not reintroduce the errors through the coordinate generator. Ultimately, the coordinate generator should be interactive with the viscous flow solver in order to properly handle the viscous flow problem under consideration with a minimum number of mesh points.

The difficulty of determining a good coordinate system mainly arises due to the complicated structure of high Reynolds number separated flows and the severe scaling laws which arise at and downstream of separation. Stewartson [1] and others have determined the asymptotic structure near separation for several problems and it is a challenging numerical problem to develop coordinate generators which honor the scalings indicated by the asymptotic theory. It is therefore important that a coordinate generation technique be flexible enough to align coordinate lines in arbitrary directions so that the scale laws can be accommodated in the most convenient manner by stretching normal to the proper coordinate lines.

Many authors have been developing coordinate generators, most of which solve elliptic boundary value problems in the manner originated by Thompson and coworkers, i.e. see Thompson, Thames and Masten [2,3,4] for example.

These methods are based on the numerical solution of partial differential equations and therefore have some difficulty with outer boundary conditions, determination of proper forcing functions for the right hand sides of the differential equations, and numerical inaccuracy at corners and other discontinuities.

Some of the most flexible and accurate methods of potential flow solution are based on the numerical solution of the integral equations resulting from source, vortex and doublet distributions on the body surface, see for example Hess and Smith [5] and Bristow [6]. These methods are attractive since they involve only the solution of boundary integrals requiring no outer boundary conditions, and allow for the removal of singularities resulting from corners or lower order discontinuities. If the solution in the interior of the flow field is desired, it can be obtained by integration. In a sense, mapping with Schwarz-Christoffel transformations can be viewed as a coordinate generation method which is done in the same spirit, i.e. it is a boundary integral method therefore requiring no outer boundary conditions, and singularities can easily be removed. After this initial mapping of the boundary has been performed, coordinates interior to the flow field can be determined by integration in arbitrary directions and therefore virtually any coordinate system desired can be accurately determined. It is the purpose of this paper to show how this can be done.

The Schwarz-Christoffel transformation for polygons is well known and can be found in any standard text on complex variables, see Carrier, Krook, and Pearson [7] for example. The extension to curved surfaces is less well known and is absent from most texts. A very extensive treatment of the extension to curved surfaces is contained in Woods [8]. We will develop this extension in Section 1 in a perhaps less complicated manner than that contained in Woods.

In this paper we will develop a numerical coordinate generator and potential flow solver based on the Schwarz-Christoffel transformation for curved surfaces. Several authors have dealt with numerical integration of the Schwarz-Christoffel transformation for polygons, i.e. the case where the sides are made up of straight lines. Trefethen [9] has presented a method for handling this case along with a review of other work in the area. The most closely related work to the present is that of Anderson [10] who has used straight elements to do curved channel mappings. Almost no work has been done on numerical integration of the transformation for curved surfaces. In fact, the only work in this area to the author's knowledge is that contained in Woods [8].

We will first discuss the Schwarz-Christoffel transformation and its generalization. Next we will show how polygons can be numerically mapped accurately by removing singularities which occur at corners. We will then develop a numerical method for handling curved elements and present some airfoil and cylinder example solutions. The case of flows involving free streamlines will be addressed next and we will show how the singularity at separation can be removed and how a special downstream asymptotic element can be developed. Finally, we will explore the extension to more complicated situations, including flows with circulation and channel mapping.

All of the mappings done in this study for external flows will first map the body to a stagnation plane. This is perhaps the most natural plane for doing viscous flow computations since the inviscid protion of the flow takes on a simple form which produces high accuracy in numerical computations. In addition, the boundary-layer equations reduce to self similar form in this plane when similarity conditions are satisfied. Finally, there is a relationship between this plane and an optimal coordinate plane, see Davis [11]. Other mapping planes will be explored as a two step mapping where a new t plane is related to the ζ plane.

2. THE SCHWARZ-CHRISTOFFEL THEOREM AND ITS GENERALIZATION TO CURVED SURFACES

The Schwarz-Christoffel Theorem is normally written in a form which is applicable only to polygons or surfaces which consist of straight line segments, see Carrier, Krook, and Pearson [7], p. 136, for example. However, the theorem can be extended to general surfaces which may be curved or made up of curved portions. One is referred to Woods [8], Chapter 5, for the generalizations of the theorem. Since the generalizations we wish to use here are easy to obtain, the following section is devoted to their development.

The most familiar form of the Schwarz-Christoffel Theorem is the one which is applicable to polygons and is given by

$$\frac{dz}{d\zeta} = f(\zeta) = M \prod_{i=1}^{n} (\zeta - a_i)^{\alpha_i/\pi} \tag{2.1}$$

where the geometry is as shown in Figure 1. Integration of the above equation yields

$$z = M \int_{o}^{\zeta} \prod_{i=1}^{n} (\zeta - a_i)^{\alpha_i/\pi} \, d\zeta + N \tag{2.2}$$

which maps the upper half of the ζ plane into the interior of the polygon shown in the z plane. The quantity n relates to the number of corners, M and N are complex constants to be determined, and the a_i's are real constants also to be determined. Normally the α_i angles are taken to be positive (in radians) for a counterclockwise rotation when the polygon is circled in a counterclockwise sense. However, here we choose to change the signs normally used and consider the α_i's to the positive for a clockwise rotation. Hence, all of the α_i angles shown in Figure 1 would be negative for the present sign convention.

Equation (2.1) can obviously be rewritten as follows for reasons which will be apparent later

$$\frac{dz}{d\zeta} = M \exp\left[\ln \prod_{i=1}^{n} (\zeta - a_i)^{\alpha_i/\pi}\right] = M \exp\left[\frac{1}{\pi} \sum_{i=1}^{n} \alpha_i \ln(\zeta - a_i)\right] . \tag{2.3}$$

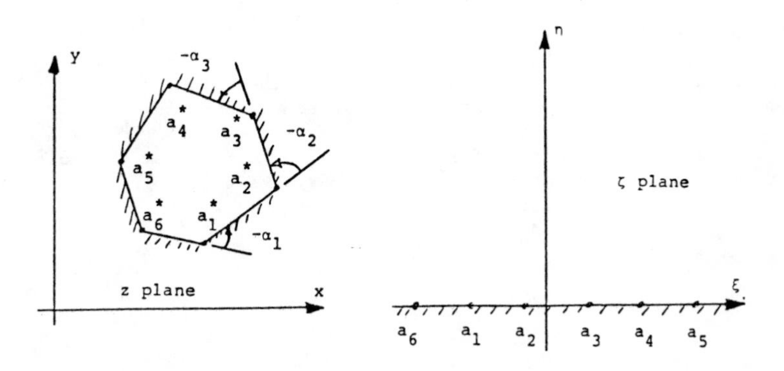

Figure 1. Polygon Mapping

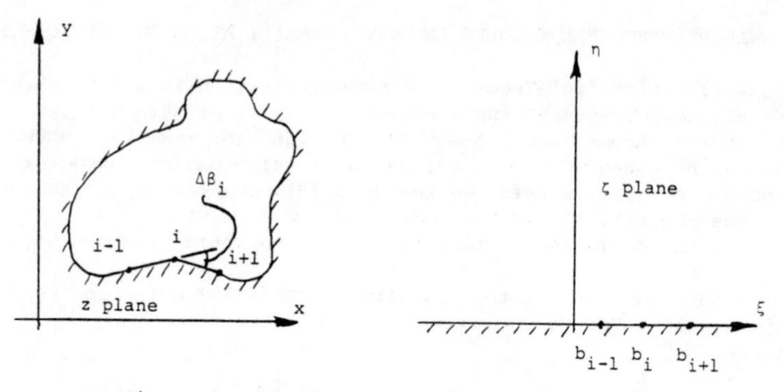

Figure 2. Mapping of Continuous Curves.

Now consider a continuous curve as shown in Figure 2 rather than a polygon. Visualize the curve as being made up of a large number of straight line segments with the turning angles α_i replaced by $\Delta\beta_i$ and the locations a_i replaced by b_i. In the limit as $n \to \infty$ the elements shrink to zero and the summation in Equation (2.3) is replaced with an integral. Therefore, for a continuous curve the Schwarz-Christoffel theorem becomes

$$\frac{dz}{d\zeta} = M \exp \left[\frac{1}{\pi} \int \ell n(\zeta - b)\, d\beta\right] \quad . \tag{2.4}$$

The above Equation (2.4) is general and includes (2.1) if one realizes that when a corner is encountered in the figure, say at an i location, β becomes a step function and the portion of the integral at the corner becomes

$$\int_{\beta_i^-}^{\beta_i^+} \ell n(\zeta - b)\, d\beta = \alpha_i\, \ell n(\zeta - a_i) \tag{2.5}$$

where α_i is the step in β at the location $b = a_i$, i.e. $\alpha_i = \beta_i^+ - \beta_i^-$. Therefore, if Equation (2.4) is properly applied, one can consider curves of a very general nature.

Equation (2.4) is a nonlinear differential-integral equation. The difficulty in solving it is that for a prescribed curve in the z plane, b is not known and must be determined as a part of the solution. One of the main purposes of this paper is to show how this can be done numerically.

The form (2.4) is convenient for external flows but is not the most convenient form to use for internal channel flows. We will delay the development of the most appropriate form for channel flows until later in Section 7.

3. COORDINATE GENERATION AND POTENTIAL FLOW SOLUTION FOR POLYGONS

The expression (2.1) can be used directly for external flow past a polygon. We will always map to a stagnation plane since this is perhaps the most natural one for doing viscous flow computations as we discussed in the Introduction. Here we are assuming that in addition to finding the potential flow solution, we wish to use the mapping as a coordinate generator. For the present, we consider only symmetric flows. However, the method is extended to unsymmetric flows including those with circulation in Section 6.

For purposes of discussion, consider the mapping and potential flow solution for flow past the hexagon shown in Figure 3. Equation (2.1) becomes for this case, taking symmetry into account (i.e., $a_3 = -a_5$, $a_2 = -a_6$ and $a_1 = -a_7$), the following:

$$\frac{dz}{d\zeta} = M(\zeta+1)^{-1/3} (\zeta+a_6)^{1/3} (\zeta+a_5)^{1/3} (\zeta)^{1/3} (\zeta-a_5)^{1/3} (\zeta-a_6)^{1/3} (\zeta-1)^{-1/3} \tag{3.1a}$$

or

$$\frac{dz}{d\zeta} = M(\zeta^2-1)^{-1/3} (\zeta^2-a_6^2)^{1/3} (\zeta^2-a_5^2)^{1/3} (\zeta)^{1/3} = f(\zeta) \tag{3.1b}$$

where a_7 has been taken to be 1 without loss of generality.

In some cases, expressions of the type (3.1) can be integrated exactly. However, in most cases this must be done numerically and therefore, since we wish to consider arbitrary geometries, we will seek a general numerical integration formula which will handle all cases. In addition, we must find in the example case the values of the constants a_5, a_6, a_7, M, and N where N is given in Equation (2.2). It can easily be shown that one of the constants can be chosen arbitrarily (which has already been done), say $a_7 = 1$, N is determined by the location of the origin (in this case N = 0), and the rest must be determined as part of the solution. In the general case, three constants can be chosen arbitrarily which is equivalent to what has been done here.

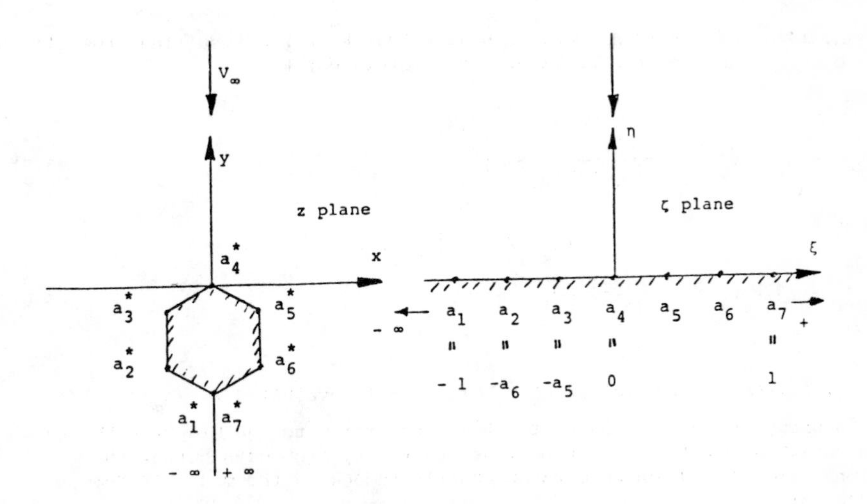

Figure 3. Hexagon Mapping.

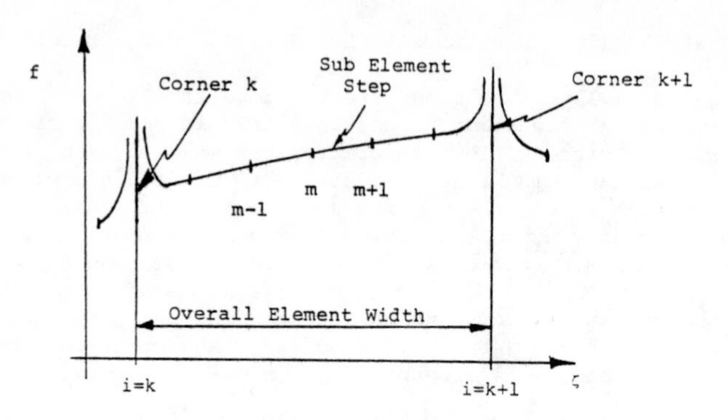

Figure 4. Composite Integration Scheme.

In attempting to numerically integrate expressions like (3.1), using say a trapezoid rule, large errors occur near corners due to the fact that $f(\zeta)$ is not analytic there. Therefore, we will develop a modified midpoint rule for complex integration which exactly integrates any nonanalytic term which occurs at a corner. We can see how to generate a general integration formula by examining two adjacent corners. In Equation (2.1) let us focus attention on any two corners say $i = k$ and $i = k+1$. At the same time let m be an index which is used for midpoint integration, see Figure 4. Let (2.1) be viewed as being made up of three factors on the right hand side, i.e.

$$\frac{dz}{d\zeta} = f_1(\zeta)\, f_2(\zeta)\, f_3(\zeta) \tag{3.2}$$

where

$$f_1(\zeta) = (\zeta - a_k)^{\alpha_k/\pi}, \tag{3.3}$$

$$f_2(\zeta) = (\zeta - a_{k+1})^{\alpha_{k+1}/\pi}, \tag{3.4}$$

and $f_3(\zeta)$ is a well behaved function near either k or k+1 resulting from all of the other corners in the problem. Therefore, near k

$$z_{m+1} - z_m = \bar{f}_2\, \bar{f}_3 \left. \frac{(\zeta - a_k)^{(\alpha_k/\pi)+1}}{(\alpha_k/\pi)+1} \right|_{\zeta_m}^{\zeta_{m+1}} \tag{3.5}$$

and near k+1,

$$z_{m+1} - z_m = \bar{f}_1\, \bar{f}_3 \left. \frac{(\zeta - a_{k+1})^{(\alpha_{k+1}/\pi)+1}}{(\alpha_{k+1}/\pi)+1} \right|_{\zeta_m}^{\zeta_{m+1}} \tag{3.6}$$

where $\bar{f}_1$, $\bar{f}_2$, and $\bar{f}_3$ are values of f_1, f_2, and f_3 evaluated at $(\zeta_m + \zeta_{m+1})/2$ which is proper for well behaved functions according to the midpoint rule. It can easily be shown (as must be true) that away from each corner the functions which have been integrated exactly reduce to the midpoint rule, i.e. near k+1

$$\left.\frac{(\zeta-a_k)^{(\alpha_k/\pi)+1}}{(\alpha_k/\pi)+1}\right|_{\zeta_m}^{\zeta_{m+1}} = \bar{f}_1\,(\zeta_{m+1} - \zeta_m) + \ldots \tag{3.7}$$

and near k

$$\left.\frac{(\zeta-a_{k+1})^{(\alpha_{k+1}/\pi)+1}}{(\alpha_{k+1}/\pi)+1}\right|_{\zeta_m}^{\zeta_{m+1}} = \bar{f}_2\,(\zeta_{m+1} - \zeta_m) + \ldots \tag{3.8}$$

Denoting

$$\Delta\zeta_m = \zeta_{m+1} - \zeta_m \tag{3.9}$$

we see that when we are away from the corners k and k+1 both integral expressions become

$$z_{m+1} - z_m = \bar{f}_1\,\bar{f}_2\,\bar{f}_3\,\Delta\zeta_m \tag{3.10}$$

as they should.

We can now develop a composite integration formula which is valid everywhere and contains the proper expressions to account for singularities at each corner. The most convenient expression to use for our case is a multiplicative composite of the type given by Van Dyke [12], p. 96. The composite is equivalent to multiplying the right hand side of Equation (3.5) by (3.6) and dividing by (3.10). This results in

$$z_{m+1} - z_m = \frac{\bar{f}_3}{\Delta\zeta_m}\;\frac{(\zeta-a_k)^{(\alpha_k/\pi)+1}}{(\alpha_k/\pi)+1}\;\left.\frac{(\zeta-a_{k+1})^{(\alpha_{k+1}/\pi)+1}}{(\alpha_{k+1}/\pi)+1}\right|_{\zeta_m}^{\zeta_{m+1}} . \tag{3.11}$$

We can now see that since $\bar{f}_3$ may include nonanalytic terms at corners other than k and k+1 we must include exact integration formulas for these effects. From the above expression (3.11) it is now easy to see how this can be done and the result is [see (2.1)]

$$z_{m+1} - z_m = \frac{M}{(\Delta\zeta_m)^{n-1}}\;\prod_{i=1}^{n}\;\left.\frac{(\zeta-a_i)^{(\alpha_i/\pi)+1}}{(\alpha_i/\pi)+1}\right|_{\zeta_m}^{\zeta_{m+1}} . \tag{3.12}$$

The above is easy to program in complex arithmetic and represents a second order accurate integration formula which is an extension of the midpoint rule for integrating the Schwarz-Christoffel transformation (2.1). It is a very convenient expression since one does not find it necessary to pay any special attention to corners if it is used. The expression can be simplified if symmetry is present and taken into account and then one needs to only sum over half of the elements.

The above integration formula (3.12) is second-order accurate even in the case where corners are present which introduce singularities. It will also be used in the cases where the body surface is curved but has corners at some specific locations. This will be done in order to properly handle the corner points. The curved element expressions for continuous segments will be developed later. Expressions similar to (3.12) could be developed to higher-order accuracy. However, since most Navier-Stokes solvers which would be used with the coordinate generator are second-order accurate, there is no reason to extend the present straight element expressions and later curved element expressions to higher order at this time.

The remaining problem in integrating (2.1) using (3.12) is that the a_i locations in the transformed plane are unknown and must be determined as part of the solution. In addition, M and N in Equation (2.2) must be determined. We will always orient the body such that $z = 0$ when $\zeta = 0$ and, therefore, $N = 0$ in all cases we will consider. The remaining constants are determined iteratively as follows:

(a) Guess values for all of the unknown constants, i.e. the a_i's and M. M is generally a complex constant which should be purely imaginary for mappings of symmetric figures which are oriented as shown in Figure 3. However, the fact that M is purely imaginary does not need to be prescribed since it will come out of the solution procedure automatically. The a_i's are picked such that $a_{i-1} < a_i < a_{i+1}$ but it is not critical that they be close to their exact final values. For the case in Figure 3, a_7 would be chosen to be 1, for example, and held at that value in subsequent iterations.

(b) Integrate Equation (2.1) using complex arithmetic using the guessed values from (a) above, using Equation (3.12). The angular rotations at the corners will be found to be correct in the physical plane, but the element lengths will be in error, i.e. the corners will not be at their proper locations in the physical z plane. The next gueses on the a_i's and M are made as follows. We will use the example case shown in Figure 3 for explanation purposes, see Figure 5. From the first iteration z_3 will not be at the proper location, say at -i. A new value of M is therefore determined such that the incorrect figure determined from (a) above is

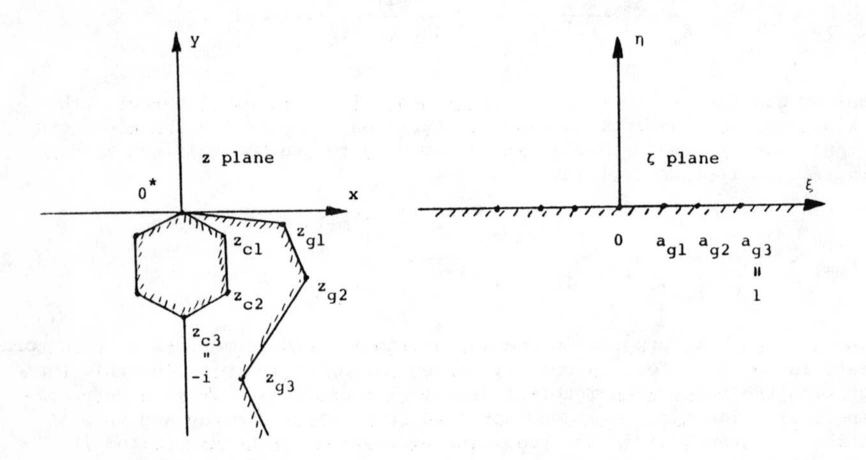

Figure 5. Location of the a_i's and Determination of M.

rotated and stretched such that z_3 is now at its proper location. The remaining points, say a_2, are now moved (because of the determination of a new M) to new locations which are still incorrect. Now it is assumed that the a_i's should be rescaled according to the scalings indicated by the errors in length of the elements in the z plane. If the subscripts c indicate correct values of g the old or guessed values, this is done as follows

$$\frac{a_{ci}-a_{ci-1}}{a_{gi}-a_{gi-1}} = k \frac{|z_{ci}-z_{ci-1}|}{|z_{gi}-z_{gi-1}|} \tag{3.13}$$

where k is determined as a scaling parameter such that a_{c3} is 1.

(c) A new iteration is performed by assuming the new values of M and a_i (step a). Step (b) is repeated and the whole process is iterated through (a) and (b) until convergence is achieved.

The above procedure converges quite rapidly and appears to do so for all cases. No under- or over-relaxation has been explored, but this might further improve the convergence rate. Typically, convergence is achieved in about ten iterations to five place accuracy in all computed results. The number of iterations has been found to be independent of the number of sides of the polygon, i.e. many sided polygon solutions converge in about ten iterations as well as polygons with few sides. Exactly the same procedure is used with curved elements and the solution converges equally as well for these cases as will be discussed later.

The main purpose of the present work is to develop a coordinate generator for arbitrary geometries. Therefore, for example, after the hexagon in Figure 3 has been mapped to the ξ axis in the ζ plane, coordinates can be generated by integrating the transformation equation in whatever direction one desires in the ζ plane. For example, Figure 6 shows the results for integrating along lines of constant ξ and η in the transformed plane for the

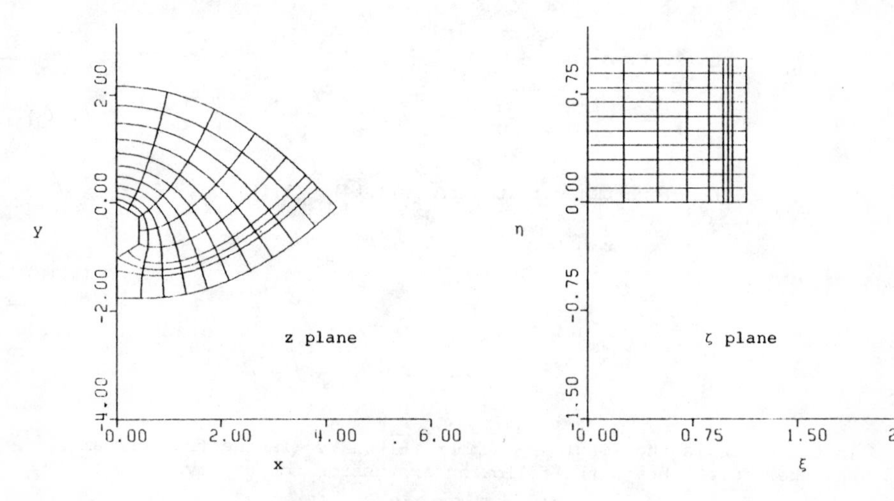

Figure 6. Hexagon.

hexagon problem. The hexagon geometry has first been mapped using the procedure discussed previously. The hexagon was subdivided into elements of equal arc length along the surface in the physical plane. In the case shown in Figure 6, each of the six sides of the hexagon was subdivided into ten elements. The coordinate curves shown for constant ξ are those which result from integrating outward from the end and midpoints of each hexagon side. The lines of constant η are those which result from a constant spacing of $\Delta\eta = 0.1$ in the transformed plane. This coordinate system is only cited as an example; one can generate virtually any one desired by prescribing different integration paths in the ζ plane. This point will be discussed later. The resulting generated coordinates can then be used with a Navier-Stokes solver, for example.

If one wishes the two dimensional incompressible potential flow solution, it can be obtained directly as follows. In the physical plane

$$u - iv = \frac{dw}{dz} = \frac{dw}{d\zeta}\frac{d\zeta}{dz} = \frac{dw/d\zeta}{dz/d\zeta} \quad . \tag{3.14}$$

In the transformed plane (see Figure 3 for example), the complex potential is the one for a stagnation point flow and is given by

$$w = \frac{a}{2}\zeta^2 \tag{3.15}$$

where a is a constant to be determined. Therefore,

$$\frac{dw}{d\zeta} = a\zeta \tag{3.16}$$

and $dz/d\zeta$ is given by Equation (2.1). For the bodies we will be considering the boundaries at $\pm\infty$ are parallel and therefore

$$\sum_{i=1}^{n} \frac{\alpha_i}{\pi} = 1 \tag{3.17}$$

Thus letting $\zeta \to \infty$ in (2.1) we obtain

$$\frac{dz}{d\zeta} \sim M\zeta \quad \text{as} \quad \zeta \to \infty \; . \tag{3.18}$$

The constant a can therefore be evaluated from (3.14), (3.16) and (3.18), (see Figure 3), as

$$i\,V_\infty = \frac{a\zeta}{M\zeta} = \frac{a}{M} \tag{3.19}$$

and therefore

$$a = i\,V_\infty M \tag{3.20}$$

where i in this case is the imaginary number $\sqrt{-1}$ and V_∞ is the free-stream speed. We, therefore, obtain the following expression for u - iv

$$u - iv = \frac{i V_\infty \zeta}{\prod_{i=1}^{n} (\zeta - a_i)^{\alpha_i/\pi}} \tag{3.21}$$

Therefore, after the transformation relationship has been integrated to give z as a function of ζ, u - iv can be obtained as a function of z from (3.21) and the pressure coefficient, etc., can be calculated in the physical plane. In the following sections, u - iv will be calculated in a similar manner when curved element expressions are used rather than the straight elements in the present case.

As a further example of the use of the method, Figure 7 shows the mapping of a six point cross. The first arm of the cross is located in the vertical direction in the physical plane between the points (0,0) and (0,-0.5). The other arms each of length 0.5, are separated by 60° angular increments in the physical plane (see Figure 7b).

This case was chosen to demonstrate the ability of the method to handle severe geometries, in this case, turning angles of 180° at the cross tips. Figure 7b shows the detail to which the mapping can be performed, even in the immediate neighborhood of the cross and in particular, its tips. Without an integration formula of the type (3.12) which handles the nonanalytic nature of the mapping exactly at corners, such a problem would be difficult to handle. It would be interesting to see how well this mapping problem could be handled with a finite difference solution to the partial differential equations.

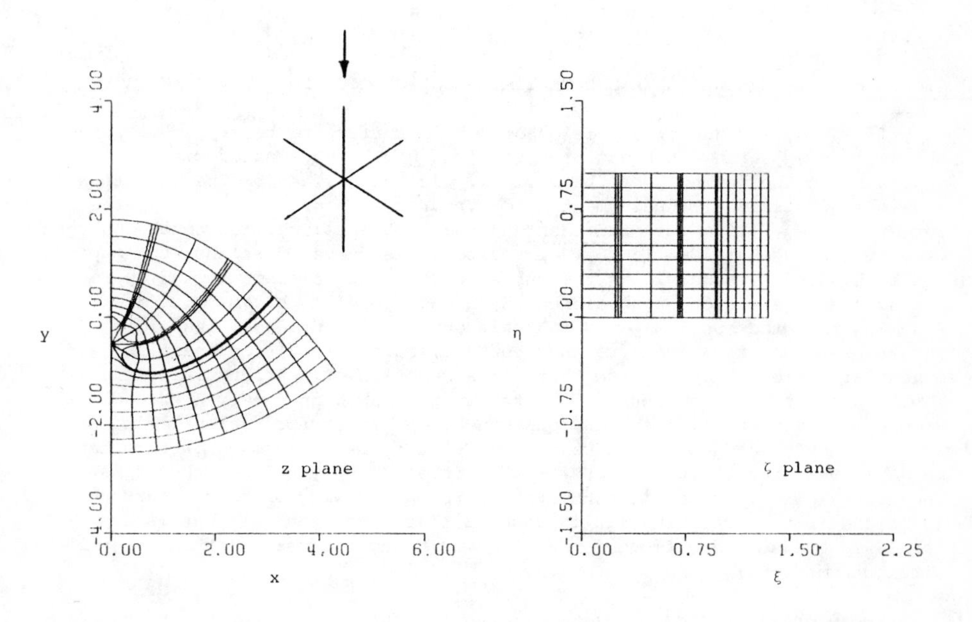

Figure 7a. Six Point Cross.

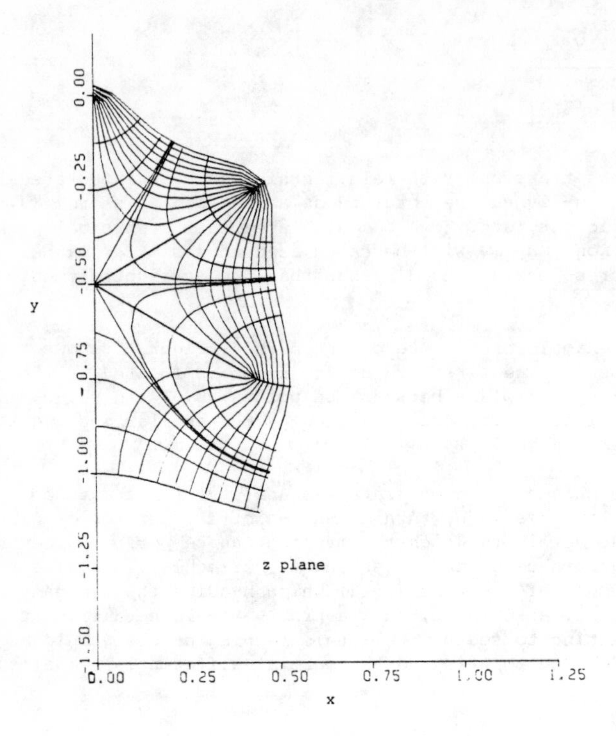

Figure 7b. Six Point Cross.

4. COORDINATE GENERATION AND POTENTIAL FLOW SOLUTION FOR CURVED SURFACES

If a body surface or a portion of a body surface to be mapped is curved, the angle β appearing in Equation (2.4) is a given function of x and y depending on the geometry of the surface. If the surface of the body totally consists of straight line segments and corners, then $d\beta = 0$ except at the corners where β is a step function and the problem simplifies to the polygon problem discussed in Section 3. If the body has curved portions, the mapping can be handled by subdividing the curved portions into elements and approximating the β variation on an element by a polynomial in b. Here we are assuming that the body shape is analytic on the elements and, therefore, care should be taken in making sure that corners, curvature discontinuities, etc. appear at element endpoints so that the assumption of a polynomial variation on the element will be accurate. Later we will show in the free streamline problem how some special expressions can be developed for regions or points where the body surface is not regular. We have already shown how this can be done for corners and the results of Section 3 apply to the present case whenever a corner occurs in connection with the mapping of curved surfaces. (Here and in the following, it is assumed that a straight element is contained in the curved element formulation as a special case with $d\beta = 0$ on a straight portion.)

Therefore, Equation (2.4) is written in general as

$$\frac{dz}{d\zeta} = M \exp\left[\frac{1}{\pi}\sum_{i=1}^{NC} \ln(\zeta-a_i)\alpha_i + \frac{1}{\pi}\sum_{m=1}^{NE}\int_{\beta_m}^{\beta_{m+1}} \ln(\zeta-b)\, d\beta\right] \tag{4.1}$$

where the NC corners have been extracted and NE represents the number of curved elements into which the surface has been subdivided.

Assuming that the body surface is analytic in the region between m and m+1, we can approximate β with a polynomial in b (see Figure 8) as

$$\beta = C_{1m} + C_{2m}b + C_{3m}b^2 \tag{4.2}$$

and therefore,

$$\frac{1}{\pi}\int_{\beta_m}^{\beta_{m+1}} \ln(\zeta-b)\, d\beta = \frac{C_{2m}}{\pi}\int_{b_m}^{b_{m+1}} \ln(\zeta-b)\, db + \frac{2C_{3m}}{\pi}\int_{b_m}^{b_{m+1}} \ln(\zeta-b) b\, db\ . \tag{4.3}$$

The integration in Equation (4.3) can be done exactly and expressed in the following form

$$\frac{1}{\pi}\int_{\beta_m}^{\beta_{m+1}} \ln(\zeta-b)\, d\beta = \ln\{[g_{2m}(b_m,b_{m+1},\zeta)]^{C_{2m}/\pi} \times [g_{3m}(b_m,b_{m+1},\zeta)]^{C_{3m}/\pi}\} \tag{4.4}$$

where

$$g_{2m} = \frac{(\zeta-b_m)^{(\zeta-b_m)}\, e^{b_m}}{(\zeta-b_{m+1})^{(\zeta-b_{m+1})}\, e^{b_{m+1}}} \tag{4.5}$$

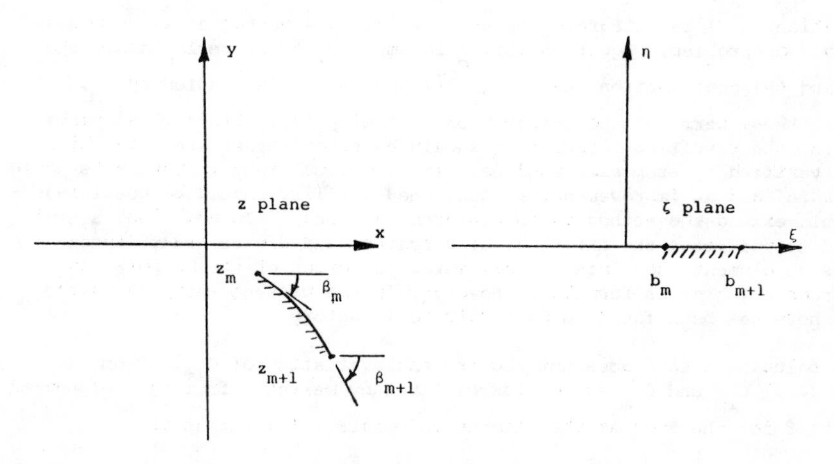

Figure 8. Curved Element Geometry.

and

$$g_{3m} = \frac{(\zeta-b_m)^{(\zeta^2-b_m^2)}\; e^{(\zeta+b_m)^2/2}}{(\zeta-b_{m+1})^{(\zeta^2-b_{m+1}^2)}\; e^{(\zeta+b_{m+1})^2/2}} \tag{4.6}$$

Higher order polynomial expressions for β will yield similar expressions for g_{4m}, g_{5m}, etc. and will appear in additional multiplying factors under the ℓn expression on the right hand side of Equation (4.4). Here we have chosen a quadradic for β since it allows us to match the polynomial with the correct body slopes at the m and m+1 locations. In addition, the numerical integration of $dz/d\zeta$ allows one to match with prescribed body points at the m and m+1 locations. Therefore, with the present formulation, one subdivides a body into a given number of elements and the element endpoints and slopes are matched with the actual body. It would, therefore, appear that if the method in its present form were integrated with the proper fourth-order accurate formulas for $dz/d\zeta$, the method would be totally fourth-order accurate. However, since the integration formula which we will use for $dz/d\zeta$ is only second-order accurate, the final results will only be second-order accurate.

From the general expression (4.1) along with the assumption of a β variation on an element of the form (4.2), Equation (4.1) can be rewritten as

$$\frac{dz}{d\zeta} = M \prod_{m=1}^{NE} (\zeta-b_m)^{\zeta_m/\pi}\; g_{2m}^{C_{2m}/\pi}\; g_{3m}^{C_{3m}/\pi} \tag{4.7}$$

where g_{2m} and g_{3m} are given by (4.5) and (4.6). For programming purposes, we have allowed for a possible corner at the end of each element (i.e. NC = NE). In an actual problem, most element endpoints will not be corners, and therefore, most of the values of the α_m's will be zero.

Equation (4.7) is integrated in exactly the same manner as was discussed in the polygon problem. Equation (3.12) is employed to properly handle the corners and the contribution due to $g_{2m}^{C_{2m}/\pi}\; g_{3m}^{C_{3m}/\pi}$ is handled by evaluating these terms at the integration step midpoint. Since these terms are smooth, the resulting integration should be second-order accurate which has been verified by step size studies. It is obvious that improvements could be made here, and an improvement, as mentioned earlier, should be possible which would extend the method to fourth-order accuracy. In addition, a study is needed to determine the influence of curvature and higher order discontinuities at element endpoints on the overall accuracy of the midpoint and higher order integration formulas. However, in the present study the method employed here has been found to be highly satisfactory.

The solution method does not require the evaluation of C_{1m}. From Equation (4.2) C_{2m} and C_{3m} are evaluated first by making β from the polynomial match with β for the body at the element endpoints. This leads to

$$\beta_{m+1} = C_{1m} + C_{2m}\, b_{m+1} + C_{3m}\, b_{m+1}^2 \tag{4.8a}$$

and

$$\beta_m = C_{1m} + C_{2m} b_m + C_{3m} b_m^2 \tag{4.8b}$$

where β_m and β_{m+1} are known quantities. Solving for C_{2m} we obtain

$$C_{2m} = \frac{\beta_{m+1} - \beta_m}{b_{m+1} - b_m} - (b_{m+1} + b_m)\, C_{3m} \tag{4.9}$$

If a midpoint rule is used in the integration, such that the g_{2m} and g_{3m} are evaluated at the element midpoints, the average angle of an element connecting the m and m+1 points in the physical plane will be denoted by θ_m and is given by

$$\theta_m = C_{1m} + C_{2m} \frac{b_m + b_{m+1}}{2} + C_{3m} \left(\frac{b_m + b_{m+1}}{2}\right)^2 . \tag{4.10}$$

This equation comes from Equation (4.2) and it can be verified that it is correct by evaluating the argument on an element in Equation (4.7). Eliminating C_{1m} and C_{2m} with the use of Equations (4.8a,b) results in

$$\theta_m = \frac{\beta_m + \beta_{m+1}}{2} - \frac{C_{3m}}{4} (b_{m+1} - b_m)^2 \tag{4.11}$$

or

$$C_{3m} = \frac{4 \left(\frac{\beta_m + \beta_{m+1}}{2} - \theta_m\right)}{(b_{m+1} - b_m)^2} \tag{4.12}$$

The quantity C_{3m} can, therefore, be viewed as an angle correction for the midpoint rule in this case. The evaluation above insures that after the mapping has been completed, the lengths of the elements in the physical plane will be correct to second-order accuracy, but the angular rotations will be exactly correct. Therefore, θ_m is an input quantity which is the angle of a straight line connecting the m and m+1 points on the body surface. Once C_{3m} is determined from (4.12), C_{2m} is determined from (4.9).

The only remaining problem is the determination of the unknown b_m points in the transformed plane corresponding to the known z_m points in the physical plane. This is done in exactly the same manner as was used in the polygon case, i.e. values of the b_m's are guessed initially and iteratively corrected by Equation (3.13) until convergence is achieved. Here a_{c_i} and a_{g_i} should be replaced with b_{cm} and b_{gm} in Equation (3.13). The convergence is as rapid as in the polygon case, typically requiring ten to fifteen iterations to achieve five place accuracy of all quantities. The rate of convergence has been found to be independent of the number of elements.

If the potential solution velocity components and pressure coefficient are desired, they are computed in exactly the same manner as in the polygon case. If the boundaries at $\pm\infty$ are parallel, rather than Equation (3.17) we obtain from (4.7)

$$\frac{1}{\pi}\sum_{m=1}^{NE}[\alpha_m + (b_{m+1} - b_m)\, C_{2m} + (b^2_{m+1} - b^2_m)\, C_{3m}] = 1 \tag{4.13}$$

and it follows from letting $\zeta \to \infty$ that

$$\frac{dz}{d\zeta} \sim M_\zeta \qquad \text{as} \quad \zeta \to \infty \tag{4.14}$$

which is the same relationship as Equation (3.18). Therefore, using Equation (4.7), we obtain

$$u - iv = \frac{i\, V_\infty \zeta}{\prod_{m=1}^{NE} (\zeta - b_m)^{\alpha_m/\pi}\, g_{2m}^{C_{2m}/\pi}\, g_{3m}^{C_{3m}/\pi}} \tag{4.15}$$

which reduces to (3.21) for polygons.

With the use of Equation (4.15) and the complex numerical integration of (4.7), we can calculate the velocity at points in the physical plane and the pressure coefficient on the body surface. Here the pressure coefficient is defined by

$$C_p = \frac{P - P_\infty}{\frac{1}{2}\rho V_\infty^2} = 1 - \frac{u^2 + v^2}{V_\infty^2} \qquad \text{at} \qquad \eta = 0 \tag{4.16}$$

As a first example case, we choose a NACA 0018 airfoil. The elements are formed by choosing locations on the airfoil surface for the endpoints of the elements. In order to provide proper resolution near the stagnation point, the element spacings were chosen such that they vary quadratically along the negative y axis. The β_m angles at the selected element endpoints are then calculated analytically with the use of the standard thickness formula for NACA four and five digit airfoils, see Abbott and von Doenhoff [13]. The results are presented in Figures 9, 10, and 11. Figure 9 shows the results for mapping to the ζ plane. Here $\Delta\eta = 0.1$ and the $\Delta\xi$'s are the ones which result from the chosen element lengths in the z plane. Figure 10 shows the pressure coefficients which result from Equations (4.15) and (4.16) for both the NACA 0018 and 0012 airfoils. These results are in agreement to the scale of the plot with those obtained by many other authors and show second-order accuracy from a step size study.

Figure 11 demonstrates the ability of the method to generate different types of coordinates. In this case, the mapping of the boundary from the z to the ζ plane is identical to Figure 9. However, in Figure 11, the coordinates are streamline and potential line coordinates. These are generated by integrating Equation (4.7) along lines for which the real or imaginary parts of w, the complex potential, are constant, i.e. since

$$w = \phi + i\psi = \frac{a}{2}\zeta^2 = \frac{a}{2}(\xi^2 - \eta^2) + ia\,\xi\eta \tag{4.17}$$

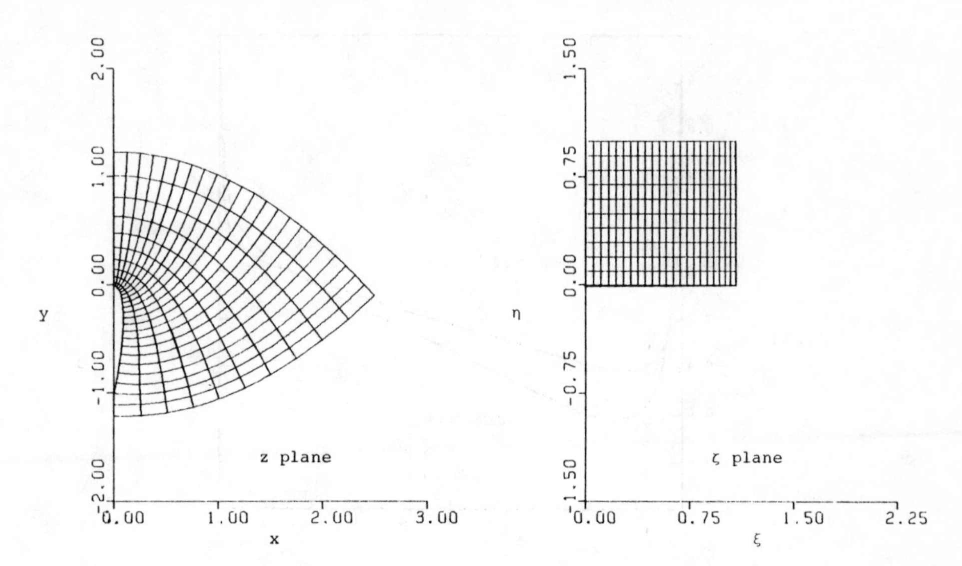

Figure 9a. NACA 0018 Airfoil.

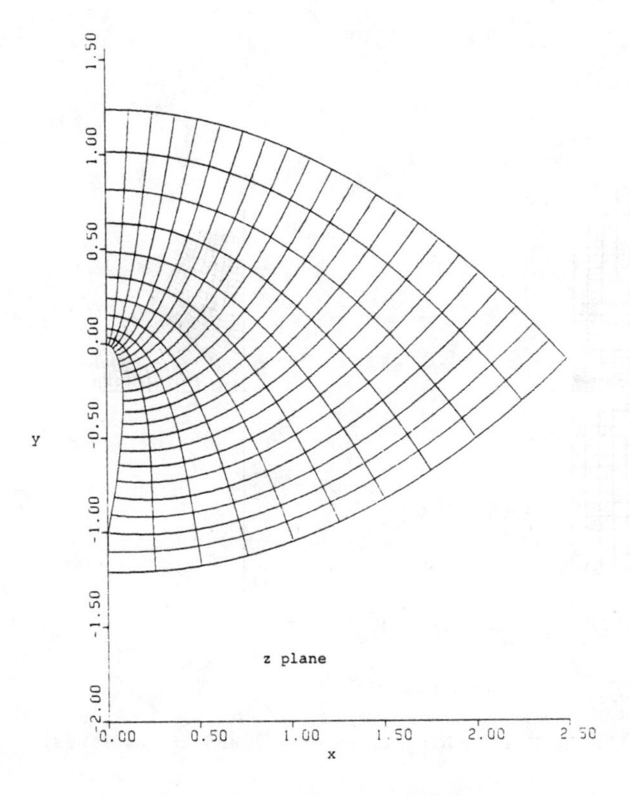

Figure 9b. NACA 0018 Airfoil.

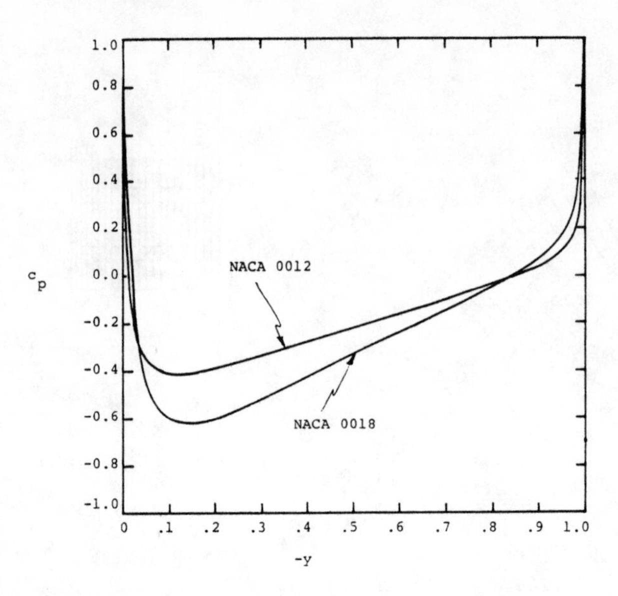

Figure 10. NACA 0012 and 0018 Airfoil Pressure Coefficients.

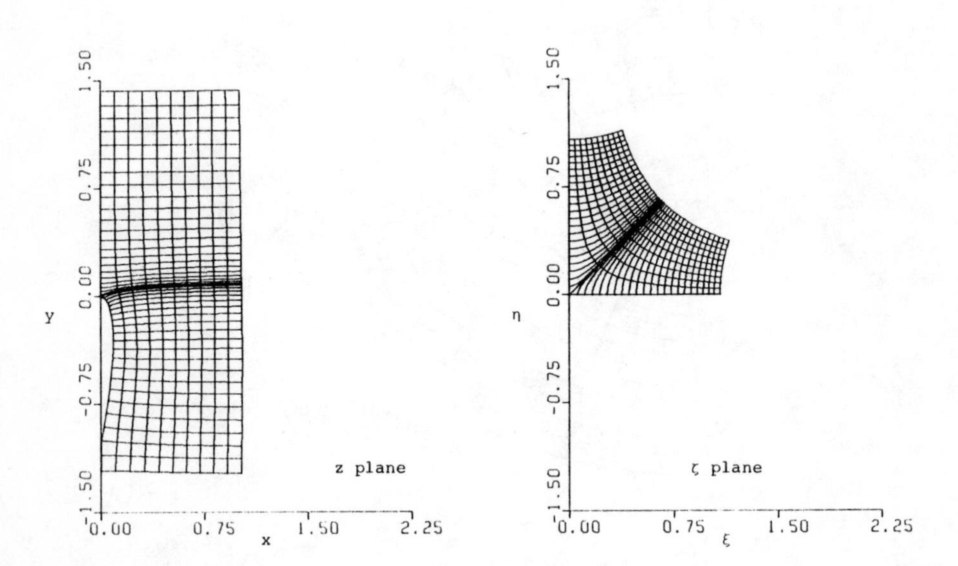

Figure 11. NACA 0018 Airfoil Potential and Stream Lines.

the streamlines are generated by integrating (4.7) along curves of constant $\xi\eta$ and the potential lines are generated by integrating (4.7) along curves of constant $\xi^2 - \eta^2$. The curves in the ζ plane are, therefore, the potential and streamlines for a stagnation point flow and the curves in the z plane are these curves mapped to the physical plane, which are the physical potential and streamlines.

Other types of mappings and coordinates are possible as well. For example, we may wish to map the original body to a cylinder and use polar coordinates in the transformed plane. This is done as a two step process by first mapping to the ζ plane as before and then relating the ζ plane to a t plane where the t plane is the plane of the circular cylinder. The mapping between the ζ and t planes is given by

$$\zeta^2 = \frac{1}{2}\left[(t + \frac{i}{2})i - \frac{i}{4}\,\frac{1}{t + \frac{i}{2}} + 1\right] \tag{4.18}$$

where the cylinder radius is 1/2 and the cylinder is located on the t axis between t = 0 and t = -i. The above expression (4.18) can be inverted to solve for t which gives for the region external to the cylinder

$$t = -i\zeta^2 + \zeta\sqrt{1 - \zeta^2} \tag{4.19}$$

Now, by introducing polar coordinates in the t plane with

$$t = -\frac{i}{2} + \rho e^{i\phi} \tag{4.20}$$

we can calculate the points on the cylinder surface (ϕ values) which correspond to the element endpoints in the z (physical plane) since the location of this position, the ζ plane, is already known. By holding these calculated values of ϕ constant and allowing ρ to vary, we can calculate from Equation (4.18) a $\Delta\zeta$ which corresponds to a given $\Delta\rho$. In addition, by holding ρ constant, we can calculate a $\Delta\zeta$ which corresponds to a given $\Delta\phi$. These values of $\Delta\zeta$ are used in Equation (4.7) to perform numerical integration.

Figure 12a shows the mapping of an example case to the ζ plane and the generation of the coordinate system which results from integration along lines of constant ξ and η. Figure 12b shows the same mapping, but this time coordinates are produced by integrating along lines of constant ρ and ϕ in the t plane. For reference purposes, Figure 12c shows the curves in the ζ plane which result from the integration path to produce the polar coordinate system.

Therefore, once the mapping to the ζ plane has been accomplished, we have at our disposal the possibility of many types of coordinate systems. We can stretch or shrink the mesh in particular regions by varying the integration step size in the transformed plane. We can even distort the mesh in the transformed plane and, as long as the integration paths in the transformed plane are orthogonal, due to the nature of conformal mappings, the coordinate system in the physical plane will also be orthogonal.

Blottner and Ellis [19] have developed a simple numerical technique for generating curves which are orthogonal to another given set of prescribed curves. This method has been used with success to produce the interesting, but poor, orthogonal coordinate system shown in Figure 12d. This example is given only to show how a distorted orthogonal coordinate system can be generated. No claim is made that this is a good coordinate system for doing numerical computations.

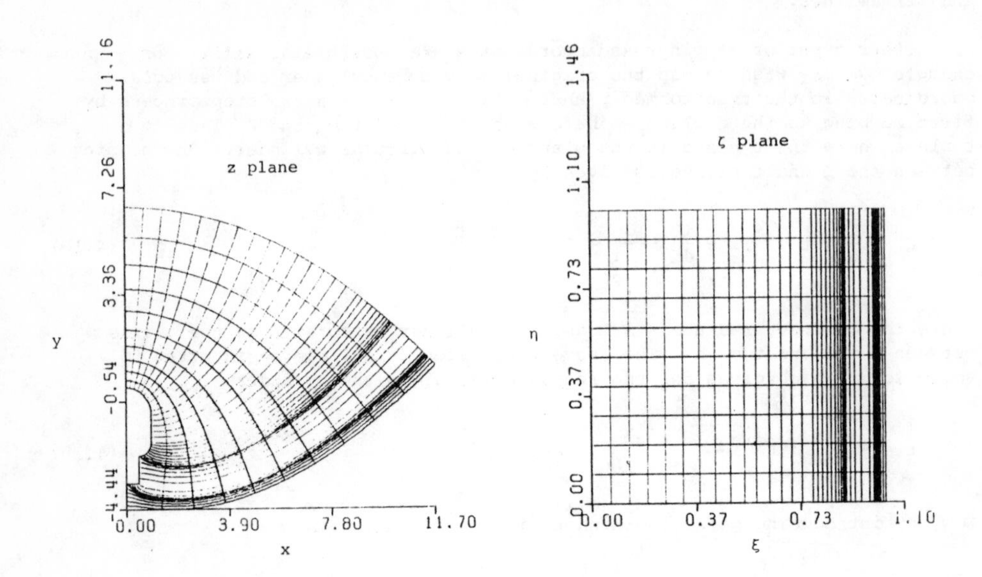

Figure 12a. Mapping to Parabolic Coordinates.

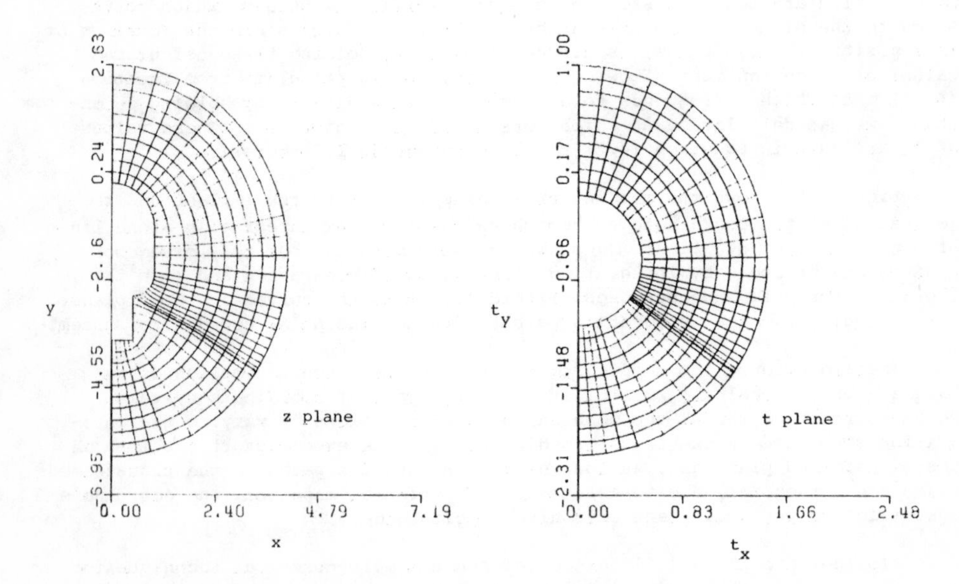

Figure 12b. Mapping to Polar Coordinates.

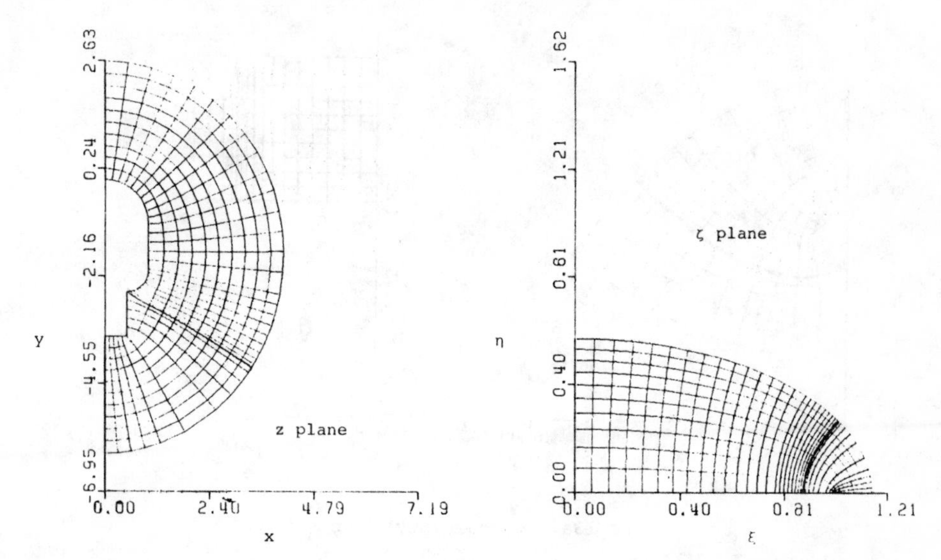

Figure 12c. Relationship Between z and ζ Planes for Figure 12b.

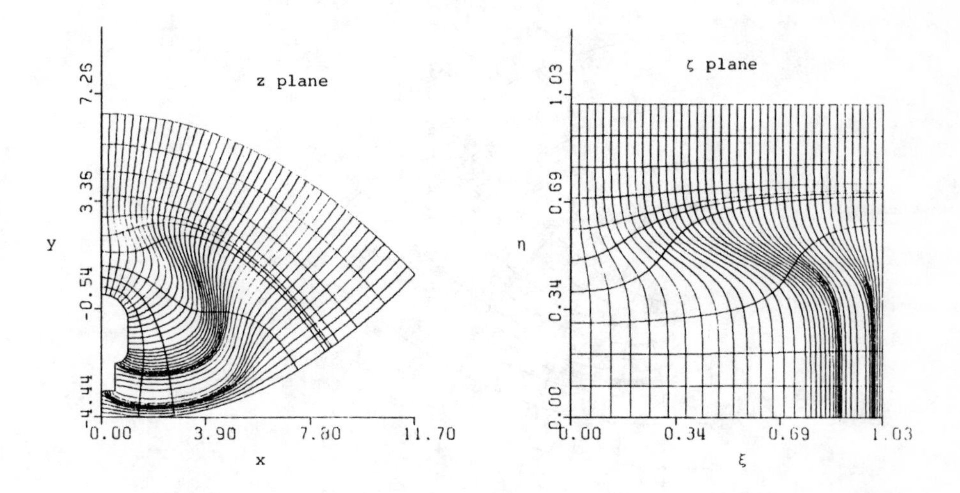

Figure 12d. A Simple Modification of the Coordinates.

Figures 13 through 17 are results for various cylinder problems. Figure 13 is a mapping of a complete circular cylinder, Figure 14 is the physical plane result for the mapping of a half cylinder, and finally, Figure 15 is the physical plane result for the mapping of the half cylinder with a constant thickness afterbody. These mappings are done to demonstrate that the method has no difficulty, with smooth surfaces, corners, or curvature discontonuities. The results in Figure 16 show the second-order accuracy of the method, i.e. the maximum velocity on the cylinder in Figure 13 goes to its exact value of 2 with an error of order $1/N^2$ where N is the total number of elements on the cylinder.

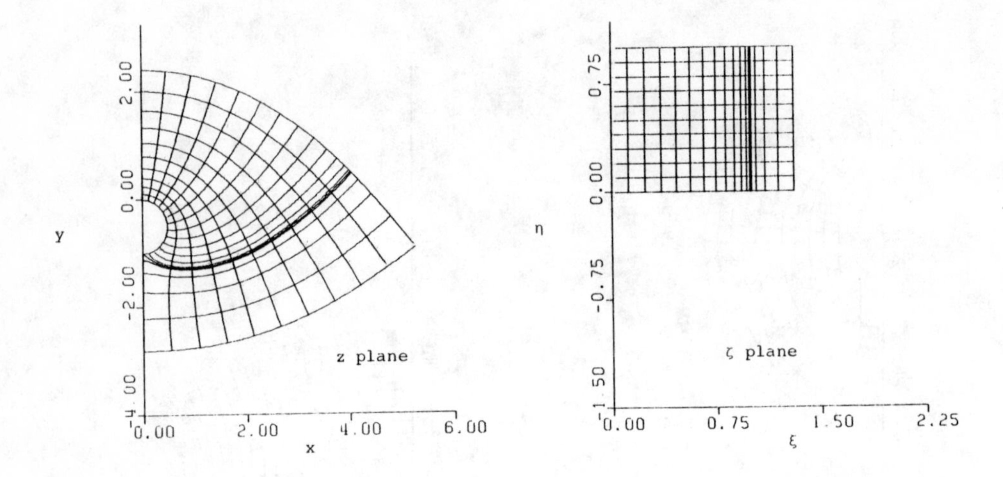

Figure 13a. Circular Cylinder.

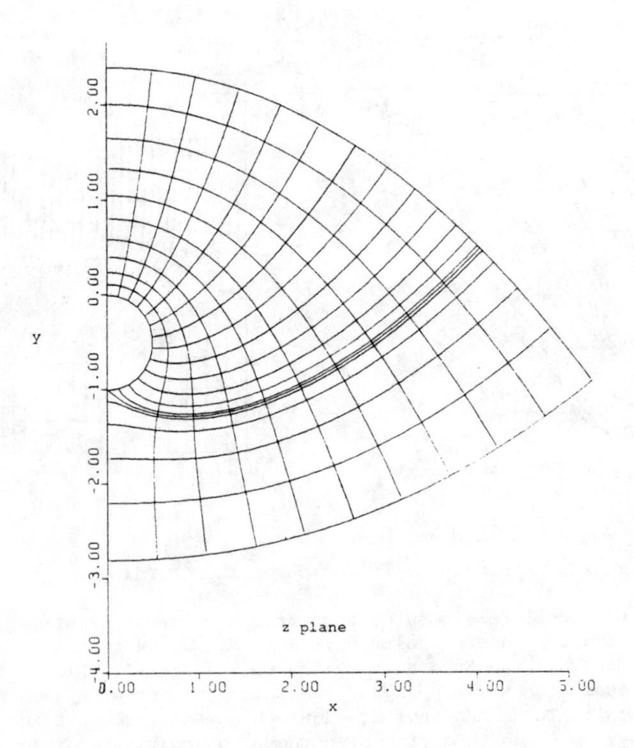

Figure 13b. Circular Cylinder.

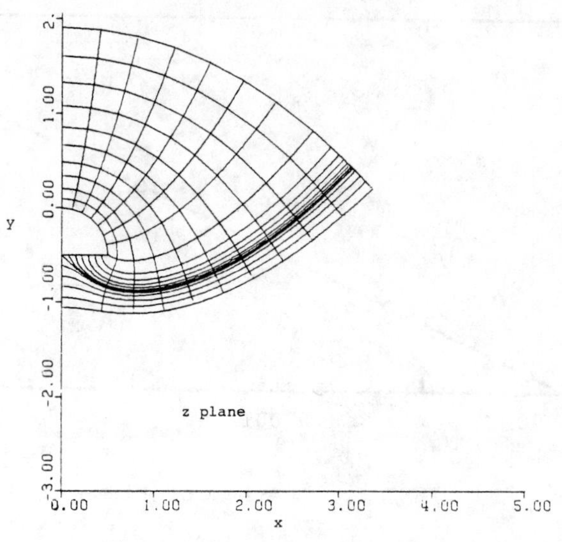

Figure 14. Truncated Cylinder.

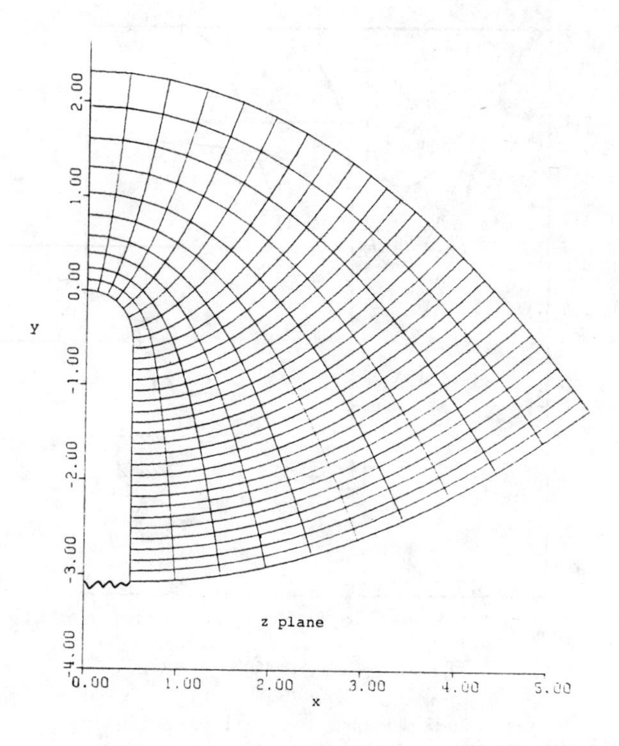

Figure 15. Cylindrically Blunted Plate.

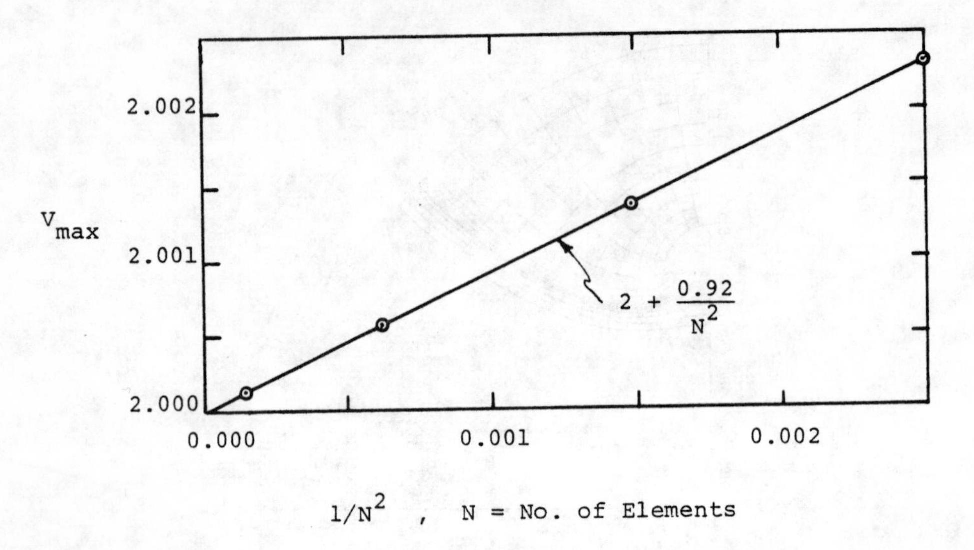

Figure 16. Cylinder Maximum Velocity Vs. Step Size Squared.

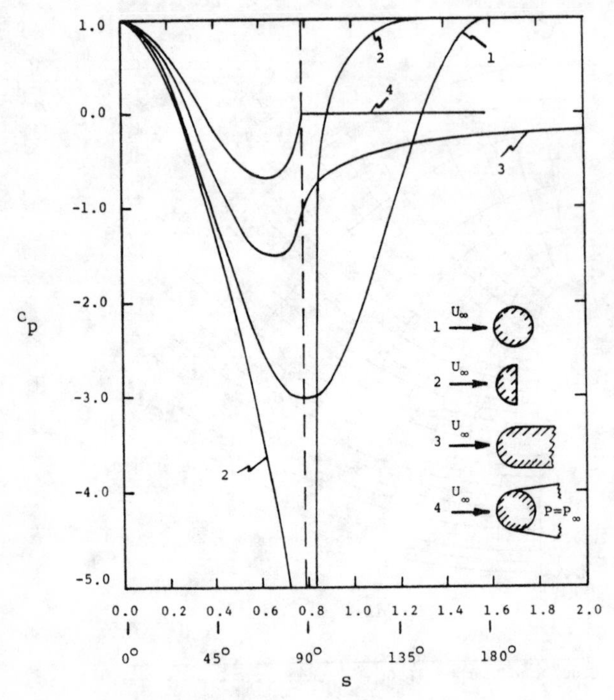

Figure 17. Pressure Coefficients for Various Cylinder and Afterbody Geometries.

Other step size studies not shown demonstrate second-order accuracy at all locations on the cylinder.

Figure 17 shows the pressure coefficient results from all of the cylinder calculations. In addition, the pressure coefficient is shown for a free streamline calculation for a flow which separates at 90° from the stagnation point. This calculation will be discussed in the next section.

The pressure coefficient results are as one would expect. The half cylinder shows (curve 2) a singularity at the corner as it should, the cylindrically blunted plate (curve 3) indicates a possible curvature discontinuity at 90°, and the free streamline calculation indicates a slope discontinuity at 90°.

5. FLOWS WITH FREE STREAMLINES

Recent asymptotic results by Sychev [14], Messiter and Enlow [15], and Smith [16] have shown that the Kirchoff [17] free streamline theory is consistent with triple-deck theory and that in the limit as Re $\rightarrow \infty$ for steady laminar flow, an asymptotic model can be constructed which is consistent. It, therefore, appears that the calculation of flows with free streamlines is important in the theory of separated flows and will become even more important as the asymptotic models become more elaborate to include wake pressure variation, etc. It also seems that the incorporation of free streamlines into the calculation of coordinate systems for high Reynolds number Navier-Stokes solvers can lead to coordinate systems which are more appropriate for high Reynolds number flows than the ones presently being used. If the free streamline model is correct, a coordinate system based on it should provide the proper alignment of coordinate curves for doing the proper scalings and mesh refinements required for accurate numerical solutions.

We will not go into the details of free streamline theory here, but refer one to Woods [8] for an extensive treatment of the subject. We will assume that the wake pressure behind the body, say a circular cylinder, is constant and is given by the free stream pressure P_∞. Therefore, from Bernoulli's equation, the velocity on the free streamline at the edge of the dead water region is V_∞, the free stream velocity. As mentioned previously, free streamline models have been constructed to incorporate a wake pressure variation. Even though we will not include this pressure variation, the pressure method can easily be extended to include it.

Therefore, we must mathematically solve a problem where we know the shape of a portion of the $\psi = 0$ surface (the body surface) but do not know the velocity on the surface, and for another portion of the $\psi = 0$ surface (the free streamline) we know the velocity but do not know the shape of the surface.

For the most part, flows with free streamlines can be handled with the same method as was discussed in the previous section. The body surface is mapped in exactly the same manner. The free streamlines are mapped to the ξ axis of the ζ plane as if the free streamlines were part of the body surface, but their location in the physical plane is unknown and must be determined as part of the solution. Since the free streamlines extend to infinity, it is useful to develop an asymptotic element such that beyond a prescribed point the slow decay of β towards its value at infinity can be accounted for by the use of the asymptotic element. This region is shown as region 2 in Figure 18. In addition, it can be shown that the curvature of a free streamline at the separation point is generally infinite, but this infinity can be analytically

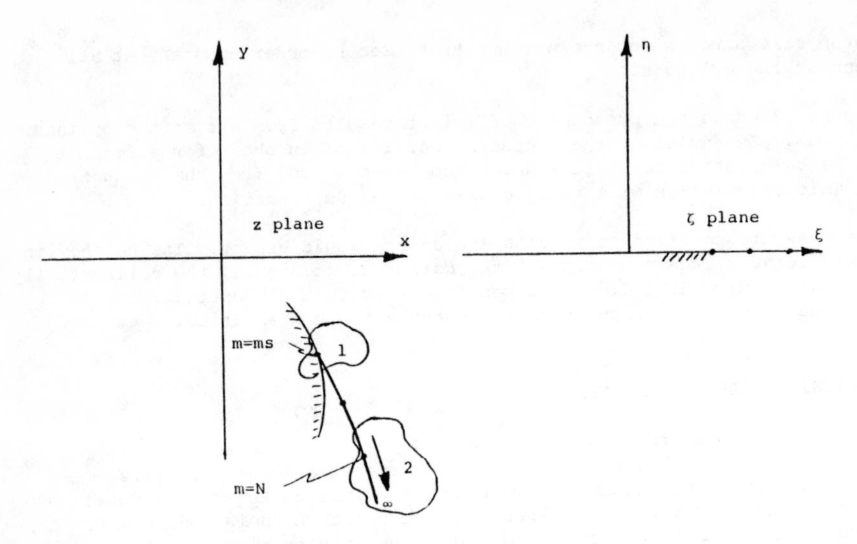

Figure 18. Free Streamline Geometry.

integrated out such that it does not introduce large errors into the numerical solution. While this effect turns out to be less severe than a corner, it is important to properly account for it in essentially the same manner as was done in the corner case. This is shown as region 1 in Figure 18.

It can be shown that the free streamline decays toward infinity like a parabola. This can be shown to imply that

$$\beta = \frac{\pi}{2} + \frac{C_{2N}}{b} \qquad \text{as} \qquad b \to \infty \tag{5.1}$$

where C_{2N} is a constant to be determined as part of the solution. Therefore, it is assumed that the location of the N point is far downstream and that the expression (5.1) is accurate enough to account for the variation of β from the N location to infinity. Therefore, we must evaluate

$$\frac{1}{\pi} \int_{\beta_N}^{\pi/2} \ell n\ (\zeta - b)\ d\beta = -\frac{C_{2N}}{\pi} \int_{b_N}^{\infty} \ell n\ (\zeta - b)\ \frac{db}{b^2} \tag{5.2}$$

and incorporate the result into our integration formula (4.7). The above Equation (5.2) results in

$$\frac{1}{\pi} \int_{\beta_N}^{\pi/2} \ell n\ (\zeta - b)\ d\beta = \ell n\ [g_{2N}\ (b_N, \zeta)]^{C_{2N}/\pi} \tag{5.3}$$

where

$$g_{2N}(b_N, \zeta) = \left[\frac{1}{(\zeta - b_N)^{(\zeta - b_N)} \ (-b_N)^{b_N}} \right]^{1/\zeta b_N} \tag{5.4}$$

On the other side of the body there is another free streamline extending to negative infinity. A downstream asymptotic element extending from $-b_N$ to $-\infty$ can, therefore, be easily incorporated into the symmetric case and the resulting expressions are similar to (5.3) and (5.4).

The new element expressions alter Equation (4.7) by only a multiplication by the new $g^{C/\pi}$ expressions. This is one of the nice features of the present method, that modification and improvement of the method to incorporate asymptotic elements, higher order accurace, etc. can be accomplished by multiplication of Equation (4.7) by such terms and, therefore, involve only minor program changes.

The remaining element expression to be developed is the one which takes care of the infinite curvature discontinuity which occurs at a free streamline separation point. This is shown as region 1 in Figure 18. Figure 19 shows this region of the body surface magnified. The quantities $\bar{s}$ and $\bar{n}$ are the arc length and normal coordinates measured from the separation point. It can be shown, see Smith [16] for example, that the shape of the free streamline near separation is of the form

$$\bar{n} \cong k\,\bar{s}^{3/2} \tag{5.5a}$$

and therefore

$$\frac{d\bar{n}}{ds} \cong \frac{3}{2}\,k\,\bar{s}^{1/2} \tag{5.5b}$$

and

$$\frac{d^2\bar{n}}{ds^2} \cong \frac{3}{4}\,k\,\bar{s}^{1/2} \tag{5.5c}$$

Using the expression (3.14) for $u - iv$ along with (3.15), (3.16), and (3.19), which are all general, we have

$$\frac{dz}{d\zeta} = \frac{i\,M\,\zeta}{u - iv} \tag{5.6}$$

Along the free streamline (see Figure 19)

$$|dz| = ds\ , \qquad |d\zeta| = db\ , \tag{5.7a,b}$$

$$|\zeta| = b\ , \qquad |u-iv| = V_\infty\ . \tag{5.7c,d}$$

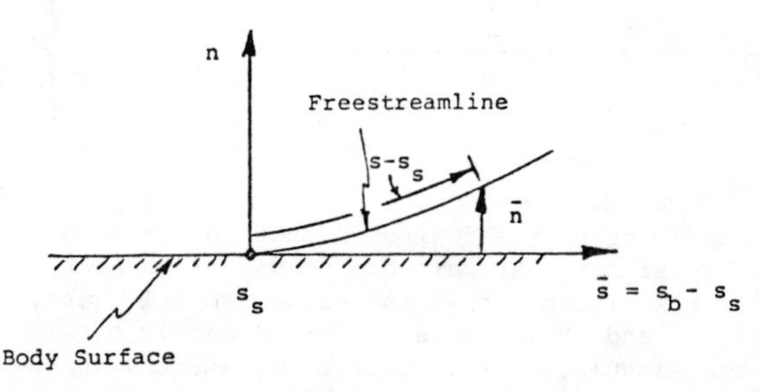

Figure 19. Free Streamline Geometry Near Separation Point.

Therefore, taking the magnitude of (5.6), we obtain

$$\frac{ds}{db} = \frac{|M|}{V_\infty} b \qquad \text{at} \qquad \eta = 0 \tag{5.8}$$

and integrating

$$s - s_s = \frac{|M|}{V_\infty} \left(\frac{b^2}{2} - \frac{b_s^2}{2}\right) \tag{5.9}$$

where $s - s_s$ is the arc length along the free streamline. Near s_s, $s - s_s$ can be replaced by $s_b - s_s$ and since dn/ds is the slope of the free streamline measured from the body surface, we can easily show that

$$\beta = \beta_s + C_{4ms} (b - b_s)^{1/2} \tag{5.10}$$

for the immediate region downstream of separation. It is, therefore, clear that the expression (4.2) used alone in this region will be inaccurate. We, therefore, develop an expression to accommodate the variation shown in Equation (5.10). In the present method we have superimposed the variation indicated by (5.10) over the free streamline from b_s to b_N, where b_N is the last point before the asymptotic element. This amounts to modifying Equation (4.2) to the form

$$\beta = C_{1m} + C_{2m} b + C_{3m} b^2 + C_{4ms} (b-b_s)^{1/2} \tag{5.11}$$

where C_{4ms} is one constant which is the same for all elements and must be determined as part of the solution. We, therefore, must now develop an expression for

$$\frac{C_{4ms}}{\pi} \int_{b_s}^{b_N} \ln(\zeta-b)\, d(b-b_s)^{1/2} = \ln \left[g_{2ms}(b_N, b_s, \zeta)\right]^{C_{4ms}/\pi} \tag{5.12}$$

Integration gives

$$g_{2ms} = \left\{ \frac{\left[\dfrac{(b_N-b_s)^{1/2} + (\zeta-b_s)^{1/2}}{(b_N-b_s)^{1/2} - (\zeta-b_s)^{1/2}}\right]^{(\zeta-b_s)^{1/2}} (\zeta-b_N)^{(b_N-b_s)^{1/2}}}{e^{2(b_N-b_s)^{1/2}}} \right\} e^{-i\pi(\zeta-b_s)^{1/2}} \tag{5.13}$$

Equation (5.13) has been written in a form such that the term $e^{-i\pi(\zeta-b_s)^{1/2}}$ which is responsible for the singular part of the variation near $\zeta = b_s$, can be integrated as a factor while the other factor is properly behaved. Therefore, incorporating (5.12) and (5.13) into (4.7), we merely obtain a new multiplying factor. The singular part is extracted, exactly in the same manner as was done in the case of corners, and integrated exactly, i.e. from (4.7) near b_s

$$\frac{dz}{d\zeta} = f_1 \; e^{-iC_{4ms}(\zeta-b_s)^{1/2}} \tag{5.14}$$

where f_1 is all of the remaining part of the coefficient which is well behaved near b_s. Integrating between m and m+1, as in the case of a polygon corner, we obtain

$$z_{m+1} - z_m = \bar{f}_1 \frac{2}{C_{4ms}^2} \{1 + iC_{4ms} (\zeta-b_s)^{1/2}\} \; e^{-iC_{4ms}(\zeta-b_s)^{1/2}} \Bigg|_{\zeta_m}^{\zeta_{m+1}} \tag{5.15}$$

where $\bar{f}_1$ is f_1 evaluated at $(\zeta_m + \zeta_{m+1})/2$ as in the case of the polygon. We can now work the above Equation (5.15) into a composite integration formula in exactly the same manner as was used in Equation (3.12) where the above expression mearly appears as a new multiplying factor. The evaluation of (5.15) near the separation point, i.e. let m be at separation and m+1 a small distance away, will exactly reproduce the behavior given by (5.5a-c) once an expansion of (5.15) is made in terms of $(\zeta-b_s)^{1/2}$.

On the other side of the body, there is another free streamline which can be handled in exactly the same manner. In the case of symmetry, which we are presently considering, all expressions on the left side of the body can be combined with the equivalent expression at the symmetrical location on the right to form new element expressions which allow one to then only consider the right side of the plane. We have done this in developing all solutions discussed in this paper.

The body is handled exactly as before, i.e. body points and slopes are prescribed, the b_m locations are guessed and subsequent iterations correct the b_m's in exactly the same manner as before.

The wake portion of the flow is handled as follows, see Figure 20. For the present, we have been handling the wake with the values of the C_{3m}'s set equal to zero. Even though this amounts to having lower order accuracy on the wake, the results have been found to be excellent. First,

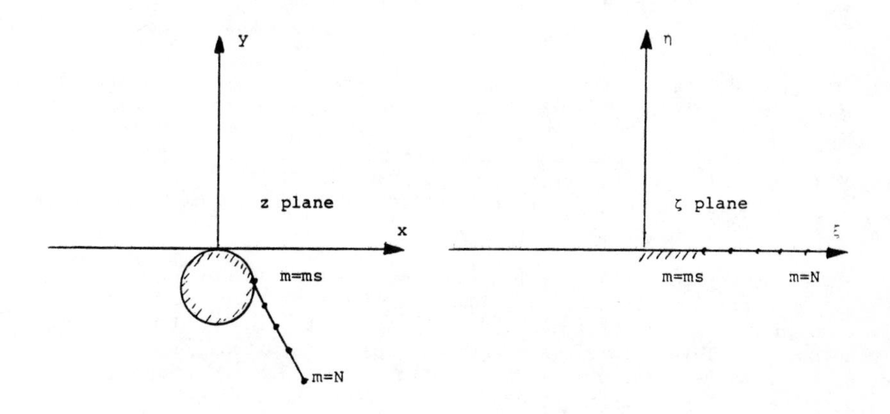

Figure 20. Wake Solution Points.

$$\frac{dw}{dz} = u - iv = V e^{-i\theta} = \frac{dw/d\zeta}{dz/d\zeta} = \frac{a\zeta}{dz/d\zeta} \tag{5.16}$$

This results in

$$\frac{dz}{d\zeta} = \frac{a\zeta}{V} e^{i\theta} \quad . \tag{5.17}$$

On the free streamline $V = V_\infty$ and $dz/d\zeta$ is the expression (4.7) which now includes the new special element expressions for the separation singularity and asymptotic behavior near infinity. Letting $f_1(\zeta)$ be the part of $dz/d\zeta$ which arises from the body elements, we obtain

$$f_1(\zeta) \prod_{m=ms}^{m=N+1} [g_{2m}]^{C_{2m}/\pi} = \frac{a\zeta}{V_\infty} e^{i\theta} \quad . \tag{5.18}$$

We pick the free streamline element endpoints at known locations in the ζ plane, i.e. b_{ms}, b_{ms+1}, etc. are specified, and, therefore, the values of g_{2m} are known. Taking the absolute value of (5.18) (remembering that the absolute value of a product is the product of the absolute values) and the log of the result, we obtain a set of simultaneous equations for the C_{2m}'s on the wake. The values of ζ at which the equations are satisfied are at the wake element endpoints.

At this point, there is one more equation than unknown, and the remaining equation is obtained from a condition like (4.13), i.e. that the total turning angle is π as the mapping proceeds from $-\infty$ to $+\infty$. This results in the following equation

$$\frac{1}{\pi} \sum_{m=ms}^{N-1} [\alpha_m + (b_{m+1} - b_m) C_{2m} + (b_{m+1}^2 - b_m^2) C_{3m}] - \frac{C_{2N}}{b_N}$$

$$+ C_{4ms} (b_N - b_s)^{1/2} = \frac{1}{2} - \frac{\beta_s}{\pi} \tag{5.19}$$

where C_{2N} and C_{4ms} are contained in (5.18) as two of the C_{2m} constants. Note that the C_{3m} constants in (5.19) for the wake elements are zero at present.

In the present method, the b_m's in the wake are held fixed and, therefore, the coefficient matrix for the simultaneous equations does not change as the body element locations are iterated. The coefficient matrix, therefore, only needs to be inverted once and each subsequent iteration step requires only a matrix multiplication for the wake portion. The method, therefore, converges as rapidly and does not require significantly more computer time than a case which does not involve free streamlines.

We must prescribe the separation location in order to obtain a solution, the true separation location being a function of Reynolds number and determined through interaction of the boundary layer and the inviscid flow such that the boundary-layer flow passes through separation without a boundary-layer singularity, see Smith [16] for example. The present method, therefore, can provide the inviscid solution which can be used in connection with an interacting boundary-layer or triple-deck solver to do separated flows.

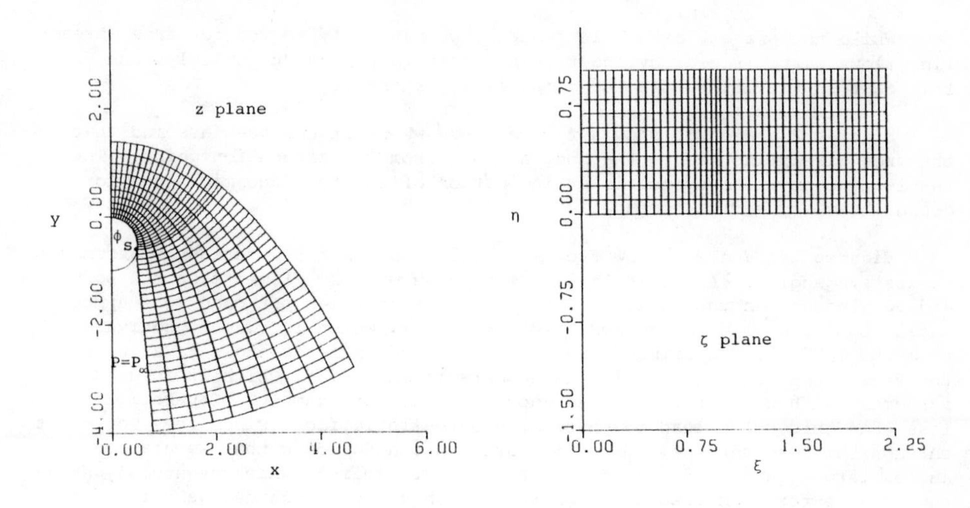

Figure 21a. Cylinder with Free Streamlines, $\phi_s = 100°$.

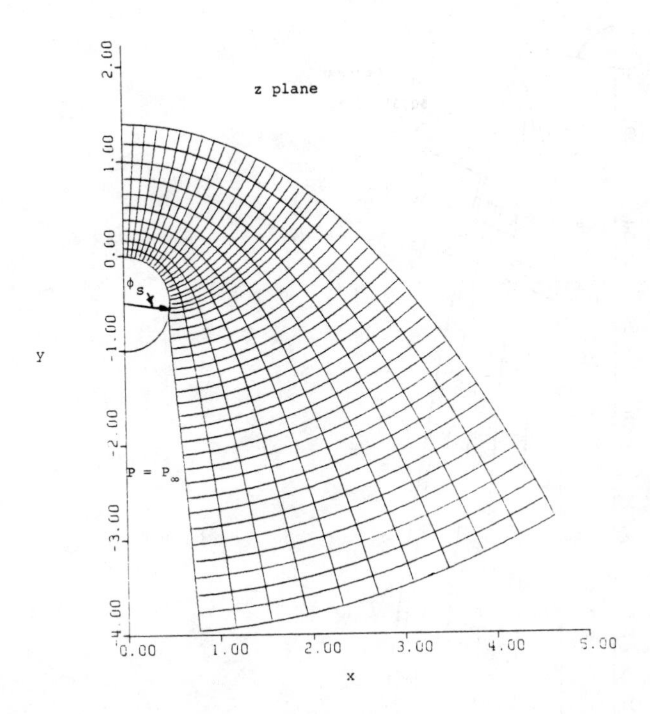

Figure 21b. Cylinder with Free Streamlines, $\phi_s = 100°$.

While the present method is general, we have only solved for free streamline flows past circular cylinders since there are previous calculations for this case in the literature for comparison purposes.

Figures 21a and 21b show a typical case where it has been assumed that the free streamline leaves the body at 100° from the stagnation point. The resulting coordinate system is an indication of the type one might use for doing the viscous flow computations.

Figures 22a and 22b show free streamline shapes for a series of assumed separation angles. All calculations were done with 20 elements on the body and 50 elements on the wake. Typical computation times per case were approximately 10 CPU seconds on an Amdahl/470 computer, which is roughly equivalent to a CDC 6600. By experimentation, it was found that the angle ϕ_s at which the free streamline singularity disappears is at 55.73°, which is in good agreement with other authors, see Woods [8, 18] for example. Separation angles less than this value lead to physically unrealistic flows since they would cut through the body surface. Experimentation has also found that separation angles larger than $\phi_s \cong 123.5°$ lead to free streamlines which eventually cross the -y axes far downstream and, therefore, should be discarded as being physically unrealistic. These results are also in agreement with Woods [8, 18].

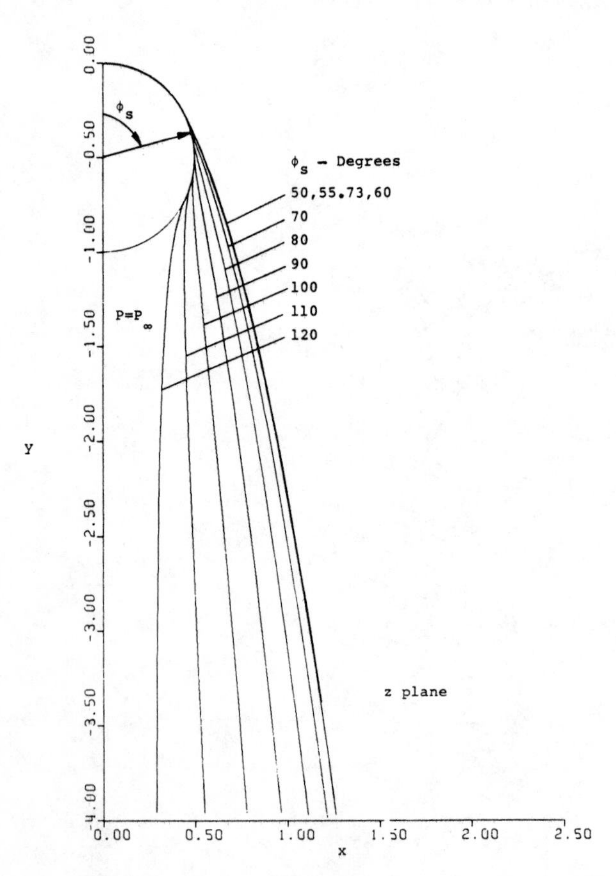

Figure 22a. Cylinders with Free Streamlines.

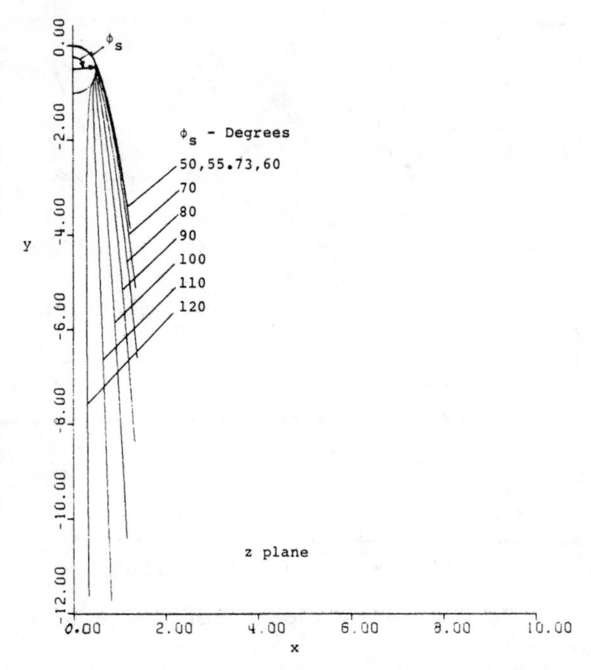

Figure 22b. Cylinders with Free Streamlines.

The pressure coefficients for the free streamline flows are calculated in exactly the same manner as before, except that the new singular and asymptotic element expressions have been incorporated. Figure 23 shows the results for the cylinder which are in excellent agreement with Woods [8, 18].

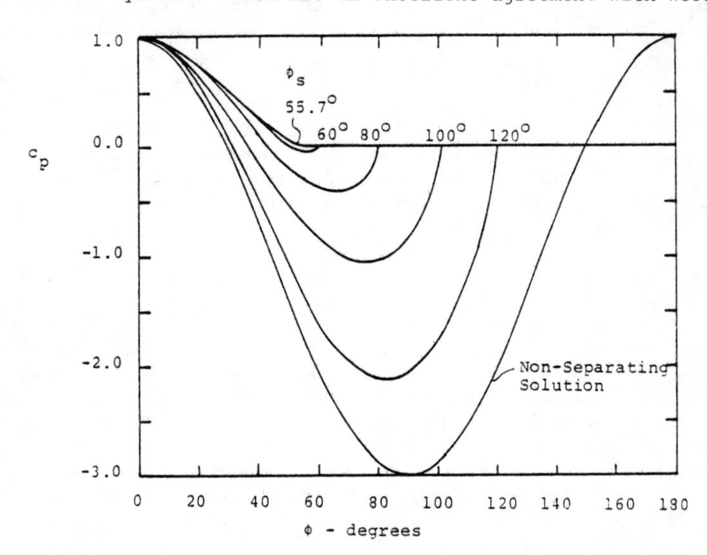

Figure 23. Pressure Coefficients for Free Streamlines Flow Past Circular Cylinder.

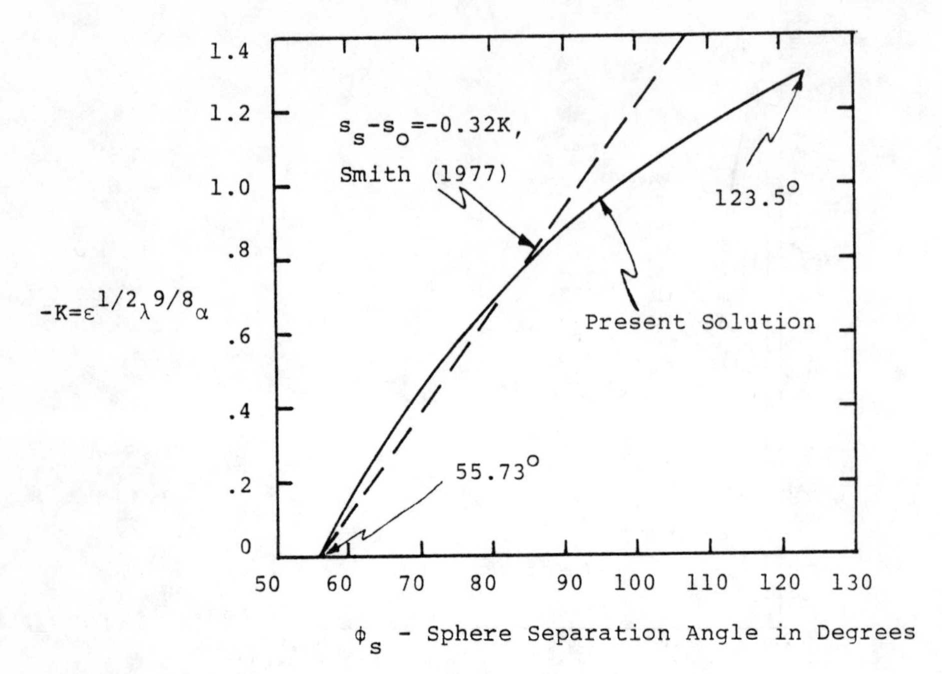

Figure 24. Comparison of Present Solution With Triple-Deck Parameters of Smith (1977).

Finally, Figure 24 shows the relationship between the sphere separation angle and the coefficient of the singular term expressed in terms of Smith's [16] triple-deck constants. The curve $s_s - s_o = -0.32K$, which is the approximate expression used in some of Smith's [16] work, is shown for comparison.

6. FLOWS WITH CIRCULATION

With modification, the present method can be used to handle the case of flows with circulation. Only a brief discussion of how this is done will be given, see Woods [8], p. 303, for a more detailed discussion of mapping of flows with circulation.

Consider the airfoil at angle of attack as shown in Figure 25. It is assumed in this figure that the Kutta condition is satisfied. The curve $B\,E_\infty$ and $B'E'_\infty$ is the trailing edge streamline. We wish to map this body and trailing edge streamline to the same ζ plane as before. This is a more difficult case since we do not have symmetry and we do not know the shape of the trailing edge streamline. Consider the complex potential plane w shown in the figure. A closed path C around the airfoil in the physical plane produces the path C in the w plane which jumps by the amount Γ (the circulation) across the trailing edge streamline. Since we intend to use the same mapping technique as before, the relationship between the w plane and the ζ plane is given by

$$w = \frac{a}{2}\zeta^2 \qquad (6.1)$$

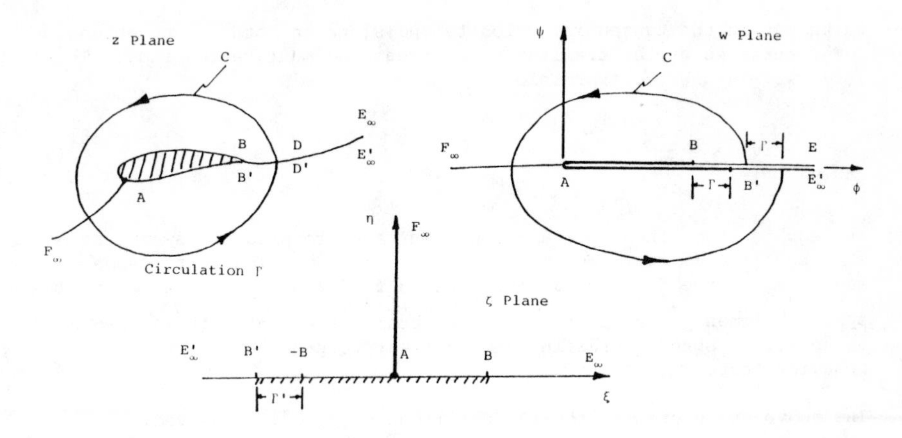

Figure 25. The Mapping with Circulation Present.

and since on the body surface or trailing streamline ϕ is given by

$$\phi = \frac{a}{2} \xi^2 \tag{6.2}$$

we have for corresponding upper and lower points on the trailing streamline

$$\phi_\ell = \phi_u + \Gamma \tag{6.3}$$

which implies

$$a \frac{\xi_\ell^2}{2} = a \frac{\xi_u^2}{2} + \Gamma \tag{6.4}$$

Therefore, once we pick the location of the trailing edge in the ζ plane on the upper surface of the airfoil, the location of the trailing edge on the lower surface is determined if Γ is specified. The problem with circulation is, therefore, handled as follows:

(a) Element locations on the body surface are picked in the physical plane. The same type of iterative method as before is used to find the locations in the ζ plane (the b_m values on the body surface). The trailing edge location for ξ_u is prescribed and the trailing edge value for ξ_ℓ is determined by prescribing Γ.

(b) We do not know the location of the trailing edge streamline in the physical plane and, therefore, we pick nodal points in the ζ plane (b_m values) and calculate the location in the physical plane by a method similar to that used in the free-streamline case. We only need to pick the upper values ξ_u since the equivalent location in the physical plane of ξ_ℓ is given by Equation (6.4).

(c) We know that the tangential velocity and slope on both the upper and lower surfaces of the trailing edge streamline must be the same. Therefore, we can use the expression

$$u - iv = \frac{a\zeta}{\frac{dz}{d\zeta}} \tag{6.5}$$

and equate it on the upper and lower surfaces to produce an expression like Equation (5.18). We solve the resulting equation in the same manner as in the free-streamline case to find the C_{2m}'s. A downstream special element to reflect the proper behavior at infinity is also used along with a special trailing edge element to produce the proper behavior there.

The above procedure is iterated until convergence is achieved.

Figure 26 shows the result for a NACA 0018 airfoil at angle of attack. It should be noted that the method puts distortion in the coordinate system to account for the circulation. With these results, we could produce a polar type of distorted coordinate system similar to the one developed in the symmetric case without circulation.

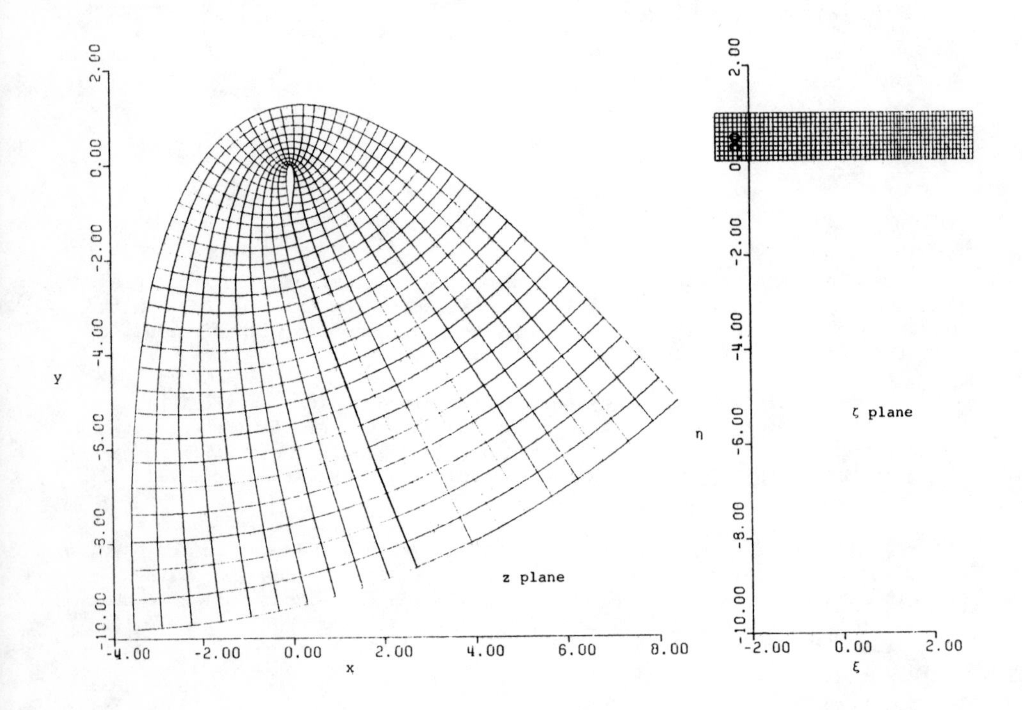

Figure 26. A NACA 0018 Airfoil Mapping with Circulation.

7. CHANNEL FLOWS

We will first demonstrate how the Schwarz-Christoffel Theorem for polygons given by Equation (2.1) is used to develop a two step numerical mapping method. This method resembles somewhat the method of O. Anderson [10]. In the last part of this section, we will also show how a one step method can be developed which will map general shaped channels, including curved sections, with higher-order accuracy.

The first method for channel mapping uses the same polygon integration formula given by Equation (3.12) as was used for external flows. In fact, the computer program written for external flows is easily modified to do channel flows. In the channel case, we take the mapping one step further and map the ζ plane to a t plane which consists of a channel with parallel walls with unit distance between them. This second step of the mapping is given by

$$t = -\frac{1}{\pi}\ \ell n\zeta + i \tag{7.1}$$

As an example of how a typical mapping is performed numerically, let us consider the simple case shown in Figure 27. Letting $\zeta \to \infty$ we find that from Equation (2.1)

$$\frac{dz}{d\zeta} \sim \frac{M}{\zeta} \qquad \text{as} \qquad \zeta \to \infty \tag{7.2}$$

for this case. The velocity in the parallel wall channel is taken to be unity and, therefore, the complex potential w is t. Therefore, from Equations (7.1) and (7.2) and letting the upstream velocity in the channel be 1, we find that

$$u - iv = i = \frac{dw}{dz} = \frac{dw}{dt}\frac{dt}{d\zeta}\frac{d\zeta}{dz} = -\frac{1}{\pi M} \tag{7.3}$$

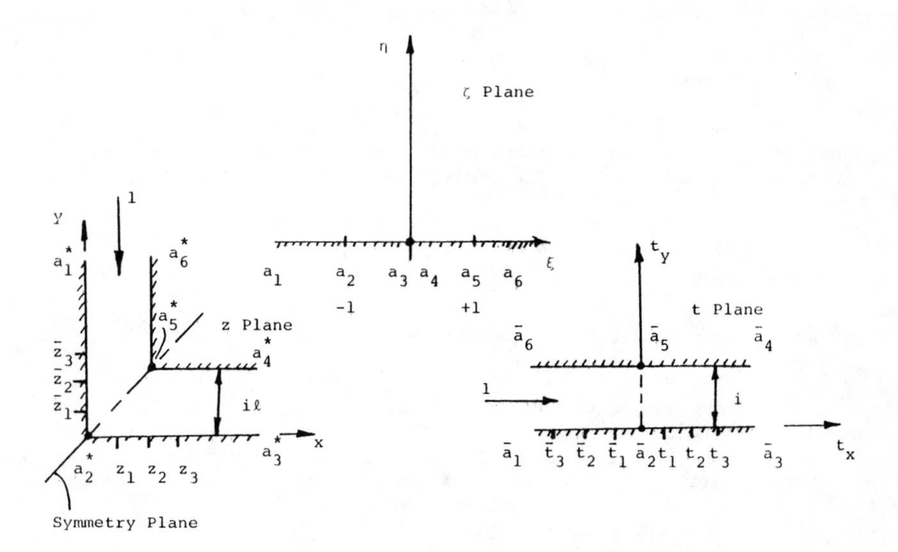

Figure 27. Channel Mapping with a Plane of Symmetry.

and therefore

$$M = \frac{i}{\pi} \tag{7.4}$$

for this case. Other cases are done in exactly the same manner but M may be different.

Next we observe that this problem has the plane of symmetry shown and that, therefore, a_2^* and a_5^* map to $\bar{a}_2$ and $\bar{a}_5$ in the t plane.

Now we divide the channel into elements with nodal locations $z_1, z_2, \ldots z_N$ to the right of the symmetry plane shown. We place elements above the symmetry plane which have nodes at the locations $\bar{z}_1, \bar{z}_2, \ldots \bar{z}_N$ which are the reflection of the z_i locations. Due to symmetry, these element endpoints must also be symmetric about the t_y axis in the t plane. Examination of Equation (7.1) reveals that this symmetry implies that the locations of the element end points on the ξ axis in the ζ plane satisfy

$$\bar{\xi}_i = \frac{1}{\xi_i} \tag{7.5}$$

This conclusion is true for any problem for which the upstream and downstream flows in a channel are symmetric about some plane.

Therefore, this and other problems of this type are solved numerically in almost exactly the same manner as the external flow case as follows:

(a) Guess values for the locations in the ζ plane for the element endpoints for the region to the right of the symmetry plane ($-1 \leq \xi_i \leq +1$). Use Equation (7.5) to determine the locations for the elements to the other side of the symmetry plane. M is determined from Equation (7.4) and Equation (3.12) is used for integration, where the range of integration is from $\xi = -1$ to $\xi = +1$. The resulting z_i values will not be at the proper locations. Call these values z_{gi}.

(b) For the guessed ξ_i values in part (a), calculate the equivalent t_{xg} location from Equation (7.1) and predict new t_{xc} locations from an equation similar to (3.13) but now written in the t plane, i.e.

$$\frac{t_{xci} - t_{xci-1}}{t_{xgi} - t_{xgi-1}} = \frac{z_{ci} - z_{ci-1}}{z_{gi} - z_{gi-1}} \tag{7.6}$$

(c) Convert these calculated t_{xci} locations to ξ_i locations and repeat the process.

This method appears to always converge and converges at about the same rate as was found in the external flow case (approximately ten iterations independent of the number of elements). It is important that the interpolation be done in the t plane for physically obvious reasons. Attempts to iterate for the element endpoints in the ξ plane resulted in either a much slower convergence rate or even divergence in some cases.

Figure 28 shows the result of this mapping. The elements were taken to be equally spaced on both surfaces in the z plane in order to perform the

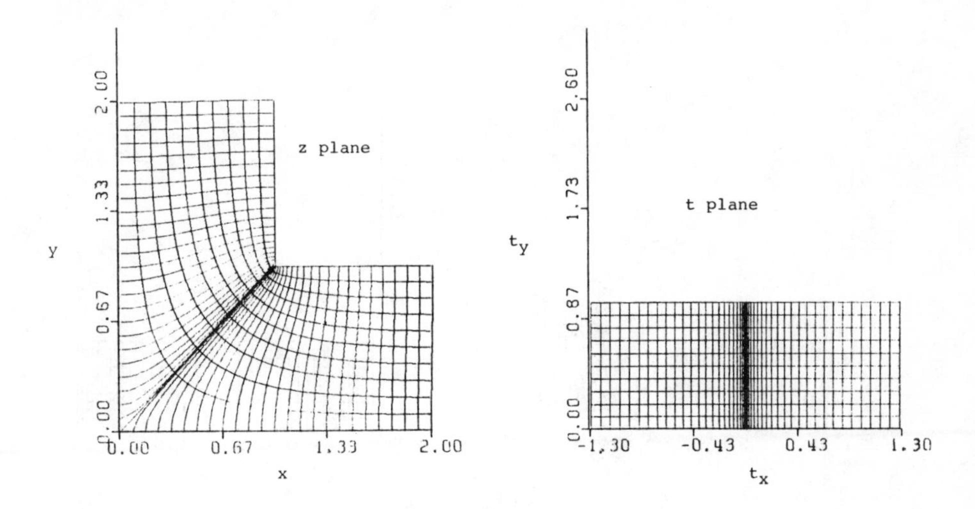

Figure 28. A Corner Mapping.

mapping. The coordinates are generated by starting in the t plane on the bottom surface at the element endpoints and integrating with t_x held constant to the upper surface, i.e. by using the expression

$$\zeta = -e^{-\pi t} \tag{7.7}$$

where t_x is held constant at the element endpoint value and t_y is incremented to calculate $\Delta\zeta$. Equation (3.12) is used to perform the integration.

Figure 29 shows another mapping which was performed in exactly the same manner. Here the symmetry plane extends from the top of the cylinder to the upper wall.

If the potential flow complex velocity is desired, it can be computed numerically from

$$u - iv = \frac{dt}{dz} \tag{7.8}$$

In both cases, the wall tangential velocities show excellent agreement with more exact or analytical results.

Figure 30 shows a problem with a different type of symmetry. In this case, the symmetry is about the centerline of the channel. Again, the same type of mapping is performed with the channel inlet placed on the t_y axis in the t plane. If we subdivide the upper and lower surfaces in the z plane with the same element spacings, we only need to integrate in the region from 0 to +1 on the ξ axis in the ζ plane and then reflect the element endpoints about $\xi = 0$ as the iteration process proceeds since the mapping in this case is symmetric about the η axis in the ζ plane. The interpolation is again performed in the t plane to update the guessed element endpoints in the

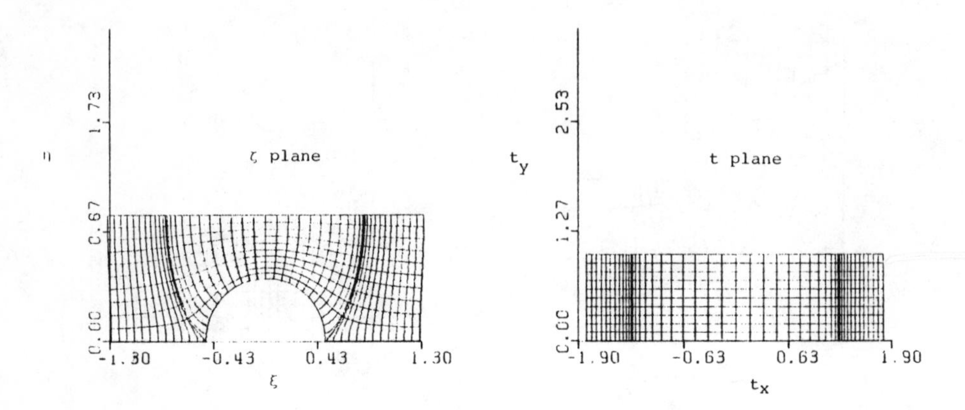

Figure 29. A Channel Mapping with a Half Cylinder on Boundary.

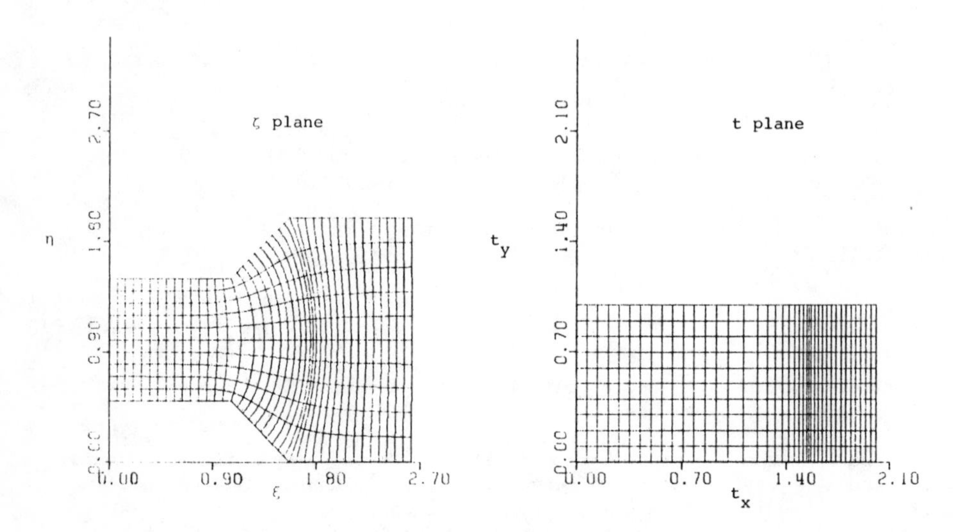

Figure 30. A Mapping of a Divergent Channel.

ζ plane. Convergence rate is again about the same as in previous cases. The coordinate lines in Figure 30 are again generated by integrating vertically from the element endpoints in the t plane.

In problems with symmetry of the types discussed above, the numerical problem becomes much simpler than for a problem which has no symmetry. This occurs since, in both cases, we know the location of the points in the z plane which map to $t = 0 + i0$ and $t = 0 + i1$ in the t plane. In problems with no symmetry, we must be sure to take an initial plane far enough upstream so that the flow is essentially parallel in the physical plane and then assume that the upper and lower points at this location map to $t = 0 + i0$ and $t = 0 + i1$ in the t plane. This results in a small error which diminishes as the location of the initial station is moved further upstream. Figure 31 shows a result of this type of mapping performed by O. Anderson of United Technologies Research Center using the same computer program as was used in the previous cases. A method to improve the results would be to assume that symmetry exists about the initial plane even though it does not. This error would not be serious as long as the initial station (assumed symmetry plane) is far enough upstream of the region of interest.

We can now develop a one step method for channel mapping and map directly from the z to the t plane. This is done as follows.

First the arbitrary channel (see Figure 32) with straight segments shown in the physical plane is ampped to the ξ axis in a ζ plane. Next the parallel wall channel, shown in the t plane of Figure 32, is mapped to the same ξ axis in the ζ plane by Equation (7.1). The ζ plane is next eliminated to allow for a direct mapping from the z to the ζ plane and the result after lengthy algebra is

$$\frac{dz}{dt} = K \exp[(\phi-\delta)\frac{t}{2h}] \frac{\prod_{m=1}^{M}\left[\sinh\frac{\pi}{2h}(t-a_m)\right]^{\alpha_m/\pi}}{\prod_{n=1}^{N}\left[\cosh\frac{\pi}{2h}(t-a_n)\right]^{\alpha_n/\pi}} \tag{7.9}$$

where the m subscripts refer to the lower wall and the n subscripts to the upper wall.

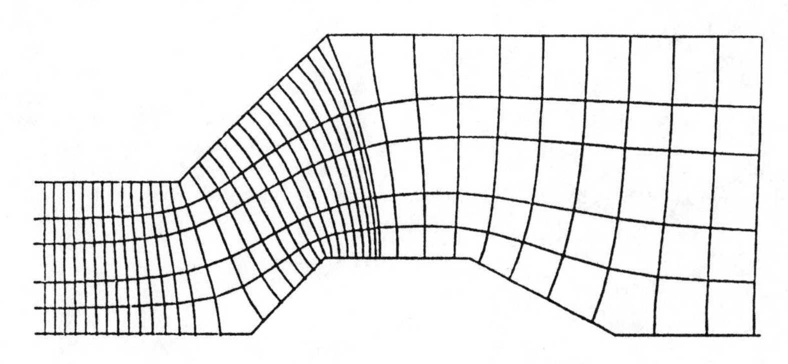

z plane

Figure 31. A Multiple Corner Channel Mapping, after O. Anderson.

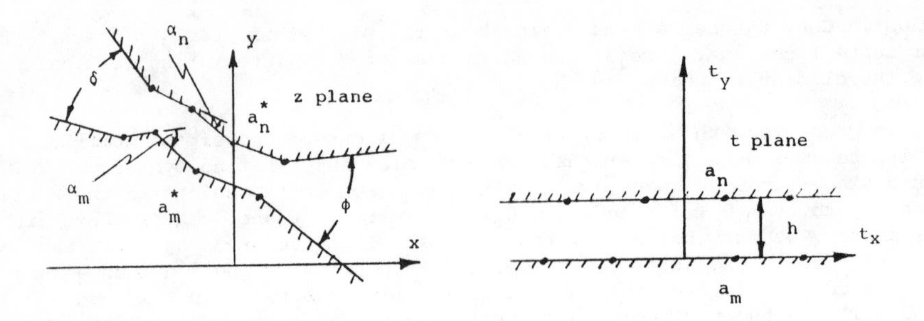

Figure 32. Channel Mapping.

Rewriting the π product expressions in exactly the same manner as was done in Equation (2.3), a limit process can be performed to extend (7.9) to curved boundaries. The resulting expression is

$$\frac{dz}{dt} = K \exp\left[(\phi-\delta)\frac{t}{2h}\right] \frac{\exp\left[\frac{1}{\pi}\int \ell n\{\sinh\left[\frac{\pi}{2h}(t-b)\right]\}\, d\beta_1\right]}{\exp\left[\frac{1}{\pi}\int \ell n \cosh\left[\frac{\pi}{2h}(t-b)\right]\}\, d\beta_2\right]} . \tag{7.10}$$

We can now repeat all of the previous analysis of this paper to produce equivalent expressions for (7.9) and (7.10), i.e. corner integration expressions, curved element expressions, etc. Some of this work is reported on in Sridhar and Davis [20].

In addition to the normal type channel flows, it should be possible to calculate internal free streamline flows along the same lines as was done in Section 5 for external flows.

8. CONCLUSION

A technique has been developed for numerical coordinate generation using complex variable theory. The method shows promise in that it overcomes several of the difficulties associated with other methods. The method is also used as a potential flow solver.

ACKNOWLEDGEMENT

This research was supported by the Office of Naval Research under Contract N00014-76-C-0364. The author is indebted to M. Napolitano and Y. Lin for valuable discussions and help in some of the calculations.

REFERENCES

1. Stewartson, K., "Multistructured Boundary Layers on Flat Plates and Related Bodies," in Advances in Applied Mechanics, Vol. 14, p. 145, Academic Press, 1974.

2. Thompson, J.F., Thames, F.C. and Masten, C.W., "Automatic Numerical Generation of Body-Fitted Curvilinear Coordinate Systems for Field Containing any Number of Arbitrary Two-Dimensional Bodies," Journal of Computational Physics, Vol. 15, p. 299, 1974.

3. Thompson, J.F., Thames, F.C. and Masten, C.W., "Boundary-Fitted Curvilinear Coordinate Systems for Solution of Partial Differential Equations on Fields Containing any Number of Arbitrary Two-Dimensional Bodies," NASA CR-2729, 1977.

4. Thompson, J.F., Thames, F.C. and Masten, C.W., "TOMCAT - A Code for Numerical Generation of Boundary-Fitted Curvilinear Coordinate Systems on Fields Containing any Number of Arbitrary Two-Dimensional Bodies," J. Comput. Phys., Vol. 24, No. 3, 1977.

5. Hess, J.L. and Smith, A.M.O., "Calculation of Potential Flow About Arbitrary Bodies," Progress in Aeronautical Sciences, Vol. 8, Pergamon Press, 1966.

6. Bristow, D.R., "Recent Improvements in Surface Singularity Methods for the Flow Field Analysis about Two-Dimensional Airfoils," AIAA 3rd Computational Fluid Dynamics Conference, A Collection of Technical Papers, 1977.

7. Carrier, G.F., Krook, M. and Pearson, C.E., Functions of A Complex Variable, McGraw-Hill Book Company, 1966.

8. Woods, L.C., The Theory of Subsonic Plane Flow, Cambridge University Press, 1961.

9. Trefethen, L.N., "Numerical Computation of the Schwarz-Christoffel Transformation," Stanford University Computer Science Report, STAN-CS-79-710, 1979.

10. Anderson, O.L., "Calculation of Internal Viscous Flows in Axisymmetric Ducts at Moderate to High Reynolds Numbers," to appear in Computers and Fluids, 1979.

11. Davis, R.T., "A Study of Optimal Coordinates in the Solution of the Navier-Stokes Equations," Dept. of Aerospace Engineering, University of Cincinnati, Report AFL 74-12-14, 1974.

12. Van Dyke, M., Perturbation Methods in Fluid Mechanics, The Parabolic Press, 1975.

13. Abbott, I.H. and von Doenhoff, A.E., Theory of Wing Sections, Including a Summary of Airfoil Data, Dover, 1959.

14. Sychev, V.V., "On Laminar Separation," Mekhanika Zhidkosti i Gaza, No. 3, p. 47, 1972.

15. Messiter, A.F. and Enlow, R.L., "A Model for Laminar Boundary Layer Flow Near a Separation Point," SIAM Journal, Vol. 25, p. 655, 1973.

16. Smith, F.T., "The Laminar Separation of an Incompressible Fluid Streaming Past a Smooth Surface," Proc. R. Soc. Lond., A, 356, p. 443, 1977.

17. Kirchoff, G., "Zur Theorie freir Flüssigkeitsstrahlen," J. Reine Angew. Math., 70, p. 289, 1969.

18. Woods, L.C., "Two-Dimensional Flow of a Compressible Fluid Past Given Curved Obstacles with Infinite Wakes," Proc. Roy. Soc. Lond., A, 227, p. 367, 1955.

19. Blottner, F.G. and Ellis, M.A., "Finite-Difference Solution of the Incompressible Three-Dimensional Boundary Layer Equations for a Blunt Body," Computers and Fluids, Vol. 1, pp. 133-138, 1973.

20. Sridhar, K.P. and Davis, R.T., "A Schwarz-Christoffel Method for Generating Internal Flow Grids," ASME/AIAA Conference on Computers in Flow Predictions and Fluid Dynamic Measurements, Nov. 15-20, 1981.

Introduction to Multi-grid Methods for the Numerical Solution of Boundary Value Problems

WOLFGANG HACKBUSCH

LIST OF SYMBOLS

B	bilinear form
C	generic constants with different values at different places
$C_0(\nu), C_1$	special bounds
f	right hand side of the boundary value problem
f_h, f_ℓ	right hand side of discrete problems
$\mathcal{G}_h, \mathcal{G}_\ell$	smoother of the multi-grid method
G_h, G_ℓ	iteration matrix of $\mathcal{G}_h, \mathcal{G}_\ell$, respectively
h, h_ℓ	grid size, discretization parameter
$\mathcal{H}_h, \mathcal{H}_\ell$	finite element subspace
$H^m(\Omega)$	Sobolev space of order m
ℓ	level number of multi-grid method
L_h, L_ℓ	matrix of discrete problem
m	2m is order of boundary value problem
M_h, M_ℓ	iteration matrix of multi-grid method
n	dimension of $\Omega < \mathbb{R}^n$
N, N_h, N_ℓ	number of unknowns
p	prolongation, interpolation
r	restriction
u	unknown function of boundary value problem
u_h, u_ℓ	discrete grid function, solution of the discrete problem
V_h, V_ℓ	linear space of discrete grid functions
γ	parameter of multi-grid method
Γ	boundary of domain Ω
ν	number of smoothing iterations

$\nu_m(h)$	upper bound of ν
$\partial/\partial n$	normal derivative
Ω	domain of the boundary value problem
Ω_h, Ω_ℓ	grid of size h, h_ℓ

1. INTRODUCTION

1.1 Problems under Consideration

In this paper we describe multi-grid algorithms for solving discretized boundary value problems. In the first subsection we give some examples of various boundary value problems. The discretization by difference schemes and finite element methods is recalled in the following sections 1.1.2 and 1.1.3.

1.1.1 Boundary value problems

Let $\Omega \subset \mathcal{R}^n$ $(n \geqslant 1)$ denote a domain in the n-dimensional space. A boundary value problem consists of (i) an elliptic partial differential equation for an unknown function (or vector) u in Ω and (ii) boundary conditions.

An example for an elliptic partial differential equation is Helmholtz' equation:

$$- \Delta u(x) + c^2 u(x) = g(x) \qquad (x \in \Omega), \tag{1.1}$$

where $\Delta = \Sigma_{i=1}^{n} \partial^2/\partial x_i$ is the Laplacian operator (other notations: ∇^2, div grad). The general linear elliptic equation of second order with variable coefficients is

$$\sum_{i=1}^{n} \sum_{j=1}^{n} a_{ij}(x) \frac{\partial^2 u}{\partial x_i \partial x_j} + \sum_{i=1}^{n} a_i(x) \frac{\partial u}{\partial x_i} + a(x)u(x) = g(x) \tag{1.2}$$

with

$$\sum_{i,j=1}^{n} a_{ij}(x) \xi_i \xi_j \neq 0 \text{ for all } x \in \Omega, \ 0 \neq \xi = (\xi_1, \ldots, \xi_n) \in \mathcal{R}^n.$$

A special elliptic equation of fourth order is the biharmonic equation :

$$\Delta^2 u(x) = g(x) \qquad (x \in \Omega). \tag{1.3}$$

As an example for a nonlinear equation we mention the equation for minimal surface

$$(1+u_{x_2}^2) u_{x_1 x_1} - 2 u_{x_1} u_{x_2} u_{x_1 x_2} + (1+u_{x_1}^2) u_{x_2 x_2} = 0 \qquad (x \in \Omega) \tag{1.4}$$

An elliptic equation of order 2m must be combined with m boundary conditions. Possible boundary conditions are Dirichlet values

$$u(x) = \phi(x) \qquad \text{for} \quad x \in \Gamma, \tag{1.5}$$

where $\Gamma = \partial\Omega$ denotes the boundary of the domain Ω, or the Neumann condition :

$$\frac{\partial u(x)}{\partial n} = \phi(x) \qquad \text{on } \Gamma. \tag{1.6}$$

$\partial u/\partial n = \underline{n}\cdot\text{grad } u$ denotes the derivative with respect to the outer normal direction $\underline{n}$. A generalization of (1.6) is the mixed condition :

$$\alpha(x)\ u(x) + \frac{\partial u(x)}{\partial n} = \phi(x) \qquad \text{on } \Gamma. \tag{1.7}$$

Boundary value problems can be formulated in a weak form. For instance, Helmholtz' equation (1.1) with Neumann condition (1.6) is equivalent to

$$B(u,v) = (f,v) \qquad \text{for all } v \in H^1(\Omega), \tag{1.8}$$

where B is the bilinear form

$$B(u,v) = \int_\Omega [(\text{grad } u)\cdot(\text{grad } v) + c^2uv]\ dx$$

and

$$(f,v) = \int_\Omega g(x)v(x)dx + \int_\Gamma \phi(x)v(x)d\Gamma.$$

$H^1(\Omega)$ is the Sobolev space of first order. Functions $v \in H^1(\Omega)$ and their first derivatives are square-integrable.

If the bilinear form B is symmetric and definite, i.e.,

$$B(u,v) = B(v,u), \qquad B(u,u) > 0 \text{ for } u \neq 0,$$

then the solution of the boundary value problem and of its weak formulation (1.8) is the unique minimizer of the variational problem

$$J(u) := B(u,u) - 2(f,u) = \text{minimum} \tag{1.9}$$

Formally, we represent a (linear) boundary value problem by

$$Lu = f. \tag{1.10}$$

Further boundary value problems are elliptic systems of partial differential equations as Stokes' problem and the steady Navier-Stokes equations. These problems are mentioned in §10. Also elliptic eigenvalue problems (compare with §11) belong to the class of problems that can be treated by multi-grid methods.

1.1.2 Difference schemes

For the numerical solution of boundary value problems we need some discretization method. Replacing derivatives by differences of step size h we obtain a difference scheme for the unknowns

$u_h(x)$, $x \in \Omega_h$, where Ω_h denotes the grid

$$\Omega_h = \{x \in \Omega : x = (\nu_1 h, \ldots, \nu_n h) \text{ with } \nu_1, \ldots, \nu_n \text{ integers}\}$$

of size h. For example, Helmholtz' equation (1.1) with n=2 can be discretized by the usual five-point scheme:

$$h^{-2}[4u(x) - u(x+he_1) - u(x-he_1) - u(x+he_2) - u(x-he_2)]$$
$$+ c^2 u(x) = f(x) \qquad (x \in \Omega_h), \tag{1.11}$$

where $e_1 = \binom{1}{0}$, $e_2 = \binom{0}{1}$ are the unit vectors with respect to x_1- and x_2-direction. The scheme (1.11) can only be used for specially shaped domains Ω : All neighbours $x \pm he_i$ (i=1,2) must belong either to Ω_h or to the boundary Γ. In the latter case the value of u is given by the boundary condition (1.5).

Usually, we must modify the difference equations at points $x \in \Omega_h$ near to the boundary. A first possibility is the Shortley-Weller scheme (cf. Shortley & Weller 1938, Meis & Marcowitz 1978, pp. 205). Interpolation at points near to the boundary is proposed by Collatz, 1966. For difference schemes in arbitrary domains in the case of mixed boundary conditions compare, for example, Bramble and Hubbard 1965. For difference schemes derived from the weak formulation (1.8) we refer to Oganesjan and Ruchovec 1979.

In the linear case we represent the system of linear equations by

$$L_h u_h = f_h. \tag{1.12}$$

$u_h = \{u_h(x)\}_{x \in \Omega_h}$ is called a grid function. The right-hand side f_h depends on the right-hand side of the differential equation and on the boundary values.

1.1.3 Finite element methods

The finite element technique is well known. The crux of the method is the approximation of functions by means of a finite dimensional subspace $H_h \subset H^1(\Omega)$*. For example, we may triangulate Ω and define H_h by all continuous functions that are linear on each triangle. The maximal diameter h of the triangles is used as discretization parameter.

The boundary value problem must be represented by its weak formulation (1.8). Then the finite element solution $\tilde{u}_h \in H_h$ is the unique solution of

$$B(\tilde{u}_h, \tilde{u}_h) = (f, \tilde{v}_h) \qquad \text{for all } \tilde{v}_h \in H_h. \tag{1.13}$$

Under suitable conditions we have the error estimate

$$\| \tilde{u}_h - u \|_{L^2(\Omega)} \leq C\ h^2 \| u \|_{H^2(\Omega)} \leq C'\ h^2\ \| f \|_{L^2(\Omega)}, \tag{1.14}$$

---

* Only for equations of second order. $H^m(\Omega)$ is case of order 2m.

where C and C' remain constant for $h \to 0$, that means, for refining the triangulation. For the technique and theory of finite elements we refer to Aziz 1972, Ciarlet 1978, Strang and Fix 1972, Thomasset 1980, Zienkiewicz 1977.

Using a suitable basis $\{e_{h1}, e_{h2}, \ldots, e_{hN}\}$ of the finite dimensional subspace H_h, we can represent the solution $\tilde{u}_h$ of problem (1.13) by

$$\tilde{u}_h = \sum_{i=1}^{N} u_{hi} e_{hi} \qquad (e_{hi} \in H_h, u_{hi} \in \mathbb{R}).$$

The coefficient vector

$$u_h := (u_{h_1}, u_{h_2}, \ldots, u_{hN}) \in \mathbb{R}^N$$

is the solution of the system of linear equations

$$L_h u_h = f_h, \tag{1.15}$$

where

$$L_{h,ij} = B(e_{hi}, e_{hj}), \qquad f_{h,j} = (f, e_{hj}).$$

The essential advantage of the finite element method (compared with the more general Ritz-Galerkin method) is the fact that L_h is a sparse matrix.

Finally we remark that often the coefficient u_{hj} of $\tilde{u}_h$ is the value of $\tilde{u}_h$ at some nodal point $x^{(j)} \in \Omega$. Moreover, in special cases the matrices L_h of the difference scheme (1.12) and of the finite element equation (1.15) are identical.

1.2 Multi-Grid Method

Multi-grid methods can be used for solving the finite difference scheme (1.12) as well as the finite element equations (1.15). The underlying boundary value problem may have constant coefficients as (1.1) or variable coefficients as (1.2), it may be definite or not, it may be of second or higher order as (1.3). The boundary conditions may be of any kind : (1.5), (1.6) or (1.7). The nonlinear multi-grid method can be applied also to nonlinear problems (see §8). Special multi-grid techniques for eigenvalue problems are described in §11.

The characteristic feature of the multi-grid method is the combination of a smoothing step and a coarse-grid correction. During the smoothing step the defect (= residuum) is not necessarily decreased but "smoothed". By the following correction step the discrete solution is improved by means of an auxiliary equation in a coarser grid*. It results an iterative process, that, usually, is very fast and effective.

---

* We say "coarser grid" also in cases, where "coarser finite element subspace" is meant.

A multi-grid method for Poisson's equation [(1.1) with c=0 and (1.5)] in a square is first described by Fedorenko 1961. Bachvalov 1966 discusses the variable coefficient case. The first multi-grid solution for finite element equations is due to Astrachancev 1971. Several numerical results and additional multi-grid techniques are described by Brandt 1977. Further numerical results are published by Wesseling 1977, 1980b, Hackbusch 1978, Foerster & Stüben & Trottenberg 1980.

Convergence proofs for multi-grid iterations are much more difficult than for other iterative processes as, for example, SOR or ADI (cf. Varga 1962). Such proofs for various situations are given by Fedorenko 1964, Bachvalov 1966, Astrachancev 1971, Nicolaides 1977, Wesseling 1980a, Bank & Dupont 1981. A very general proof of convergence is formulated by Hackbusch 1980a, 1981b.

We introduce the multi-grid algorithm in §2 for a very simple boundary value problem, in order to explain the behaviour of the smoothing step and the coarse-grid correction. Moreover, we restrict our considerations to the case of two grids (two-grid method). The general multi-grid iteration is defined in §3. Usually, the multi-grid iteration is initialized in a special manner. This nested iteration technique is described in §6. The multi-grid algorithm is not uniquely defined. The "smoother" applied in the smoothing step can be chosen suitably. The smoothing techniques are discussed in §4.

2. LINEAR TWO-GRID METHOD

2.1 One Dimensional Model Problem

We consider the differential equation

$$-u''(x)=f(x) \quad \text{for} \quad x\in\Omega:=(0,1) \qquad u(0)=u(1)=0 \tag{2.1}$$

The usual difference equation approximating (2.1) reads as

$$L_h\, u_h = f_h \tag{2.2}$$

with $u_h = (u_h(h), u_h(2h), \ldots, u_h(1-h))^T$, $f_h = (f(h), \ldots, f(1-h))^T$ and

$$L_h = h^{-2} \begin{bmatrix} 2 & -1 & & & 0 \\ -1 & 2 & -1 & & \\ & \ddots & \ddots & \ddots & \\ & & -1 & 2 & -1 \\ 0 & & & -1 & 2 \end{bmatrix}$$

Also the finite element method with linear elements leads to equation (2.2) with (2.3).

The Jacobi iteration (Gesamtschrittverfahren, cf. Varga 1962) is defined as

$$u_h^{i+1} = D_h^{-1}\,(f_h+B_h u_h^i) = u_h^i + \frac{h^2}{2}\,(f_h-L_h u_h^i) \tag{2.4}$$

where $L_h = D_h - B_h$, D_h diagonal part of L_h, u_h^i = i-th iterate. A very simple "smoother" is the damped Jacobi iteration. Damping by a factor $\omega = 1/2$ yields the iteration

$$u_h^{i+1} = u_h^i + \frac{1}{4} h^2 (f_h - L_h u_h^i) \tag{2.5}$$

Denote the exact discrete solution of (2.2) by u_h and the error of u_h^i by

$$e_h^i := u_h^i - u_h.$$

From (2.5 we conclude)

$$e_h^{i+1} = G_h e_h^i \qquad \text{with} \qquad G_h = I - \frac{h^2}{4} L_h. \tag{2.6}$$

The eigenfunctions of L_h and G_h are the sine-functions

$$s_{h\mu} = (\sin \mu\pi x)_{x=h,2h,\ldots,1-h} \text{ for } 1 \leq \mu \leq N := \frac{1}{h} - 1. \tag{2.7}$$

A simple calculation shows that the eigenvalues of L_h and G_h are

$$L_h s_{h\mu} = \frac{4}{h^2} \sin^2\left(\frac{\mu\pi h}{2}\right) s_{h\mu}, \; G_h s_{h\mu} = \cos^2\left(\frac{\mu\pi h}{2}\right) s_{h\mu}. \tag{2.8}$$

Since the argument $\frac{\mu\pi h}{2}$ varies in the interval $(0, \frac{\pi}{2})$ we obtain the following result :

The rate of convergence of the iteration (2.5) is

$$\cos^2 \frac{\pi h}{2} = 1 - \frac{\pi^2 h^2}{8} + O(h^4).$$

This very slow rate is caused by the slowest frequency $\mu = 1$. On the other hand we have fast convergence with respect to the high frequencies: If $\mu \approx N = h^{-1} - 1$ the eigenvalue $\cos^2(\mu\pi h/2)$ is much less than unity.

The initial error e_h^0 is a sum of all eigenfunctions $s_{h\mu}$. While the slow frequencies $\mu \approx 1$ remain almost unchanged, the high frequencies nearly vanish by few iterations. Hence, the slow frequencies dominate in the error e_h^i of the i-th iterate for $i>0$. Therefore, e_h^i is a smooth grid function (oscillations would require high sine-frequencies). We summarize :

By few iterations of the "smoother" (2.5) the error e_h^i is not decreased remarkably, but the error becomes smoother.

Assume that $\tilde{u}_h$ is an approximation to the exact solution u_h with smooth error:

$$e_h := \tilde{u}_h - u_h. \tag{2.9}$$

e_h satisfies the equation

$$L_h e_h = L_h \tilde{u}_h - f_h =: d_h. \tag{2.10}$$

d_h is called the defect of $\tilde{u}_h$. Equation (2.10) differs from the original equation (2.2) only by its right-hand side.

Since e_h is smooth, it should be possible to approximate e_h by a coarser grid function e_{2h}, solution of

$$L_{2h} e_{2h} = d_{2h} \qquad (L_{2h} \text{ defined as } L_h \text{ in } (2.3)), \tag{2.11}$$

where d_{2h} approximates d_h. For example, the restriction

$$r: d_h \to d_{2h}$$

may be defined by

$$d_{2h}(x) = (rd_h)(x) = \frac{1}{4} d_h(x-h) + \frac{1}{2} d_n(x) + \frac{1}{4} d_n(x+h) \tag{2.12}$$

for x = multiple of 2h. Hence, r is the rectangular matrix

$$r = \frac{1}{4} \begin{bmatrix} 1 & 2 & 1 & & & & & \\ & & 1 & 2 & 1 & & & \\ & & & & & \ddots & & \\ & & & & & 1 & 2 & 1 \end{bmatrix} \tag{2.13}$$

The result $e_{2h} = L_{2h}^{-1} d_{2h}$ of (2.11) is defined on the coarser grid. Therefore we have to interpolate (prolongate) in a suitable way. A simple prolongation

$$p: e_{2h} \to pe_{2h}$$

is the piece-wise linear interpolation

$$(pe_{2h})(x) = \begin{cases} e_{2h}(x) & \text{if } x \text{ multiple of } 2h, \\ \frac{1}{2}[e_{2h}(x-h) + e_{2h}(x+h)] & \text{otherwise.} \end{cases} \tag{2.14}$$

p is represented by the rectangular matrix

$$p = \frac{1}{2} \begin{bmatrix} 1 & & \\ 2 & & \\ 1 & 1 & \\ & 2 & \\ & 1 & \\ & & \ddots \end{bmatrix} \tag{2.15}$$

Now we choose pe_{2h} as approximation to e_h. Since $u_h = \tilde{u}_h - e_h$, we obtain $\hat{u}_h := \tilde{u}_h - pe_{2h}$ as a new approximation to u_h. The arguments used above show that the smallness of $\|\hat{u}_h - u_h\|$ depends on the smoothness of the error e_h. Thus, the coarse-grid correction must be preceded by a smoothing step.

The combination of the smoothing step and the coarse-grid correction yields the following two-grid iteration:

u_h^i: given i-th iterate; (2.16a)

$\tilde{u}_h$: result of ν applications of the smoothing procedure (2.5): (2.16b)

$$\tilde{u}_h^0 := u_h^i, \quad \tilde{u}_h^{j+1} := \tilde{u}_h^j + \frac{h^2}{4}(f_h - L_h \tilde{u}_h^j) \text{ for } j=0,1,\ldots,\nu-1,$$

$$\tilde{u}_h := \tilde{u}_h^\nu;$$

$$d_h := L_h \tilde{u}_h - f_h \quad \text{(calculation of the defect);} \qquad (2.16c_1)$$

$$d_{2h} := r d_h \quad \text{(restriction of the defect);} \qquad (2.16c_2)$$

$$e_{2h} := L_{2h}^{-1} d_{2h} \quad \text{(solution of coarse-grid eq.);} \qquad (2.16c_3)$$

$$u_h^{i+1} := \tilde{u}_h - p e_{2h} \quad \text{(correction of } \tilde{u}_h) \qquad (2.16c_4)$$

Equations $(2.16c_1)$ to $(2.16c_4)$ yield

$$u_h^{i+1} := \tilde{u}_h - p\, L_{2h}^{-1}\, r\, (L_h \tilde{u}_h - f_h).$$

2.2 Convergence for the Model Problem

The two-grid iteration (2.16) is of the form

$$u_h^{i+1} = M_h u_h^i + N_h f_h \qquad (2.17)$$

with

$$M_h = (I - pL_{2h}^{-1} r L_h)\, G_h^\nu \qquad (I = \text{identity}) \qquad (2.18)$$

and $N_h = (I-M_h)L_h^{-1}$. The iterates u_h^i converge to u_h, if the spectral radius of M_h is smaller than unity (cf. Varga 1962). A sufficient condition is $\|M_h\|<1$ ($\|\cdot\|$ spectral norm) as can be seen directly from

$$\|u_h^{i+1} - u_h\| \leq \|M_h\|\ \|u_h^i - u_h\|.$$

The eigenfunctions $s_{h\mu}$ and $s_{h\mu'}$ from (2.7) with $\mu' = \frac{1}{h} - \mu$ are spanning a subspace that is invariant under M_h:

$$M_h\,(\alpha s_{h\mu} + \beta s_{h\mu'}) = \gamma s_{h\mu} + \delta s_{h\mu'} \quad \text{for} \quad 1 \leqslant \mu < 1/(2h). \tag{2.19}$$

The coefficients satisfy

$$\begin{bmatrix} \gamma \\ \delta \end{bmatrix} = M_{h\mu} \begin{bmatrix} \alpha \\ \beta \end{bmatrix}, \text{ where } M_{h\mu} = \begin{bmatrix} s^2 & c^2 \\ s^2 & c^2 \end{bmatrix} \begin{bmatrix} c^2 & 0 \\ 0 & s^2 \end{bmatrix}^{\nu} \tag{2.20}$$

with

$$s^2 = \sin^2 \frac{\mu\pi h}{2}, \quad c^2 = \cos^2 \frac{\mu\pi h}{2} = 1 - s^2.$$

The first matrix $\begin{bmatrix} s^2 & c^2 \\ s^2 & c^2 \end{bmatrix}$ illustrates that the coarse-grid correction reduces in particular the lower frequencies (note that $\mu < \mu'$, $s^2 < 1/2 < c^2$). The second matrix corresponds to the ν smoothing iterations eliminating the high frequencies. From (2.19), (2.20) one concludes

$$\| M_h \| = \max_{\mu} \| M_{h\mu} \| = \sup_{0<\theta<1/2} \sqrt{2[\sigma^2(1-\sigma)^{2\nu} + (1-\sigma)^2 \sigma^{2\nu}]} =: m_\nu . \tag{2.21}$$

The estimate is independent of h. The asymptotic dependence of m_ν on ν (number of smoothing iterations in (2.16b)) is

$$m_\nu \approx \sqrt{2}/(e\nu).$$

The first values are

$$m_1 = 1/2, \; m_2 = 1/4, \; m_3 < 0.1502$$

(cf. Hackbusch 1976, 1978).

We shall see that the convergence rates can still be improved by better smoothers. But it is typical that the rates are bounded independently of h. Also the behaviour of m_ν for $\nu \to \infty$ can be proved in very general situations.

2.3 General Linear Two-grid Iteration

Let (2.2), $L_h u_h = f_h$, be the system of linear equations obtained from a general boundary value problem by a difference scheme (cf. (1.12)) or a finite element method (cf. (1.15)).

We denote the "smoother" by the symbol $\mathcal{G}_h(u_h, f_h)$. One step of the smoothing procedure is

$$v_h \to \mathcal{G}_h(v_h, f_h)$$

where

$$\mathcal{G}_h(u_h,f_h) = G_h u_h + H_h f_h, \qquad H_h L_h = I-G_h. \tag{2.22}$$

If the differential equation is a scalar elliptic equation of order 2m, a possible choice of $\mathcal{G}_h$ is the following generalization of (2.4):

$$\mathcal{G}_h(u_h,f_h) = u_h-\omega_h h^{2m}(L_h u_h-f_h), \tag{2.23}$$

where $\omega_h \approx 1/\rho(h^{2m}L_h)$, ρ: spectral radius. Other smoothers are discussed in §4.

Let $h'>h$ be a second coarser grid size. We need a coarse-grid analogue of (2.2) :

$$L_{h'} u_{h'} = f_{h'}. \tag{2.24}$$

Assume, for example, that $L_{h'}$ is obtained in the same way as L_h from (2.2). Define

V_h = linear space of grid functions of step size h.

$V_{h'}$ = linear space of grid functions of step size h'.

In the case of a finite element discretization, we use the term "grid function" instead of "coefficient vector".

Assume that we are able to define a suitable prolongation

$$p: V_{h'} \to V_h \tag{2.25}$$

and a suitable restriction

$$r: V_h \to V_{h'}. \tag{2.26}$$

Usually, the choice of p and r is no severe problem. Consider first the case of a difference scheme with h'=2h. If the elliptic equation is of second order, we may choose p as piece-wise linear interpolation*. In the two dimensional case this prolongation p acts on a unit vector of $V_{h'}$ (h'=2h) as follows:

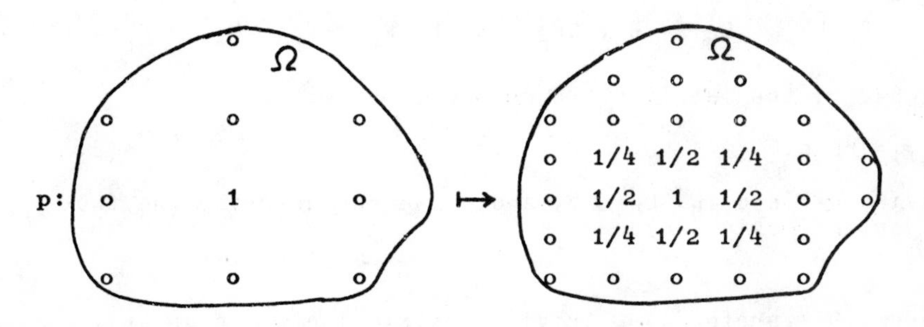

* Other prolongations are proposed by Wesseling 1980b. For higher order equations the order of interpolation must be increased.

A good choice of the restriction r is*

$$r = p^*, \tag{2.27}$$

where p^* denotes the transposed mapping with respect to the scalar products

$$\langle u_h, v_h \rangle_h = h^{-n} \sum_{x \in \Omega_h} u_h(x) v_h(x), \quad \langle u_{h'}, v_{h'} \rangle_{h'} = h'^{-n} \sum_{x \in \Omega_{h'}} u_{h'}(x) v_{h'}(x)$$

of V_h and $V_{h'}$. Equation (2.27) reads as

$$\langle r u_h, v_{h'} \rangle_{h'} = \langle u_h, p v_{h'} \rangle_h .$$

Note that p and r from (2.13) and (2.15) satisfy (2.27).

In the case of a finite element method there is a canonical choice of p and r if we have**

$$\mathcal{H}_{h'} \subset \mathcal{H}_h, \tag{2.28}$$

that means, $\mathcal{H}_h$ is a genuine refinement of $\mathcal{H}_{h'}$. Let $u_{h'} \in V_{h'}$ be the coefficient vector of the function $\tilde{u}_{h'} \in \mathcal{H}_{h'}$. Since, by (2.28), $\tilde{u}_{h'} \in \mathcal{H}_h$, it has a coefficient vector $u_h \in V_h$. We define $pu_{h'}$ by u_h. If we denote the mappings $u_{h'} \to \tilde{u}_{h'}$ and $u_h \to \tilde{u}_h$ by $P_{h'}$ and P_h, respectively, the prolongation p is uniquely defined by $P_h p = P_{h'}$:

$$\begin{array}{ccc} V_{h'} & \xrightarrow{\quad p \quad} & V_h \\ \downarrow P_{h'} & & \downarrow P_h \\ \mathcal{H}_{h'} & \xrightarrow{\text{inclusion}} & \mathcal{H}_h \end{array}$$

The canonical choice of the restriction r is given by (2.27). Note that a finite element discretization (1.15) of the boundary value problem (1.10) satisfies

$$L_h = R_h L P_h, \quad L_{h'} = R_{h'} L P_{h'} \quad \text{with} \quad R_h = P_h^*, \quad R_{h'} = P_{h'}^*. \tag{2.29}$$

Since $P_h p = P_{h'}$, the matrices are connected by

$$L_{h'} = r L_h p. \tag{2.30}$$

Having defined G_h, $L_{h'}$, p, and r, we can describe the two-grid algorithm as in §2.1.

---

* Brandt 1977 prefers the trivial restriction $ru_h = u_h|_{\Omega h'}$.

** If (2.28) is not satisfied (cf. Bank & Sherman 1981), p must be defined by interpolation as in the case of finite differences.

Two-grid iteration for solving $L_h u_h = f_h$:

$$u_h^i \quad : \text{ given the i-th iterate;} \tag{2.31a}$$

$$\tilde{u}_h \quad : \text{ result of } \nu \text{ smoothing iterations by } \mathcal{G}_h: \tag{2.31b}$$

$$\tilde{u}_h^0 := u_h^i, \quad \tilde{u}_h^{j+1} := \mathcal{G}_h(\tilde{u}_h^j, f_h) \quad \text{for } j=0,\ldots,\nu-1,$$

$$\tilde{u}_h := \tilde{u}_h^\nu;$$

$$u_h^{i+1} := \tilde{u}_h - p\, L_{h'}^{-1}\, r\, (L_h \tilde{u}_h - f_h). \tag{2.31c}$$

2.4 Convergence of the Two-Grid Iteration

As in section 2.2 we want to ensure that the iterates u_h^i of (2.31) converge to the solution u_h of (2.2). It suffices to show that the iteration matrix M_h of the two-grid iteration (cf. (2.17)) has a sufficiently small norm.

The smoothness of a vector v_h can be measured by $\|L_h v_h\|$. Therefore, the smoothing property of the smoother $\mathcal{G}_h$ with iteration matrix G_h (cf. (2.22)) can be expressed by

$$\|L_h G_h^\nu\| \leq C_0(\nu)\, h^{-2m} \qquad \text{for all } \nu \leq \nu_m(h), \tag{2.32}$$

where

$$\lim_{\nu\to\infty} C_0(\nu) = 0, \qquad \lim_{h\to 0} \nu_m(h) = \infty,$$

and 2m = order of the elliptic equation. The smoother of §2.1 satisfies (2.32) with $C_0(\nu) = 4/(\nu+1)$, $2m=2$, $\nu_m(h) = \infty$. The inequality (2.32) can also be proved for the smoother (2.23).

The approximation property

$$\|L_h^{-1} - p\, L_{h'}^{-1}\, r\| \leq C_1\, h^{2m} \tag{2.33}$$

is a natural estimate for finite element equations. The argumentation is as follows. Let $f_h \in V_h$ be arbitrary and choose f with $R_h f = f_h$. Let u, u_h, $u_{h'}$ be the solutions of $Lu=f$, $L_h u_h = f_h$, $L_{h'} u_{h'} = f_{h'}$, respectively. The "coefficient vector" u_h describes the finite element solution $u_{\mathcal{H}_h} = P_h u_h \in \mathcal{H}_h$. The latter equation defines a prolongation $P_h : V_h \to \mathcal{H}_h$. By (1.14) (2m=2) we have

$$\|u - P_h u_h\|_{L^2(\Omega)} \leq C h^{2m} \|f\|_{L^2(\Omega)}, \quad \|u - P_{h'} u_{h'}\|_{L^2(\Omega)} \leq C h^{2m} \|f\|_{L^2(\Omega)}$$

If $h'/h \leq$ const, these inequalities yield

$$\|u_h - p u_{h'}\| \leq C \|P_h u_h - P_{h'} u_{h'}\|_{L^2(\Omega)} \leq C' h^{2m} \|f\|_{L^2(\Omega)} \leq C'' h^{2m} \|f_h\|$$

Since $u_h = L_h^{-1} f_h$ and $u_{h'} = L_{h'}^{-1} r f_h$, (2.33) is proved.

THEOREM. Suppose (2.32) and (2.33). Let ν_{min} and h_{max} be numbers such that $C_0(\nu) < 1/C_1$ for $\nu \geq \nu_{min}$ and $\nu m(h) \geq \nu_{min}$ for $h \leq h_{max}$. Then the two-grid iteration converges. The iteration matrix M_h (cf. (2.17)) satisfies

$$\| M_h \| \leq C_0(\nu)\, C_1 < 1. \tag{2.34}$$

Proof: $\| M_h \| = \| (I - pL_{h'}^{-1} r L_h) G_h^{\nu} \| \leq \| L_h^{-1} - pL_{h'}^{-1} r \| \; \| L_h G_h^{\nu} \| \leq C_0(\nu) C_1$.

For precise proofs of (2.32), (2.33) and generalizations compare Hackbusch 1980a, 1980d, 1981b.

3. LINEAR MULTI-GRID METHOD

3.1 Notations

In §2 we approximated the solution of $L_h u_h = f_h$ by means of one coarser equation $L_{h'} u_{h'} = d_{h'}$ that was solved exactly (cf. (2.16c$_3$)). Now we introduce several coarser grid sizes. Suppose that h is one of the members of the sequence

$$h_0 > h_1 > h_2 > \ldots > h_{\ell-1} > h_\ell > \ldots \qquad (\ell\text{: level number}). \tag{3.1}$$

Usually, one chooses h_0 as coarse as possible and

$$h_\ell := h_0/2^\ell \qquad (\ell = 0, 1, 2, \ldots). \tag{3.2}$$

In order to avoid the notations L_{h_ℓ}, u_{h_ℓ}, etc, we replace the subscript h_ℓ by the level number ℓ:

$$L_\ell = L_{h_\ell}, \quad u_\ell = u_{h_\ell}, \quad \Omega_\ell = \Omega_{h_\ell}, \quad \text{etc.}$$

The levels ℓ and $\ell-1$ are connected by the restriction $r = r_\ell$: $V_\ell \to V_{\ell-1}$ and the prolongation $p = p_\ell$: $V_{\ell-1} \to V_\ell$. In the following we omit all subscripts of r and p.

Assume that we are interested in the solution of

$$L_\ell\, u_\ell = f_\ell, \tag{3.3}$$

where ℓ is fixed by $h_\ell = h$. The coarse-grid matrices L_k $(0 \leq k < \ell)$ obtained by the same discretization can be used for the multi-grid algorithm defined below. But another good choice of these matrices is the recursive definition by*

$$L_k = r\, L_{k+1}\, p \qquad (k = \ell-1, \ell-2, \ldots, 1, 0). \tag{3.4}$$

---

* It is very important that the sparsity pattern of L_k is almost the same as of L_ℓ (cf. Hackbusch 1978, Wesseling 1980b).

This "Galerkin approximation"* is used by Hackbusch 1976, 1977, 1978 and Wesseling 1977, 1980a, 1980b. Note that (3.4) is satisfied for the one-dimensional model problem of §2.1 (3.4) is also fulfilled for finite element equations because of (2.30)

3.2 Multi-Grid Algorithm

The auxiliary equation ($2.16c_3$) of the two-grid iteration is of the same form as the original equation. Therefore, the exact solution ($2.16c_3$) can be replaced by few iterations of the same method for the lower level. The recursive application yields the following algorithms for $\ell = 1,2,...$

Level $\ell=1$ (two-grid algorithm) : $L_\ell u_\ell = f_\ell$ can be solved by the two-grid iteration of §2 with $h=h_1$, $h'=h_0$.

Level $\ell=2$ (three-grid algorithm) : Apply the two grid iteration for the levels ℓ and $\ell-1$ ($h=h_\ell$, $h'=h_{\ell-1}$). But approximate the solution of the auxiliary equation $L_{\ell-1}v_{\ell-1}=d_{\ell-1}$ as follows: Start with $v^0_{\ell-1}:=0$ and apply γ iterations of the two-grid algorithm of level $\ell-1=1$. Set $v_{\ell-1}:=v^\gamma_{\ell-1}$.

Level $\ell=3$ (four-grid algorithm) : Apply the two-grid iterations for the levels ℓ and $\ell-1$. Approximate $v_{\ell-1}=L^{-1}_{\ell-1}d_{\ell-1}$ by iterations of the three-grid algorithm of level $\ell-1=2$ starting with $v^0_{\ell-1}=0$.

Because of the recursive definition the multi-grid iteration can be described most conveniently by an ALGOL program. The following procedure m g m (multi-grid-method) performs only one iteration of the multi-grid algorithm at level ℓ.

Multi-grid iteration for solving $L_\ell u_\ell = f_\ell$ $(\ell \geq 0)$

```
Procedure mgm (ℓ, u,f);
comment ℓ = actual level number
        u = u_ℓ^i as input,  u = u_ℓ^(i+1)  as output,
        f = f_ℓ is right-hand side of L_ℓ u_ℓ = f_ℓ ;
if ℓ = 0 then u := L_0^(-1)*f  else                      (3.5a)
begin integer j; array v,d;
   for j:=1 step 1 until ν do u :=G_ℓ(u,f);              (3.5b)
   d := r*(L_ℓ*u-f);                                    (3.5c)
   v := 0                                                (3.5d)
   for j := 1 step 1 until γ do mgm (ℓ-1,v,d);          (3.5e)
   u := u-p*v                                            (3.5f)
end multi-grid method;
```

The statement (3.5b) performs the smoothing step consisting of ν iterations of the smoother G_ℓ as in (2.26b). Vector d of (3.5c) is the restricted defect (cf. ($2.16c_{1,2}$)). The lines (3.5d,e) replace the statement $v:=L^{-1}_{\ell-1}d$ of the two-grid iteration.

---

* It is very important that the sparsity pattern of L_k is almost the same as of L_ℓ (cf. Hackbusch 1978, Wesseling 1980b)

replace the statement $v := L_{\ell-1}^{-1} d$ of the two-grid iteration. $v = v_{\ell-1}^{\gamma}$ is the result of γ iterations of the multi-grid method mgm of level $\ell-1$ starting with $v_{\ell-1}^{0} = 0$. The correction (3.5f) is the same as in (2.16c$_4$).

The algorithm (3.5) contains the parameters ν and γ. The usual values of γ are

$$\gamma = 1 \quad \text{or} \quad \gamma = 2 \tag{3.6}$$

For the choice of ν compare section 5. In a more complicated program the parameter ν and γ can be determined a-posteriori by checking the smoothness of u and the smallness of $\|v - L_{\ell-1}^{-1} d\|$, respectively

A flow chart for i iterations of the multi-grid algorithm of level ℓ is given below.

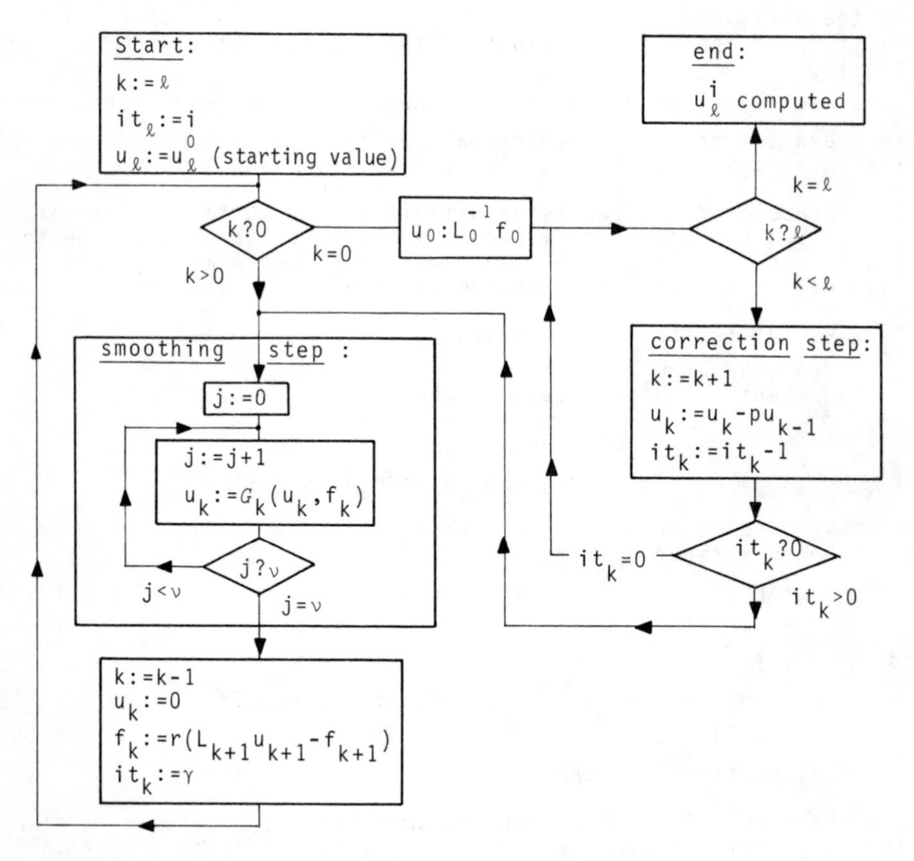

FLOW-CHART (solution of $L_\ell u_\ell = f_\ell$ by i multi-grid iterations

For readers familiar with FORTRAN but not with ALGOL we give the following (non-recursive) FORTRAN version of the multi-grid iteration. We assume that we have the following subroutines :

```
SUBROUTINE g(ℓ,u,f)                  performing:
DIMENSION u(*,*), f(*,*)
.                                    u(ℓ,*) := G_ℓ(u(ℓ,*), f(ℓ,*))
.
.

SUBROUTINE defect (ℓ,u,f)
DIMENSION u(*,*), f(*,*)             f(ℓ-1,*) := r(L_ℓ u(ℓ,*) -f(ℓ,*))
.
.
.

SUBROUTINE p(ℓ,u)
DIMENSION u(*,*)                     u(ℓ,*) :=  u(ℓ,*) -pu(ℓ-1,*)
.
.
.
```

Subroutine mgm(i,ℓ,u,f) performs i iterations of the multigrid iteration. u(ℓ,*) contains the starting value u_ℓ^0 in the beginning and the result u_ℓ^1 after execution. The right-hand side of equation $L_\ell u_\ell = f_\ell$ is about to be stored in f(ℓ,*).

```
      SUBROUTINE mgm(i,ℓ,u,f)
      DIMENSION u(*,*),f(*,*),it(*)
      k=ℓ
      it(k)=i
C     beginning of the iteration at level k
10    IF (k.EQ.1) GOTO 40
C     smoothing step
20    DO 30 j=1,ν
30    CALL g(k,u,f)
      CALL defect(k,u,f)
      k=k-1
      u(k,*)=0
      it(k)=γ
      GOTO 10
40    u(1,*)=L_1^-1*f(1,*)
50    IF (k.EQ.ℓ) RETURN
C     correction step
      k=k+1
      CALL p(k,u)
      it(k)=it(k)-1
      IF (it(k).EQ.0) GOTO 50
      GOTO 20
      END
```

According to the FORTRAN convention for indices, $\ell=1$ is the lowest level.

Finally we remark that the smoothing step and the coarse-grid correction may be interchanged. That means, the statement (3.5b) of the ALGOL program may be executed after statement (3.5f).

3.3 Computational Work

Let

$$N_\ell = \dim V_\ell$$

be the number of unknowns of equation $L_\ell u_\ell = f_\ell$. By the sparsity of the matrix L_ℓ the partial steps of the multi-grid iteration take the following number of arithmetic operations :

statement	number of operations
$u_\ell := G_\ell(u_\ell, f_\ell)$	$\leqslant C_G N_\ell$
$d_{\ell-1} := r(L_\ell u_\ell - f_\ell)$	$\leqslant C_D N_\ell$
$u_\ell := u_\ell - p * v_{\ell-1}$	$\leqslant C_C N_\ell$
$u_0 := L_0^{-1} * f_0$	$\leqslant C_0$

(3.6)

The constants C_G, C_D, C_C are independent of ℓ. Assume

$$N_{\ell-1} / N_\ell \leqslant C_H \qquad \text{for all } \ell = 1,2... \tag{3.7}$$

Usually, the choice (3.2) of the sequence $\{h_\ell\}$ implies

$$C_H = 2^{-n}$$

where n is the dimension of the problem ($\Omega \subset \mathbb{R}^n$). The following note shows that the computational work of one multi-grid iteration is proportional to the number of unknowns.

> NOTE : Assume $\eta := \gamma C_H < 1$. Then the number of operations required by one multi-grid iteration at level ℓ (i.e., by one call of mgm(ℓ,u,f), cf. (3.5)) is bounded by $C_\ell N_\ell$, where
>
> $$C_\ell \leqslant \frac{\nu C_G + C_D + C_C}{1-\eta} + C_0 \eta^\ell / \gamma$$

In the standard case of $n=\gamma=2$ and (3.2) we have $\eta = 1/2 < 1$.

We need some additional storage to treat the auxiliary equations at the levels $k=0,1,...,\ell-1$. The storage for u_ℓ and f_ℓ requires $2N_\ell$ units. Thanks to (3.7) one needs

$$\text{total storage} = 2 \sum_{k=0}^{\ell} N_k \leqslant 2N_\ell/(1-C_H) \quad \text{units.}$$

Since $C_H = 2^{-1} \ll 1$, the additional storage is only a small percentage of the total one.

3.4 Convergence of the Multi-Grid Iteration

A precise formulation and a proof of the following theorem can be found in Hackbusch 1981b.

THEOREM : Suppose (2.32), (2.33), $\gamma \geq 2$, and some technical conditions on p and $h_{\ell-1}/h_\ell$. Assume that h_0 is sufficiently small: $h_0 \leq h_{max}$, and that ν is sufficiently large: $\nu \geq \nu_{min}$. Then the multi-grid iteration converges at all levels ℓ. The iteration matrix M_ℓ satisfies

$$\|M_\ell\| \leq C \cdot C_1 \cdot C_0(\nu) < 1. \tag{3.8}$$

The first part of (3.8) is equivalent to

$$\|u_\ell^{i+1} - u_\ell\| \leq C \cdot C_1 \cdot C_0(\nu) \|u_\ell^i - u_\ell\|. \tag{3.8'}$$

The restriction $h_0 \leq h_{max}$ can be omitted if (2.32) holds with $\nu_m(h) = \infty$. Although the theorem requires $\nu \geq \nu_{min}$, the multi-grid iteration converges for most of the boundary value problems with

$$\nu = 2. \tag{3.9}$$

Experiments showed that $\gamma = 1$ yields worse convergence than $\gamma = 2$, but often $\gamma = 1$ will be preferred because of its less computational work (cf. §3.3).

In order to give an idea of the fastness of the convergence we shall report some numerical results in §5.

4. SMOOTHING TECHNIQUES

4.1 Damped Jacobi Iteration

The explanation of the smoothing step in §2.1 showed the important role of the smoother $\mathcal{G}_h$. In general, not any smoother can be used for any boundary value problem.

In §2 we defined the damped Jacobi iteration as smoother :

$$\mathcal{G}_h(u_h, f_h) = u_h - \omega_h h^{2m} (L_h u_h - f_h) \tag{4.1}$$

(cf. (2.23)). The iteration matrix of $\mathcal{G}_h$ is $G_h = I - \omega_h h^{2m} L_h$. As in the case of (2.9) one proves :

NOTE : Let $\mathcal{G}_h$ be defined by (4.1). If L_h is symmetric and positive definite and bounded by $\|L_h\| \leq C_L h^{-2m}$, and if ω_h is chosen so that $\omega_h h^{2m} \|L_h\| \in [\eta, 1]$, $\eta > 0$, then the smoothing property (2.32) holds with

$$C_0(\nu) = 1/[\omega_h \cdot (\nu+1)] \leq C_L/[\eta(\nu+1)]$$

for all h. Hence, $\nu_m(h) = \infty$ can be chosen.

By the last remark, the multi-grid iteration works without restriction on the step size. On the other hand we have:

> If L_h is symmetric and indefinite (i.e., if L_h has positive and negative eigenvalues), the inequality (2.32) cannot be valid for all ν (i.e., with $\nu_m(h)=\infty$).

Therefore, in the latter case we have a restriction $h \leqslant h_{max}$. The numerical results of §5.2 reflect this fact.

Also for nonsymmetric matrices L_h one can prove the smoothing property (2.32) (with $\nu_m<\infty$), if the underlying scalar differential equation is elliptic (cf. Hackbusch 1981b).

An improvement of (4.1) is the following choice of G_h. Define

$$G_h(\omega,u_h f_h) = u_h - \omega h^{2m}(L_h u_h - f_h) \qquad (\omega \text{ real}) \tag{4.2}$$

and denote the iteration matrix of (4.2) by

$$G_h(\omega) = I - \omega h^{2m} L_h \tag{4.3}$$

Choosing a suitable sequence

$$\omega_1, \omega_2, \ldots, \omega_\nu, \tag{4.4}$$

we can replace the smoothing step (2.31b) or (3.5b) by

$$\text{for } j := 1 \text{ step } 1 \text{ until } \nu \text{ do } \quad u := G_\ell(\omega_j, u, f). \tag{4.5}$$

For instance, the parameters ω_j are to be chosen so that

$$\| L_h G_h(\omega_\nu) G_h(\omega_{\nu-1}) \cdot \ldots \cdot G_h(\omega_1) \| = \text{minimum}. \tag{4.6}$$

4.2 Relaxation

Usually, the relaxation method (Gauss-Seidel iteration) is a more effective smoother than Jacobi's iteration. It is defined by

$$G_h(u_h, f_h) = (D_h - R_h)^{-1}(S_h u_h + f_h), \tag{4.7}$$

where

$$L_h = D_h - R_h - S_h \qquad \text{with} \tag{4.8}$$

- D_h (block-) diagonal matrix,
- R_h strictly lower triangular matrix,
- S_h strictly upper triangular matrix.

Note that the splitting of L_h into D_h, R_h, S_h depends on the ordering of the grid points $x^i \in \Omega_h$. For a two dimensional grid there are the following possibilities :

lexicographic ordering: $x_2^{i+1} > x_2^i$ or

$$(x_2^{i+1} = x_2^i \text{ and } x_1^{i+1} > x_1^i) \quad \text{for } x^i = (x_1^i, x_2^i) \in \Omega_h; \tag{4.9a}$$

chequer-board ordering (red-black ordering):

$$x^i \ (1 \leq i \leq N_{red}) \text{ if and only if } (x_1^i + x_2^i)/h \text{ even}^*; \tag{4.9b}$$

four-coloured ordering:

$$x^i \ (1 \leq i \leq N_1) \ : \ x_1^i/h \text{ odd}, \ x_2^i/h \text{ even},$$

$$x^i \ (N_1 < i \leq N_2) \ : \ x_1^i/h \text{ even}, \ x_2^i/h \text{ odd},$$

$$x^i \ (N_2 < i \leq N_3) \ : \ x_1^i/h \text{ and } x_2^i/h \text{ odd},$$

$$x^i \ (N_3 < i \leq N) \ : \ x_1^i/h \text{ and } x_2^i/h \text{ even}^*. \tag{4.9c}$$

The lexicographic ordering (4.9a) admits a simple local Fourier analysis**. Chequer-board ordering (4.9b) is a good choice in the case of a five-point scheme [e.g. for (1.11)], while the third ordering (4.9c) should be used for nine-point schemes. In the latter cases (4.9b,c) defines a 2-cyclic splitting that satisfies the smoothing property (2.32) (cf. Hackbusch 1980a).

It must be emphasized that computational work can be saved by using the ordering (4.9b,c). Consider the case of a five-point scheme. After the smoothing step the defect $L_h u_h - f_h$ vanishes at all "black" points x^i $(i > N_{red})$. Therefore, the calculation of the restricted defect $r(L_h u_h - f_h)$ requires the computation of $(L_h u_h - f_h)(x^i)$ only at the "red" points. A similar argument holds for (4.9c) in the case of a nine-point formula.

4.3 Anisotropic problems

Consider the model problem

$$a\, u_{x_1 x_1} + b\, u_{x_2 x_2} = f \quad \text{in } \Omega, \quad 0 < a \ll b, \tag{4.10}$$

with some boundary conditions. In this case Jacobi's iteration of §4.1 and all relaxation methods mentioned in §4.2 are bad smoothers. The reason is that errors of high frequencies like $\sin \nu x_1 \ \sin \mu x_2$

---

* Ordering of points with same "colour" is arbitrary.

** Local Fourier analysis gives a good insight into the behaviour of the various parts of the multi-grid algorithm. This technique named "local mode analysis" is described in detail by Brandt 1977.

(ν large, μ small) are poorly reduced. This is an immediate result of the local mode analysis (see footnote preceding page**).

Nevertheless, block-relaxation methods can be used successfully. In the case of $a \ll b$ one must apply x_2-line relaxation (for $a \gg b > 0$ x_1-line-relaxation). That means, the points $x^i = (x_1^i, x_2^i) \in \Omega_h$ with same x_1^i belong to one block. The matrix D_h of the splitting (4.8) is block diagonal. Each block is tridiagonal if L_h is a nine-point formula. Again, we can choose different orderings of the blocks:

<u>lexicographic</u> <u>ordering</u>: x_1-component increases; (4.11a)

<u>red</u>-<u>black</u> <u>ordering</u>: first blocks with "x_1/h odd", then blocks with "x_1/h even". (4.11b)

Ordering (4.11b) is recommended for five-or nine-point schemes.

If it is not known in before, whether $a \ll b$ or $b \ll a$, one can use a combination of x_1- and x_2-line relaxation. However, a better smoother seems to be the incomplete LU-decomposition.

4.4 Incomplete LU-Decomposition

Recently, very good results are obtained by an incomplete LU-decomposition for smoothing (cf. Wesseling 1980b, Wesseling & Sonnefeld 1980, Hemker 1980a, Mol 1980)) The exact LU-decomposition

$$L_h = R_h S_h \quad (R_h \text{ lower, } S_h \text{ upper triangular matrix})$$

cannot be used (for $n \geq 2$), since R_h and S_h have not the same sparsity pattern as L_h. However, we can obtain

$$L_h = R_h S_h - C_h \quad (R_h \text{ lower, } S_h \text{ upper triangular matrix}) \tag{4.12}$$

with

$$L_{h,ij} = 0 \text{ implies } R_{h,ij} = S_{h,i,} = 0, \tag{4.13}$$

where i,j run from 1 to N = number of grid points of Ω_h.

The following program performs the incomplete LU-decomposition satisfying (4.12) and (4.13). After the execution the triangular matrices R_h and S_h are both stored on the matrix $L = L_h$. The diagonal elements are identical: $R_{h,ii} = S_{h,ii}$.

```
for k := 1 step 1 until N do
begin L_kk := sqrt(L_kk);
    for j ∈ {j>k  and  L_kj=0}  do  L_kj := L_kj/L_kk;
    for i ∈ {i>k  and  L_ik=0}  do  L_ik := L_ik/L_kk
    for (i,j)∈{(i,j)  with  L_ij≠0,  i>k, L_ik≠0,  j>k, L_kj≠0} do
        L_ij := L_ij-L_ik L_kj
end incomplete LU-decomposition:
```

The results of Wesseling 1980b show that smoothing by incomplete LU-decomposition, namely

$$G_h(u_h,f_h) = S_h^{-1} R_h^{-1} (f_h+C_h u_h), \tag{4.14}$$

is well-suited not only for usual problems but also for anisotropic problems. The rate of convergence depends on the ordering of the grid points, but in any case one obtains very good results.

5. NUMERICAL RESULTS

5.1 Helmholtz' Equation

The considerations of §2 and §3 showed that the rate of convergence of the multi-grid iteration is bounded by some $C(\nu)$, where $C(\nu)$ does not depend on the grid size h, but tends to zero for $\nu\to\infty$. The following examples shall show that, usually, good convergence can be obtained for small ν (here $\nu = 2$).

We consider Helmholtz' equation (1.1):

$$-\Delta u + c^2 u = -4+c^2(x^2+y^2) \qquad \text{in } \Omega \tag{5.1}$$

with Dirichlet boundary values

$$u = x^2+y^2 \qquad \text{on } \Gamma. \tag{5.2}$$

The solution is $u(x,y) = x^2+y^2$. Note that the smoothness of the solution does not influence the rate of convergence.

The boundary value problem (5.1), (5.2) is discretized by the five-point formula with Shortley-Welier's modification at points near the boundary.

The details of the used multi-grid iteration are the following:

- prolongation p: piece-wise linear interpolation*
- restriction r : $r = p^*$, cf. (2.27)
- coarser grid matrices L_k defined by (3.4)
- smoother G_h: point-wise or line-relaxation with ordering (4.9c) or (4.11b), respectively
- number of smoothing iterations** $\nu := 2$
- $\gamma = 1$

The FORTRAN program is contained in the report of Hackbusch 1976.

---

* modification for $c^2\neq 0$: compare Hackbusch 1978.

** more precisely: $\nu=1+3/4$ (point-wise), $\nu=3/2$ (line wise relaxation) steps of the smoother after coarse grid correction, $\nu=1/4$ ($\nu=1/2$, resp.) before correction. Note that smoothing step and coarse-grid correction are interchanged (cf. end of §3.2). Compare Hackbusch 1978.

First we give some results for Poisson's equation [$c^2=0$ in (5.1)] in a circle of radius 1/2.

μ	1	2	3	4	5	6	7
error $\Vert u^{\mu}_{1/64}-u_{1/64}\Vert$	$6.8_{10}-1$	$4.0_{10}-2$	$3.4_{10}-4$	$1.5_{10}-5$	$2.4_{10}-7$	$1.0_{10}-8$	$3.5_{10}-10$
ratio		0.060	0.008	0.045	0.016	0.043	0.034

Table 1 : Equation (5.1), $c^2=0$, Ω=circle, h=1/64, point-wise relax.

Table 1 shows the errors of seven iterations of the multi-grid algorithm solving the difference equations for the grid size h=1/64. The sequence (3.1) of the step sizes is

$$h_0 = \frac{1}{2},\ h_1 = \frac{1}{4},\ h_2 = \frac{1}{8},\ h_3 = \frac{1}{16},\ h_4 = \frac{1}{32},\ h_5 = h = \frac{1}{64}.$$

The averaged rate of convergence is

$$\rho_{1/64} = \left[\Vert u^7_{1/64}-u_{1/64}\Vert / \Vert u^3_{1/64}-u_{1/64}\Vert\right]^{1/4} = 0.032.$$

The computational work of one multi-grid iteration (cf. §3.3) corresponds to nearly 5 relaxation iterations. Thus, the rate corresponding to the work of one relaxation iteration is about 1/2.

Table 2 confirms the fact that the rates are bounded independently of the grid size.

1/h	8	12	16	24	32	48	64	96	128
rate	.014	.014	.019	.018	.021	.023	.023	.025	.027

Table 2 : rates of convergence* (Eq. (5.1), $c^2=0$, Ω=circle, point-wise relaxation).

The results of Hackbusch 1978 show that the rate of convergence depends on the shape of the domain (more precisely : on the regularity of the boundary value problem). The rate deteriorates in the case of a re-entrant boundary. In the latter case line-relaxation is superior to point-wise relaxation. As example we compare the results for a circle with those for a cracked region (worst case).

---

* These rates are obtained for the Mehrstellenverfahren (special nine-point formula of fourth order, cf. Collatz 1966).

domain	smoothing by point-wise relaxation	smoothing by line-wise relaxation
circle :	0.030 to 0.035	0.030 to 0.034
cracked circle :	0.113 to 0.147	0.053 to 0.079

Table 3 : rates of convergence (Eq. (5.1), $c^2=0$)

The rates depend only very weakly on the parameter c^2 of Helmholtz' equation (5.1). Even for negative c^2 good results can be obtained.

c^2	-20^{**}	-15	-10	-5	0	10	100	10000
rate	0.26	.038	.032	.031	.030	.029	.027	.025

Table 4 : rates of convergence (Eq. (5.1), $h=1/64$, Ω=circle, point-wise relaxation)

If we replace the boundary condition (5.2) by other conditions we obtain similar results. Table 5 shows the rates of convergence for various boundary conditions in the case of Poisson's equation.

prescribed boundary values	u (top), u (left), u (right), u (bottom)	$\partial u/\partial n$ (top), $\frac{\partial u}{\partial n}$ (left), $\frac{\partial u}{\partial n}$ (right), $\partial u/\partial n$ (bottom)	$u+\frac{\partial u}{\partial n}$ (left), u (right)	per. (top), per. (left), per. (right), per. (bottom)
rate of convergence	0.038	0.023	0.038	0.032

Table 5 : rates of convergence (Eq. (5.1), $c^2=0$, $h=1/64$, Ω=unit square, point-wise relaxation)

In the case of Neumann or periodic conditions the solution is determined up to a constant. Nevertheless, the multi-grid iteration** works as well as for other boundary conditions.

---

* The first eigenvalue of (5.1) for this Ω is $c^2 \approx -23.14$.

** The equation $L_0 u_0 = f_0$ at the lowest level $\ell=0$ must be completed by an additional equation (e.g. $u_0(P)=0$ for some $P \in \Omega_0$).

An ALGOL and FORTRAN program for the multi-grid solution of the general linear second order equation (1.2) in a rectangle subject to the boundary conditions mentioned above is contained in the report of Hackbusch 1977.

5.2 Indefinite Problems

The convergence rates for boundary value problems with variable coefficients are very similar to the results of §5.1. More interesting examples are indefinite boundary value problems (i.e., problems with positive and negative eigenvalues). The equation

$$a_{11}u_{x_1x_1}+a_{22}u_{x_2x_2}+a_2u_{x_2}+(\lambda+a)u = f \quad \text{in} \quad \Omega=(0,1)\times(0,1)$$
$$a_{11}=-e^{x_1}, \quad a_{22}=-e^{-x_2}, \quad a_2=\sin\pi x_1, \quad a=\pi^2(a_{11}+a_{22})-\pi\cos\pi x_1, \quad f=1 \tag{5.3}$$

with

$$u = 0 \quad \text{on} \quad \Gamma \tag{5.4}$$

has the first eigenvalue $\lambda=0$. Thus, the problem (5.3), (5.4) is indefinite for $\lambda<0$.

Note that the smoothers of §4.1 and §4.2 are divergent for indefinite problems. Therefore, the smoothing property (2.32) cannot hold for all ν (i.e., for $\nu_m(h)=\infty$). The coarsest grid size h_0 must be sufficiently small (cf. §3.4). The following table 6 contains the rates of convergence of the multi-grid iteration for the difference scheme with grid size h=1/64 under varying values of λ. We used the two different sequences

$$h_0 = 1/4, \quad h_1 = 1/8, \quad h_2 = 1/16, \quad h_3 = 1/32, \quad h_4 = 1/64, \tag{5.5a}$$

$$h_0 = 1/8, \quad h_1 = 1/16, \quad h_2 = 1/32, \quad h_4 = 1/64 \tag{5.5b}$$

λ	5	2	1	-1	-2	-3	-5	-8	-10	-15
sequence (5.5a)	.12	.37	.64	(3.7)	.95	.55	.29	.15	.21	.55
sequence (5.5b)	.042	.14	.32	.23	.12	.083	.049	.078	.11	.34

Table 6 : rates of convergence (Eq. (5.3), (5.4), h=1/64)

6. NESTED ITERATION

The results of §5 are obtained by starting the multi-grid iteration with $u_h^0 = 0$. In this case the error $e_h^0 = u_h^0 - u_h$ is of size $O(1)^*$. We need $\mu = \text{const} \cdot \log(1/\varepsilon)$ multi-grid iterations to obtain accuracy $\varepsilon : \| e_h^\mu \| \leq \varepsilon$. Since one iteration requires $O(N)^*$ operations we obtain*:

$$\| u_h^\mu - u_h \| \leq \varepsilon \quad \text{takes} \quad \leq C \cdot N \cdot \log \frac{1}{\varepsilon} \text{ operations,} \tag{6.1}$$

where N = number of grid points (cf. §3.3). The constant C does not depend on h and ε.

The difference between the exact discrete solution u_h and and the continuous solution u is proportional to h^κ (κ: consistency order). Usually, the iteration is stopped as soon as $\varepsilon = O(h^\kappa)$, since more iterations cannot improve the difference $\| u_h^\mu - u \|$. In this case (6.1) becomes

$$\| u_h^\mu - u_h \| = O(h^\kappa) \quad \text{takes} \quad O(N \cdot \log \frac{1}{h}) \text{ operations.} \tag{6.2}$$

However, there is a better strategy for computing more accurate starting values. This "nested iteration" technique can be applied for any iterative process (cf. Kronsjö 1975). It can be combined very easily with the multi-grid algorithm. Let $L_\ell u_\ell = f_\ell$ ($\ell = 0, 1, \ldots, \ell_{max}$) be the discretizations of the boundary value problem Lu=f for the grid sizes h_ℓ. The program** is as follows :

$$
\begin{aligned}
&\tilde{u}_0 := \text{approximation of } L_0^{-1} f_0; \\
&\textbf{for } \ell := 1 \textbf{ step } 1 \textbf{ until } \ell_{max} \textbf{ do} \\
&\textbf{begin } \tilde{u}_\ell := \tilde{p} * \tilde{u}_{\ell-1}; \\
&\qquad \textbf{for } j := 1 \textbf{ step } 1 \textbf{ until } i \textbf{ do } mgm(\ell, \tilde{u}_\ell, f_\ell) \\
&\textbf{end} \quad \text{nested iteration;}
\end{aligned}
\tag{6.3}
$$

Here, $\tilde{p}$ denotes a prolongation $V_{\ell-1} \to V_\ell$. $\tilde{p}$ may coincide with the prolongation p from the multi-grid method mgm, but $\tilde{p}$ may also be an interpolation of higher consistency order***. mgm is the procedure (3.5). The program (6.3) takes

---

* Symbol $O(\phi(x))$ with some function $\phi(x)$ means $\leq \text{const.} \phi(x)$

** This technique is called "full multi-grid method" by Brandt 1977

*** At least of order κ

$$i\cdot\sum_{\ell} O(N_\ell) = O(i\cdot N_{\ell_{max}}) \qquad (N_\ell \text{ from §3.3})$$

operations. One can prove that even for small i (usually, i=1 or 2) the error $\|\tilde{u}_\ell - u_\ell\|$ is of order $O(h^\kappa)$. Therefore, we obtain

$$\|\tilde{u}_\ell - u_\ell\| \leqslant O(h_\ell^\kappa) \quad \text{takes } O(N_\ell) \text{ operations} \tag{6.4}$$

instead of (6.2). N_ℓ is the number of grid points of step size h_ℓ. For a proof compare Hackbusch 1981b.

We remark that in the case of Poisson's equation in a square, the algorithm (6.3) is even faster than direct Poisson solvers (e.g. of Buneman 1969, Schröder & Trottenberg 1973).

7. FURTHER MODIFICATIONS

7.1 Extrapolation

The nested iteration technique of §6 yields not only an approximation to u_ℓ, but also approximations to $u_{\ell-1}$, $u_{\ell-2}$, etc. In some cases discretizations admit an expansion

$$u_\ell = u + h_\ell^\kappa e_\kappa + O(h_\ell^\lambda), \quad e_\kappa \text{ independent of } h_\ell, \quad \lambda > \kappa, \tag{7.1}$$

of the global error. Under assumption (7.1), the extrapolated value

$$u_{\ell-1}^{extrap} := \alpha u_{\ell|\Omega_{\ell-1}} + \beta u_{\ell-1}, \quad \beta = 1/\left[1 - \left(\frac{h_{\ell-1}}{h_\ell}\right)^\kappa\right], \quad \alpha = 1-\beta, \tag{7.2}$$

is of accuracy $O(h_\ell^\lambda)$. Applying extrapolation (7.2) to the results $\tilde{u}_\ell$ and $\tilde{u}_{\ell-1}$ of the nested iteration, we can obtain an $O(h_\ell^\lambda)$-approximation to the continuous solution u, provided that $\tilde{u}_\ell$ and $\tilde{u}_{\ell-1}$ are computed up to an iteration error $O(h_\ell^\lambda)$.

The extrapolation (7.2) is equivalent to the solution of a discrete problem $L_{\ell-1}u_{\ell-1} = f_{\ell-1}^{extrap}$ with an extrapolated right-hand side $f_{\ell-1}^{extrap}$. The construction of $f_{\ell-1}^{extrap}$ and the multi-grid solution of the corresponding equation is called "τ-extrapolation" by Brandt 1979 (p. 67).

7.2 Multi-Grid Algorithm with Defect Correction

Even if an expansion (7.2) of the global error does not exist, we can improve the accuracy of the discrete solution u_ℓ by using a defect correction technique (cf. Brandt 1979 (p. 97), Hackbusch 1979c).

Let $h=h_{\ell_{max}}$ denote the finest step size and assume that we have two discretizations of the boundary value problem at the level ℓ_{max} :

$$L_\ell u_\ell = f_\ell \qquad (0 \leq \ell \leq \ell_{max}) \tag{7.3a}$$

is a discretization of consistency order κ. (7.3a) is used for the coarser grids, too. In addition, there must be a more accurate discretization

$$L'_{\ell_{max}} u'_{\ell_{max}} = f'_{\ell_{max}} \tag{7.3b}$$

of consistency order $\kappa' > \kappa$. For instance, in the case of Helmholtz' equation the first discretization (7.3a) may be the five point scheme, while (7.3b) is the Mehrstellenformel* of Collatz 1966. In this case we have $\kappa=2$, $\kappa'=4$. The formula (7.3b) may also contain complicated discretizations at points near the boundary. Here, it is important that the scheme L'_ℓ may be unstable or even not invertible. L_ℓ should be a very sparse matrix, whereas multiplication by L'_ℓ may require more computational work.

Now we modify the multi-grid iteration as follows :

- smoothing step : use always L_ℓ $(0 \leq \ell \leq \ell_{max})$ and f_ℓ,
- coarse-grid correction :

$$\text{use the defect } \left\{ \begin{matrix} L_\ell u_\ell - f_\ell \\ L'_\ell u_\ell - f'_\ell \end{matrix} \right\} \text{ if } \left\{ \begin{matrix} \ell < \ell_{max} \\ \ell = \ell_{max} \end{matrix} \right\}$$

This modified multi-grid iteration converges to a discrete grid function $u^*_{\ell_{max}}$, that is neither the solution of (7.3a) nor of (7.3b). But it approximates the continuous solution u better than the solution $u_{\ell_{max}}$ of (7.3a):

$$u^*_{\ell_{max}} = u + O\left(h_{\ell_{max}}^{\min(2\kappa,\kappa')}\right).$$

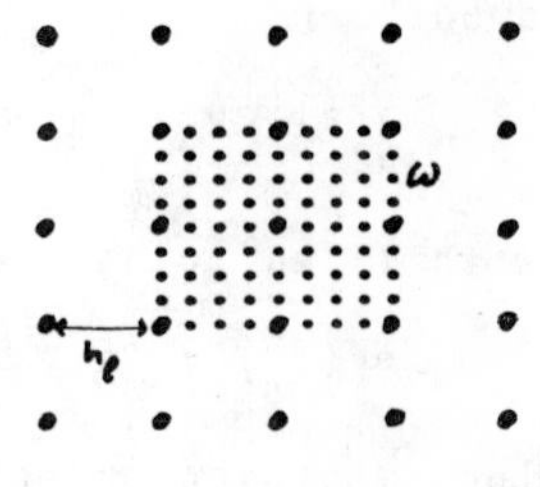

Proofs are given by Hackbusch 1981d.

The finer discretization (7.3) may also be obtained by local refinement of the grid Ω_ℓ. In a small part $\omega \subset \Omega$ of the grid the right-hand side f_ℓ of $L_\ell u_\ell = f_\ell$ is replaced by f'_ℓ, where $f'_\ell = L_\ell u'_\ell$ in ω, u'_ℓ = restriction of an accurate solution u' approximated in a grid of ω with size smaller than h_ℓ. This technique is described by Brandt 1977.

---

* Compare footnote on page 68.

8. NONLINEAR MULTI-GRID ITERATION

8.1 Nonlinear Problem and Notations

A nonlinear boundary value problem as (1.4) is denoted by

$$\mathcal{L}(u) = 0 \tag{8.1}$$

Discretizing (8.1) by a difference scheme or a finite element method we obtain a system

$$\mathcal{L}_h(u_h) = 0 \tag{8.2}$$

on nonlinear equations. Referring to a sequence (3.1) of step sizes $h_0>h_1>...>h_\ell>...$ we also write.

$$\mathcal{L}_\ell(u_\ell) = 0 \tag{8.3}$$

for (8.2) with $h = h_\ell$.

There may be more than one solution of the continuous and discrete problems (8.1), (8.3). In the following we fix one solution u^* of (8.1) and the related solution u_ℓ^* of (8.3).

In order to define the nonlinear multi-grid iteration, the system (8.3) must be generalized to

$$\mathcal{L}_\ell(u_\ell) = f_\ell, \quad \|f_\ell\| \leq \rho \qquad (\rho \text{ small}). \tag{8.4}$$

If u_ℓ^* is an isolated solution of (8.3), the inverse mapping theorem ensures the existence of a unique solution u_ℓ of (8.4) in the neighbourhood of u_ℓ^*, provided that $\rho>0$ is small enough. Formally, we write

$$u_\ell = \phi_\ell(f_\ell) \quad \text{solution of (8.4)}, \quad u_\ell^* = \phi_\ell(0). \tag{8.5}$$

8.2 Nonlinear Two-Grid Iteration

Also for the nonlinear multi-grid iteration we need a smoother

$$u_\ell^{k+1} = \mathcal{G}_\ell\left(u_\ell^k, f_\ell\right) \tag{8.6}$$

with fixed point $u_\ell = \phi_\ell(f_\ell)$:

$$\phi_\ell(f_\ell) = \mathcal{G}_\ell(\phi_\ell(f_\ell), f_\ell) \quad \text{or} \quad u_\ell = \mathcal{G}_\ell\left(u_\ell, \mathcal{L}_\ell(u_\ell)\right).$$

The smoothing property of $\mathcal{G}_\ell$ can be expressed by (2.32) with G_ℓ and L_ℓ being the derivatives

$$G_\ell = \partial\mathcal{G}_\ell(u_\ell^*, 0)/\partial u_\ell, \quad L_\ell = \partial\mathcal{L}_\ell(u_\ell^*)/\partial u_\ell \qquad \left(u_\ell^* = \phi_\ell(0)\right). \tag{8.7}$$

The nonlinear analogue of Jacobi's iteration (4.1) is

$$G_\ell(u_\ell,f_\ell) = u_\ell-\omega_\ell h_\ell^{2m}\left(\mathcal{L}_\ell(u_\ell)-f_\ell\right) \tag{8.8}$$

yielding $G_\ell = I-\omega_\ell h_\ell^{2m} L_\ell$. Smoothing by relaxation seems to be more complicated. But in many cases relaxation is as simple as in the linear case. Often, we have that the function Λ_i,

$$\left(\mathcal{L}_\ell(u_\ell)\right)(x^i) = \Lambda_i\left(u_\ell(x^i),\ldots\right) \qquad (x^i \text{ grid point, } 1\leq i\leq N) \tag{8.9}$$

is linear in its first argument. An example is the difference analogue of the nonlinear differential equation (1.4) if the first derivatives are discretized by central differences. Thus relaxation can be defined as in the linear case. But even if Λ_i is nonlinear with respect to the first argument, we can define a simple Newton-relaxation. Let $x^1,\ldots,x^{N_\ell}$ be an ordering of the grid points of Ω_ℓ and denote the derivative of Λ_i by

$$L_{ii}(u_\ell) = \partial\Lambda_i(u_\ell)/\partial u_\ell(x^i).$$

For given u_ℓ^k we define $u_\ell^{k+1} = G_\ell(u_\ell^k,f_\ell)$ by

$u := u_\ell^k;$

for $i := 1$ step 1 until N_ℓ do

$u(x^i) := u(x^i) + \left[f_\ell(x^i)-\left(\mathcal{L}_\ell(u)\right)(x^i)\right]/L_{ii}(u);$

$u_\ell^{k+1} := u;$ (8.10)

The coarse-grid correction is derived from the following argumentation. Let $\tilde{u}_{\ell-1}$ be $u_{\ell-1}^* = \phi_{\ell-1}(0)$ or an approximation to be $u_{\ell-1}^*$. Setting

$$\tilde{f}_{\ell-1} := \mathcal{L}_{\ell-1}(\tilde{u}_{\ell-1}) \tag{8.11}$$

we know that*

$$\tilde{u}_{\ell-1} = \phi_{\ell-1}(\tilde{f}_{\ell-1}) \tag{8.12}$$

---

* Here we assume $\|\tilde{f}_{\ell-1}\|\leq\rho$ (cf. (8.4)), that means, $\tilde{u}_{\ell-1}$ is suffiently close to $u_{\ell-1}^*$.

Given an arbitrary approximation $\hat{u}_\ell$ to the solution $u_\ell=\phi_\ell(f_\ell)$ of $\mathcal{L}_\ell(u_\ell)=f_\ell$, we are looking for a correction $v_\ell=\hat{u}_\ell-u_\ell$. By means of the defect $d_\ell := \mathcal{L}_\ell(\hat{u}_\ell)-f_\ell$, we obtain

$$v_\ell = \phi_\ell(f_\ell+d_\ell)-\phi_\ell(f_\ell) \approx \phi_\ell' d_\ell \qquad (\phi_\ell' \text{ derivative}^*). \tag{8.13}$$

If we have a solution $u_{\ell-1} = \phi_{\ell-1}(\tilde{f}_{\ell-1}+\sigma r d_\ell)$ of

$$\mathcal{L}_{\ell-1}(u_{\ell-1}) = \tilde{f}_{\ell-1} + \sigma r d_\ell \qquad (\sigma\neq 0 \text{ real number}),$$

we can compute

$$u_{\ell-1}-\tilde{u}_{\ell-1} = \phi_{\ell-1}(\tilde{f}_{\ell-1}+\sigma r d_\ell) - \phi_{\ell-1}(\tilde{f}_{\ell-1}) \approx \phi_{\ell-1}' \sigma r d_\ell. \tag{8.14}$$

Comparing the right-hand sides of (8.13) and (8.14), and noting that $\phi_\ell' d_\ell \approx p\phi_{\ell-1}' r d_\ell$ for smooth d_ℓ, we are led to

$$v_\ell \approx \frac{1}{\sigma} p(u_{\ell-1}-\tilde{u}_{\ell-1}) \quad \text{or} \quad u_\ell \approx \hat{u}_\ell - p(u_{\ell-1}-\tilde{u}_{\ell-1})/\sigma.$$

The complete two-grid iteration reads as follows :

u_ℓ^i given i-th iterate;

$\tilde{u}_{\ell-1}$ fixed value;

$\tilde{f}_{\ell-1} := \mathcal{L}_{\ell-1}(\tilde{u}_{\ell-1})$ satisfies $\|\tilde{f}_{\ell-1}\| \leqslant \rho/2$;

$\hat{u}_\ell$: result of ν iterations of the smoother G_ℓ (as in (2.31b)); (8.15a)

$d_{\ell-1} := r(\mathcal{L}_\ell(\hat{u}_\ell)-f_\ell)$; (8.15b)

σ such that $\|\sigma d_{\ell-1}\| \leqslant \rho/2$; (8.15c)

$u_{\ell-1}$ solution of $\mathcal{L}_{\ell-1}(u_{\ell-1}) = \tilde{f}_{\ell-1}+\sigma d_{\ell-1}$; (8.15d)

$u_\ell^{i+1} := \hat{u}_\ell - p(u_{\ell-1}-\tilde{u}_{\ell-1})/\sigma$. (8.15e)

Note that the choice of σ ensures $\|\tilde{f}_{\ell-1}+\sigma d_{\ell-1}\| \leqslant \rho$ and thereupon the solvability of step (8.15d). The algorithm of Brandt 1977 corresponds to (8.15) with $\tilde{u}_{\ell-1} := r\hat{u}_\ell$, $\sigma = -1$.

Algorithm (8.15) is a genuine generalization of the linear two-grid iteration (2.31) in the following sense. If the problem (8.3) is affine, that means $\mathcal{L}_\ell(u_\ell)=L_\ell u_\ell - f_\ell$, then (8.15) is algebraically equivalent to (2.31) for any choice of $\tilde{u}_{\ell-1}$ and $\sigma\neq 0$.

---

* $\phi_\ell' = L_\ell^{-1}$ with L_ℓ from (8.7)

8.3 Nonlinear Multi-Grid Iteration

The two-grid iteration requires the solution $u_{\ell-1} = \phi_{\ell-1}(\tilde{f}_{\ell-1}+\sigma d_{\ell-1})$ of a coarse-grid equation in step (8.15d). Thanks to the generalization (8.4) of the original problem (8.3), the auxiliary equation (8.15d) is of the same form as equation (8.4). Hence, we can apply the method recursively, obtaining the multi-grid iteration. Its ALGOL version nmgm (nonlinear multi-grid method) is as follows :

Nonlinear multi-grid iteration solving $\mathcal{L}_\ell(u_\ell) = f_\ell$

$\tilde{u}_k$ $(0 \leq k \leq \ell-1)$ are given (fixed) grid functions of level k;

$\tilde{f}_k := \mathcal{L}_k(\tilde{u}_k) \qquad (0 \leq k \leq \ell-1)$

```
procedure nmgm (ℓ,u,f);
comment ℓ : actual level number,
                  i                  i+1
          u=u_ℓ as input,   u=u_ℓ    as output,
          f=f_ℓ right-hand side of equation (8.4)
if ℓ = 0 then u := approximation of φ0(f) else              (8.16a)
begin integer j; real s; array v,d;
   for j:=1 step 1 until ν do u := G_ℓ(u,f);                (8.16b)
   d := r*(L_ℓ(u)-f);                                       (8.16c)
   if d=0 then go to end of iteration;                      (8.16d)
   s := ρ/(2*‖d‖);                                          (8.16e)
   d := f~_ℓ-1+s*d;                                         (8.16f)
   v := u~_ℓ-1;                                             (8.16g)
   for j:=1 step 1 until γ do nmgm (ℓ-1,v,d);               (8.16h)
   u := u-p*(v-u~_ℓ-1)/s                                    (8.16i)
end of iteration:
end nonlinear multi-grid method;
```

As in the linear case we need some other method for solving the nonlinear system (8.4) at the coarsest grid in step (8.16a). Often, one can use few iterations of the smoother G_0.

The exact solution of the coarse-grid equation $\mathcal{L}_{\ell-1}(v)=d$ in (8.15d) is replaced by γ iterations of nmgm at level $\ell-1$ using the starting value $\tilde{u}_{\ell-1}$. Note that $\tilde{u}_{\ell-1}$ is the exact solution if $s \cdot d=0$ [cf. (8.12)].

As in §3.2 the algorithm (8.16) can easily be rewritten in FORTRAN.

It remains the problem how to choose the values $\tilde{u}_k$ $(k<\ell)$. If we start at level ℓ we may set $\tilde{u}_{\ell-1} := ru_\ell$, where u_ℓ is the actual value before the coarse-grid correction. Then we have to compute $\tilde{f}_{\ell-1}$ for every new definition of $\tilde{u}_{\ell-1}$.

A more convenient choice is the definition of $\tilde{u}_k (0 \leqslant k<\ell)$ by the nested iteration (cf. §6). The program may be as follows

$\tilde{u}_0$:= approximation of $\phi_0(0)$;
for ℓ := 1 step 1 until ℓ_{max} do
begin $\tilde{u}_\ell := \tilde{p} * \tilde{u}_{\ell-1}$;
$\tilde{f}_{\ell-1} := \mathcal{L}_{\ell-1}(\tilde{u}_{\ell-1})$;
for j:=1 step 1 until i do nmgm $(\ell, \tilde{u}_\ell, 0)$
end nested iteration;

In (8.17) we use $f_\ell=0$, since we are interested in the solution of the original equation (8.3). The number i of iterations must be large enough to guarantee $\|\tilde{f}_k\| \leqslant \rho/2$.

For the convergence of the multi-grid iteration (8.16) we have the following result (cf. Hackbusch 1981b).

THEOREM. Suppose that (i) G_ℓ and $\mathcal{L}_\ell$ are continuously differentiable, (ii) the conditions for the linear multi-grid iteration (cf. Theorem of §3.4) are satisfied for G_ℓ and L_ℓ from (8.7), (iii) $\|\tilde{f}_k\| \leqslant \rho/2, \rho$ sufficiently small, (iv) the starting value u_ℓ^0 is sufficiently close to the discrete solution $u_\ell^* = \phi_\ell(0)$. Then the iteration $\{u_\ell^i\}$ converges to u_ℓ^* :

$$\| u_\ell^{i+1} - u_\ell^* \| \leqslant \eta \| u_\ell^i - u_\ell^* \| .$$

The rate η depends on ρ, ε, u_ℓ^0, but is independent of h_ℓ.

The condition (iv) is necessary if the nonlinear system (8.3) has more than one solution. Then a starting value u_ℓ^0 not close enough to u_ℓ^* may lead to another solution u_ℓ^{**}.

8.4 Alternative Approaches

A well-known iteration for solving the nonlinear discrete equation (8.3) is Newton's method. It requires the solution of a linear equation per iteration. Using few iterations of the linear multi-grid method we obtain a practical algorithm (cf. Hackbusch 1978). A further method for solving nonlinear equations is discussed in §12.3.

9. NONLINEAR EXAMPLE: NATURAL CONVECTION IN AN ENCLOSED CAVITY*

Let $\Omega\subset\mathbb{R}^2$ be the unit square $\{(x,z):\ 0\leq x\leq 1,\ 0\leq z\leq 1\}$. The two-dimensional flow can be described by

$$v_h \frac{\partial\zeta}{\partial x} + v_v \frac{\partial\zeta}{\partial z} = Ra\cdot Pr \frac{\partial T}{\partial x} + Pr\ \Delta\zeta$$

$$\zeta = -\ \Delta\psi$$

$$v_h \frac{\partial T}{\partial x} + v_v \frac{\partial T}{\partial z} = \Delta T$$

$$v_h = -\frac{\partial\psi}{\partial z},\quad v_v = \frac{\partial\psi}{\partial x}$$

(cf. Mallison & de Vahl Davis 1977), where v_h, v_v: horizontal and vertical velocities, T: temperature, ζ: vorticity, ψ: stream function. Eliminating v_h, v_v, ζ we obtain

$$\Delta^2\psi - Ra\frac{\partial T}{\partial x} + \frac{1}{Pr}\frac{\partial\psi}{\partial z}\frac{\partial}{\partial x}\Delta\psi - \frac{1}{Pr}\frac{\partial\psi}{\partial x}\frac{\partial}{\partial z}\Delta\psi = 0 \quad \text{in } \Omega, \tag{9.1a}$$

$$-\Delta T - \frac{\partial\psi}{\partial z}\cdot\frac{\partial T}{\partial x} + \frac{\partial\psi}{\partial x}\frac{\partial T}{\partial z} = 0 \quad \text{in } \Omega, \tag{9.1b}$$

The boundary conditions are

$$\psi = 0,\ \frac{\partial\psi}{\partial n} = 0 \quad \text{on } \Gamma, \tag{9.2a}$$

$$T = \begin{cases} 1 \text{ at } x=0,\ 0\leq z\leq 1 \\ 0 \text{ at } x=1,\ 0\leq z\leq 1 \end{cases},\quad \frac{\partial T}{\partial n} = 0 \quad \text{at } z = 0 \text{ or } 1,\ 0\leq x\leq 1.^{**} \tag{9.2b}$$

The Rayleigh and Prandtl numbers are chosen as

$$Ra = 10\ 000, \qquad Pr = 0.71$$

The system (9.1a,b) consists of a fourth order equation for ψ with large lower order part $Ra\,\frac{\partial T}{\partial x}$, and of a second order equation for T. Correspondingly, we have two boundary conditions for ψ and one for T. The first equation (9.1a) is nonlinear, the second equation (9.1b) is linear with respect to T. For the pair of unknowns we use the notation

---

* cf. I.P. Jones and C.P. Thompson: Numerical solutions for a comparison problem on natural convection in an enclosed cavity. Report AERE-R 9955, AERE Harwell, Jan. 1981.

** x=0: hot wall; x=1: cold wall

$$u = \begin{bmatrix} \psi \\ T \end{bmatrix}$$

Discretizing equation (9.1) and boundary condition (9.2) by central finite differences we obtain the system

$$\mathcal{L}_h(u_h) = 0 \tag{9.4}$$

of two nonlinear difference schemes*.

In order to apply the nonlinear multi-grid method we have to define ν, γ, p, r, G_h :

- $\nu=2$, $\gamma=2$:
- p: $\left\{\begin{matrix} \text{cubic} \\ \text{linear} \end{matrix}\right\}$ interpolation with respect to $\left\{\begin{matrix} \psi \\ T \end{matrix}\right\}$;
- r: weighted restriction (transposed mapping to linear interpolation, two-dimensional analogue of (2.13));
- G_h: point-wise relaxation (Gauss-Seidel iteration);
- $\rho = 1/1000$ (cf. (8.16e)).

Relaxation can easily be performed since the equations of (9.4) are linear with respect to the central unknown (cf. page 33). The sequency of grid sizes is

$$h_0 = 1/4, \quad h_1 = 1/8, \quad h_2 = 1/16, \quad h_3 = 1/32.$$

The starting value $\tilde{u}_0$ of the nested iteration (8.17) is computed from $\psi = 0$, $T = 1-x$ by several (40) iterations of the smoother G_0 (=relaxation). The nested iteration (8.17) with $\tilde{p} = p$ and $i = 3$ yields a sufficiently accurate result for the discrete solution u_3 of grid size $h_3=1/32$. The complete program takes about 1.1 s CPU time (Cyber 72/76 of the Rechenzentrum der Universität zu Köln).

Table 7 shows the discrete values of ψ, v_h, v_v, ζ at grid point $x=z=1/4$ for the step sizes $h_2 = 1/16$ and $h_3 = 1/32$. They are obtained by the nested iteration with i=3 (as mentioned above) and i=10. The values admit a comparison of the discretization error ($h=\frac{1}{16}$ ·/· $h=\frac{1}{32}$) and the iteration error (i=3 ·/· i=10).

---

* Δ is discretized by the five-point scheme, Δ^2 by the squared five-point scheme. The boundary condition $\partial\psi/\partial n = 0$ at $x = 0$ becomes $\psi(-h,z) = \psi(h,z)$.

	h = 1/16		h = 1/32	
	i = 3	i = 10	i = 3	i = 10
ψ	-2.4955	-2.4367	-2.3205	-2.3199
v_h	11.001	10.817	10.616	10.616
v_v	-9.1840	-8.9830	-8.8680	-8.8757
ζ	-81.073	-66.801	-66.143	-66.758

Table 7 : Results at (x,z) = (1/4, 1/4) for problem (9.1)

10. STOKES' PROBLEM AND STEADY NAVIER-STOKES EQUATIONS

10.1 Stokes' problem

Let $\Omega \subset \mathbb{R}^n$ be a n-dimensional domain (usually, n=2 or 3). Stokes' problem.

$$-\nu\Delta u_i + \partial p/\partial x_i = g_i \quad \text{in } \Omega \qquad (i=1,\ldots,n),$$

$$-\sum_{i=1}^{n} \partial u_i/\partial x_i = 0 \quad \text{in } \Omega \tag{10.1}$$

or briefly

$$-\nu\Delta \underline{u} + \text{grad} p = \underline{g}, \quad \text{div}\underline{u} = 0, \quad \text{in } \Omega \quad (\underline{u} = (u_1,\ldots,u_n)) \tag{10.1'}$$

is an example for an elliptic system of differential equations. In the case of the foregoing problem (9.1) each equation of the system is an elliptic (scalar) equation. In the present situation neither the first part of (10.1') is an elliptic equation for p, nor the second part, $\text{div}\underline{u} = 0$, is an elliptic equation for u.

The boundary value problem is complete is we add some boundary conditions, for example

$$\underline{u} = 0 \quad \text{on } \Gamma. \tag{10.2}$$

There are many papers about finite element methods for (10.1), (10.2). For instance, compare Thomasset 1980 or Girault & Raviart 1979 and the references given therein. Also difference schemes can be applied. For our purpose we assume that

$$L_\ell w_\ell = f_\ell \qquad (\ell = 0,1,\ldots) \tag{10.3}$$

is a reasonable discretization of Stokes' problem for a sequence $\{h_\ell\}$ of step sizes. The unknowns are denoted by w_ℓ, since $\underline{u}_\ell$ is the abbreviation of the discrete velocities $\underline{u}$.

The only difficulty for applying the multi-grid iteration is the right choice of the smoother G_ℓ. All smoothers of §4 do not work*. For a difference discretization of Stokes' problem (with Ω=rectangle) Brandt & Dinar 1979 proposed a generalized relaxation iteration ("distributed relaxation") for smoothing. Here, we describe an analogue of Jacobi's iteration.

Note that the differential operator L of the system (10.1) has the block structure

$$L = \begin{bmatrix} A & B \\ C & 0 \end{bmatrix}, \quad A = -\nu\Delta, \quad B = \text{grad}, \quad C = -\text{div}.$$

In our situation $A^*=A$, $B^*=C$ hold; hence, L is symmetric. Also the discrete equations (10.3) admit a partition into analogous blocks :

$$L_\ell = \begin{bmatrix} A_\ell & B_\ell \\ C_\ell & 0 \end{bmatrix}, \quad w_\ell = \begin{bmatrix} \underline{u}_\ell \\ p_\ell \end{bmatrix}, \quad f_\ell = \begin{bmatrix} \underline{f}_{1\ell} \\ f_{2\ell} \end{bmatrix},$$

where A_ℓ is the discrete analogue of A, etc.

A smoother for Stokes' problem can be defined by

$$G_\ell(w_\ell, f_\ell) = w_\ell - \omega_\ell \begin{bmatrix} h_\ell^4 A_\ell^* & h_\ell^2 C_\ell^* \\ h_\ell^2 B_\ell^* & 0 \end{bmatrix} (L_\ell w_\ell - f_\ell) \tag{10.4}$$

where

$$\omega_\ell \approx 1 \Big/ \left\| \begin{bmatrix} h_\ell^2 A_\ell & h_\ell B_\ell \\ h_\ell C_\ell & 0 \end{bmatrix} \right\|^2 . \tag{10.5}$$

Note that $\omega_\ell = O(1)$ with respect to $h_\ell \to 0$.

The multi-grid iteration (3.5) with G_ℓ from (10.4) converges as described in the theorem of §3.4. But the convergence is not uniform with respect to all components. Instead of (3.8') we have

$$h_\ell \| \underline{u}_\ell^{i+1} - \underline{u}_\ell \| + \| p_\ell^{i+1} - p_\ell \| \leq C(\nu) \left[h_\ell \| \underline{u}_\ell^{i} - \underline{u}_\ell \| + \| p_\ell^{i} - p_\ell \| \right].$$

The precise analysis of Stokes' problem and similar systems (saddle point problems) is given by Hackbusch 1980c.

10.2 Steady Navier-Stokes Equations

The multi-grid solution of Stokes' problem can easily be generalized to the nonlinear steady Navier-Stokes equations

---

* An exception is the modification (4.5) of Jacobi's iteration

$$
\begin{aligned}
-\nu\Delta u_i + \sum_{j=1}^{n} u_j \partial u_i/\partial x_j + \partial p/\partial x_i &= g_i \quad \text{in } \Omega \ (1\leq i\leq n) \\
-\operatorname{div} \underline{u} &= 0 \quad \text{in } \Omega
\end{aligned}
\tag{10.6}
$$

or

$$
\begin{aligned}
-\nu\Delta\underline{u} + (\underline{u}\cdot\nabla)\underline{u} + \operatorname{grad} p &= \underline{g} \quad \text{in } \Omega \\
-\operatorname{div} \underline{u} &= 0 \quad \text{in } \Omega
\end{aligned}
\tag{10.6'}
$$

with boundary condition (10.2).

Let

$$
\mathcal{L}_\ell(w_\ell) = 0, \quad w_\ell = \begin{bmatrix} \underline{u}_\ell \\ p_\ell \end{bmatrix} \qquad (\ell = 0,1,\ldots) \tag{10.7}
$$

denote the discrete nonlinear problem. For instance, the discretization of (10.6), (10.2) by the finite element method of Bercovier & Pironneau 1979 is analysed in Hackbusch 1980c. As in §10.1 the derivative of $\mathcal{L}_\ell$ has the following block structure :

$$
L_\ell(w_\ell) = \begin{bmatrix} A_\ell(u_\ell) & B_\ell \\ C_\ell & 0 \end{bmatrix} := \partial\mathcal{L}_\ell(w_\ell)/\partial w_\ell .
$$

A possible smoother for the discrete Navier-Stokes problem (10.7) is

$$
G_\ell(w_\ell, f_\ell) = w_\ell - \omega_\ell \begin{bmatrix} h_\ell^4 A_\ell(u_\ell) & h_\ell^2 C_\ell \\ h_\ell^2 B_\ell & 0 \end{bmatrix} \left(\mathcal{L}_\ell(w_\ell) - f_\ell\right) \tag{10.8}
$$

with ω_ℓ from (10.5). With this choice of G_ℓ the nonlinear multigrid iteration of §8 can be applied.

11. ELLIPTIC EIGENVALUE PROBLEMS

11.1 Problem

As a model problem for an elliptic eigenvalue problem we consider

$$
-\Delta u = \lambda u \quad \text{in } \Omega, \qquad u = 0 \quad \text{on } \Gamma. \tag{11.1}
$$

More complicated problems as the plate eigenvalue equation

$$
\Delta^2 u = \lambda u \quad \text{in } \Omega, \qquad u = \frac{\partial u}{\partial n} = 0 \quad \text{on } \Gamma \tag{11.2}
$$

or the Steklov eigenvalue problem

$$-\Delta u = 0 \quad \text{in } \Omega, \qquad \frac{\partial u}{\partial n} = \lambda u \quad \text{on } \Gamma \tag{11.3}$$

can be treated similarly. The discretization of (11.1) with grid size h_ℓ yields the matrix eigenvalue equation

$$L_\ell u_\ell = \lambda_\ell u_\ell \qquad (\ell = 0,1\ldots). \tag{11.4}$$

The following paragraphs describe two different multi-grid solutions of the discrete eigenvalue problem (11.4). In both cases the multi-grid iteration yields one eigenvalue λ_ℓ and the corresponding eigenfunction u_ℓ.

11.2 Direct Approach of the Eigenvalue Problem

We restrict our considerations to problem (11.4) with symmetric L_ℓ and a single eigenvalue. The multi-grid solution for the more general case is described by Astrachancev 1976 and Hackbusch 1979a.

Using the nested iteration technique of §6 we may assume that

e_k: approximate eigenfunction of level $k=0,\ldots,\ell-1$,
λ_k: approximate eigenvalue of level $k=0,1\ldots,\ell-1$

are known. We define

$L_{\lambda k} := L_k - \lambda I$ (I: identity matrix)
Q_k := orthogonal projection onto $\{e_k\}^\perp$:

$$Q_k v_k = v_k - \langle v_k, e_k\rangle_k \, e_k / \| e_k \|^2,$$

where $\langle\cdot,\cdot\rangle_k$ is the scalar product of level k. Then the iteration at level ℓ reads as follows :

u_ℓ^i : i-th iterate approximating the eigenfunction e_ℓ;

$$\lambda := \langle L_\ell u_\ell^i, u_\ell^i\rangle_\ell / \| u^i \|^2 \qquad \text{(Rayleigh quotient)}; \tag{11.5a}$$

$$\tilde{u}_\ell := (I - \omega_\ell h_\ell^{2m} L_{\lambda\ell})^\nu u_\ell^i \tag{11.5b}$$

(result of ν iterations of smoothing by Jacobi's iteration);

$$d_\ell := L_{\lambda\ell}\tilde{u}_\ell \quad \text{(defect)}; \tag{11.5c}$$

$$d_{\ell-1} := Q_{\ell-1} r d_\ell \quad \text{(restricted and projected defect)}; \tag{11.5d}$$

$v_{\ell-1}$: approximate solution of the (singular)equation (11.5e)

$$L_{\lambda_{\ell-1},\ell-1} v_{\ell-1} = d_{\ell-1}; \tag{11.5f}$$

$$u_\ell^{i+1} := \tilde{u}_\ell - pQ_{\ell-1}v_{\ell-1} \quad \text{(correction step)}. \tag{11.5g}$$

Note that the solution $Q_{\ell-1}v_{\ell-1}$ of (11.5f) is well-defined if $d_{\ell-1}\in\{e_{\ell-1}\}^\perp$. (11.5) must be completed by an algorithm performing step (11.5e). Here we shall apply a multi-grid iteration for the almost singular problem (11.5f) that is very similar to the linear multi-grid method of §3.

In order to understand the iteration (11.5), split the iterate u_ℓ^i into $v_\ell^i+\alpha e_\ell$, where $v_\ell^i \perp e_\ell$. If v_ℓ^i is sufficiently small compared with α, λ from (11.5a) is a good approximation to the eigenvalue λ_ℓ. The part αe_ℓ remains almost unchanged by the smoothing step (11.5b) and the coarse-grid correction in (11.5c-g). But the term v_ℓ^i is reduced by the usual convergence rate of the linear multi-grid iteration. A precise analysis is given by Hackbusch 1979a.

The full process of solution is described by the following three programs: the nested iteration (11.6), the eigenvalue multi-grid method emgm performing (11.5), and the singular multi-grid method smgm performing step (11.5e).

Multi-grid method for the eigenvalue problem $L_\ell u_\ell = \lambda u_\ell$

```
ẽ₀ := approximate eigenfunction at level ℓ = 0;
for k:=1 step 1 until ℓ do
begin ẽ_k := p̃*ẽ_{k-1}; λ̃_{k-1}:=<L_{k-1}ẽ_{k-1},ẽ_{k-1}>_{k-1}/‖e_{k-1}‖²;     (11.6)
      for j:=1 step 1 until i do emgm (k,ẽ_k)
end nested iteration;
```

```
procedure emgm(ℓ,u);
begin real λ; integer j; array d,v;
     λ  := <L_ℓ u,u>_ℓ/‖u‖²;
     for j:=1 step 1 until ν do u:=u-ω*h_ℓ^{2m}*L_{λℓ}*u;
     d:=Q_{ℓ-1}*r*L_{λℓ}*u;  v := 0;                                   (11.7)
     for j:=1 step 1 until γ do smgm(ℓ-1,v,d);
     u := u-p*Q_{ℓ-1}*v
end eigenvalue multi-grid iteration;
```

```
procedure smgm(ℓ,u,f);
if ℓ=0 then u := L_{λ0}^{-1}*Q₀*f else
begin integer j; array v,d;
     for j:=1step 1 until ν do u := u-ω*h_ℓ^{2m}*(L_{λ̃ℓℓ}*u-f);
     d := Q_{ℓ-1}*r*(L_{λ̃ℓℓ}*u-f); v := 0;                             (11.8)
     for j:=1 step 1 until γ do smgm (ℓ-1,v,d);
     u := Q_ℓ* (u-p*v)
end singular multi-grid iteration;
```

The projections Q_k are defined by means of $\tilde{e}_k$ computed in (11.6). Note that we need $\tilde{e}_k$ and $\tilde{\lambda}_k$ only for the foregoing levels $k<\ell$. The parameter ω of the smoothing steps is the same as in (2.23). Also other smoothers* of §4 may be used with L_ℓ and f_ℓ replaced by $L_{\lambda\ell}$ and 0, or by L_ℓ and λu_ℓ.

A simple numerical example is described by Hackbusch 1979a. The computation of 46 eigenvalues of the biharmonic equation (11.2) is reported by Hackbusch & Hofmann 1980.

11.3 Transformation into a nonlinear boundary value problem

Eigenvalue problems can be transformed into equivalent nonlinear boundary value problems. Thanks to the nonlinear multi-grid method, the resulting nonlinear problems can be solved as easily as linear problems.

Consider again the eigenvalue problem (11.4). Assume that there is only one eigenfunction e_ℓ corresponding to the eigenvalue λ_ℓ. Let $\phi_\ell : V_\ell \to C$ be some linear or nonlinear functional with $\phi_\ell(e_\ell) \neq 0$ in the linear case and $\phi_\ell'(e_\ell)e \neq 0$ in the nonlinear case. Examples are the mean value or the squared norm:

$$\phi_\ell(e_\ell) = h_\ell^n \sum_{x\in\Omega_\ell} e_\ell(x), \qquad \phi_\ell(e_\ell) = h_\ell^n \sum_{x\in\Omega_\ell} |e_\ell(x)|^2 .$$

The eigenfunction e_ℓ can be normalized by the condition

$$\lambda_\ell = \phi_\ell(e_\ell). \tag{11.9}$$

Combining this equation with (11.4) we obtain the equivalent nonlinear equation

$$\mathcal{L}_\ell(u_\ell) := L_\ell u_\ell - \phi_\ell(u_\ell)u_\ell = 0 \tag{11.10}$$

Any eigenfunction e_ℓ normalized by (11.9) is a solution of (11.10). Any solution u_ℓ of (11.10) is an eigenfunction corresponding to the eigenvalue $\lambda = \phi_\ell(u_\ell)$. If λ_ℓ is a single eigenvalue, then the solution u_ℓ of (11.10) is isolated.

The nonlinear problem (11.10) can be solved by the nonlinear multi-grid iteration of §8. Minor but useful modifications and an analysis are given by Hackbusch 1980b. Numerical examples, in particular, the solution of the Steklov eigenvalue problem (11.3) is contained in the paper mentioned above.

12. MULTI-GRID METHOD OF THE SECOND KIND

12.1 Equations of the second kind

The multi-grid method of the foregoing paragraphs (now named "multi-grid method of the first kind") has been applied to differential equations. In this section we consider abstract equations of the second kind.

---

* In addition to (2.32) we need $G_\ell e_\ell = e_\ell$ (e_ℓ : eigenfunction)

$$u = Ku + f \quad \text{(linear case)} \tag{12.1}$$

$$u = \mathcal{K}(u) \quad \text{(nonlinear case)} \tag{12.2}$$

where K ($\mathcal{K}$) is some linear (nonlinear) operator. A model problem for equation (12.1) is Fredholm's integral equation of the second kind :

$$u(x) = \int_\Omega k(x,y)\, u(y)\, dy + f(x) \qquad (x \in \Omega). \tag{12.3}$$

This integral equation is of the form (12.1) where K is the integral operator defined by

$$(Ku)(x) = \int_\Omega k(x,y)\, u(y)\, dy. \tag{12.4}$$

If k is continuously differentiable, the operator K maps continuous functions u into differentiable functions Ku. This is the essential condition we need in the following. A precise formulation of this smoothing effect of K is given by Hackbusch 1981a, 1981c.

In the nonlinear case we must assume that the derivative $K := \mathcal{K}'$ satisfies the condition mentioned above.

Let

$$u_\ell = K_\ell u_\ell + f_\ell \quad \text{or} \quad u_\ell = \mathcal{K}_\ell(u_\ell), \text{ respectively,} \tag{12.5}$$

be a discretization of (12.1/2) for step size h_ℓ, where h_ℓ is a member of the sequence (3.1). In case of problems (12.3), equation (12.5) may be obtained e.g. by some quadrature formula.

In many other applications Ku is defined implicitly as the solution of a boundary value problem or of a parabolic differential equation. Then, $K_\ell u_\ell$ may be defined to be the discrete solution corresponding to the parameter h_ℓ. An example is given in §12.3.

12.2 Algorithm

We obtain the multi-grid algorithm of the second kind from the multi-grid algorithm of §3 (linear case) and §6 (nonlinear case) by the following formal replacements :

- replace L_ℓ by $I-K_\ell$ $\quad$ ($\mathcal{L}_\ell$ by $I-\mathcal{K}_\ell$)
- replace $\mathcal{G}_\ell(u_\ell,f_\ell)$ by $K_\ell u_\ell + f_\ell$ $\quad$ (by $\mathcal{K}_\ell(u_\ell) + f_\ell$)
- set $\nu := 1$, $\gamma := 2$.

The resulting algorithm is simpler than the multi-grid method of the first kind, since one needs not look for a suitable choice of ν and γ. Nevertheless, the convergence is usually much faster than for the multi-grid method of the first kind. The rate of convergence tends to zero for $h_\ell \to 0$, whereas the rate of the first algorithm is uniformly bounded from above. More precisely, one can prove

$$\text{convergence rate} \leq C\, h_\ell^\alpha, \quad \alpha > 0,$$

where α depends on the increase of differentiability by K and on the consistency order of K_ℓ. The constant C depends on the problem (i.e., on K). Rates of the size $Ch_\ell^\alpha \leq 0.001$ or even better for a practical choice of h_ℓ are possible (compare example of the next section).

12.3 Application to Differential Equations

The multi-grid algorithm of the second kind can be applied not only to integral equations but also to boundary value problems. As a simple example consider the nonlinear boundary value problem

$$-\Delta u = f(u) \quad \text{in } \Omega, \quad u = 0 \quad \text{on } \Gamma$$

with some nonlinear right-hand side f(u) and discretize the problem by

$$-\Delta_\ell u_\ell = f(u_\ell). \tag{12.6}$$

Δ_ℓ may be the five-point scheme. Assume that linear problem $-\Delta_\ell w_\ell = g_\ell$ is easily solvable, for instance by a direct Poisson solver or by a linear multi-grid program. Define the discrete non-linear mapping K_ℓ by

$$K_\ell(u_\ell) = -\Delta_\ell^{-1} f(u_\ell).$$

Then obviously, problem (12.6) is equivalent to (12.2).

The following numerical results are cited from Hackbusch 1979b. The boundary value problem is

$$-\Delta u = e^u \text{ in } \Omega = (0,1)\times(0,1), \; u = 0 \text{ on } \Gamma. \tag{12.7}$$

We choose the sequence $h_0 = 1/2$, $h_1 = 1/4$, ..., $h_5 = 1/64$ and apply the nested iteration (8.17) with $\ell_{max} = 5$ and $i = 1$. Hence, we perform only one iteration of the nonlinear multi-grid method per level. At the mid point (1/2, 1/2) we obtain the following results.

Level ℓ	Grid Size	Result of the Nested Iteration	Exact Discrete Solution u_ℓ	Iteration Error	Discretization Error
0	1/2	0.066 819	0.066 819	-	$1.13_{10}-2$
1	1/4	0.074 715 05	0.074 716 77	$1.7_{10}-6$	$3.39_{10}-3$
2	1/8	0.077 200 481	0.077 206 541	$6.1_{10}-6$	$9.01_{10}-4$
3	1/16	0.077 872 649	0.077 874 047	$1.4_{10}-6$	$2.28_{10}-4$
4	1/32	0.078 043 724	0.078 044 063	$3.4_{10}-7$	$5.73_{10}-5$
5	1/64	0.078 086 685	0.078 086 769	$8.4_{10}-8$	$1.43_{10}-5$

Table 8 : Results of problem (12.7) at the midpoint (1/2,1/2)

The amount of computational work is mainly determined by performance of the mapping $u_\ell \to K_\ell(u_\ell) = -\Delta_\ell^{-1}\exp(u_\ell)$. For computing Δ_ℓ^{-1} we used the direct Poisson solver of Buneman 1969. The following table shows how often the Poisson solver must be performed for obtaining all results of table 8.

Step Size	1/4	1/8	1/16	1/32	1/64
Number of Calls	48	24	12	8	2

Table 9 : number of calls needed for computing $u_{1/64}$ by (8.17)

In further papers of the author one can find more applications of the multi-grid method of the second kind : applications to elliptic and integral eigenvalue problems, to optimal control problems for elliptic and parabolic equations, and to time-periodic parabolic differential equations.

References

ASTRACHANCEV, G.P. (1971): An iterative method of solving elliptic net problems. Z. vycisl. Mat. Mat. Fiz., Vol. 11, pp 439-448 (= U.S.S.R. Comp. Math. Math. Phys. Vol. 11, No. 2, pp 171-182).

ASTRACHANCEV, G.P. (1976): The iterative improvement of eigenvalues. Z. vycisl. Mat. Mat. Fiz. Vol. 16, No. 1, pp 131-139 (= U.S.S.R. Comp. Math. Math. Phys., Vol. 16, No. 1, 123-132).

AZIZ, A.K. (1972) (ed.): The mathematical foundations of the finite element method with applications to partial differential equations. Academic Press, New York.

BACHVALOV, N.S. (1966): On the convergence of a relaxation method with natural constraints on the elliptic operator. Z. vycisl. Mat. Mat. Fiz., Vol. 6, No. 5, pp 861-883 (= U.S.S.R Comp. Math. Math. Phys, Vol. 6, No. 5, 101-135).

BANK, R.E. & DUPONT, T. (1981): An optimal order process for solving elliptic finite element equations. Math. Comp. Vol. 36, pp 35-51.

BANK, R.E. & SHERMAN, A.H. (1981): An adaptive, multi-level method for elliptic boundary value problems. Computing, Vol. 26, pp 91-105.

BERCOVIER, M. & PIRONNEAU, O. (1979): Error estimates for finite element method solution of the Stokes problem in the primitive variables. Numer. Math. Vol. 33, pp 211-224.

BRAESS, D. (1981): The contraction number of a multi-grid method for solving the Poisson equation. Numer. Math. Vol. 37, pp 387-404.

BRAMBLE, J.H. & HUBBARD, B.E. (1965): Approximation of solutions of mixed boundary value problems for Poisson's equation by finite differences. J. of the ACM, Vol. 12, pp 114-123.

BRANDT, A. (1977): Multi-level adaptive solutions to boundary-value problems. Math. Comp. Vol. 31, pp 333-390.

BRANDT, A. & DINAR, N. (1979): Multigrid solutions to elliptic flow problems. In: Numerical Methods for PDEs (S. Parter, ed.). Academic Press, New York.

BUNEMAN, O. (1969): A compact non-iterative Poisson solver. SUIPR Report No. 294, Institute for Plasma Research, Stanford University, California.

CIARLET, P.G. (1978): The finite element method for elliptic problems. North-Holland, Amsterdam.

COLLATZ, L. (1966): The numerical treatment of differential equations. 3rd ed., Springer, Berlin-Heidelberg-New York.

FEDORENKO, R.P. (1961): A relaxation method for solving elliptic difference equations. Z vycisl. Mat. Mat. Fiz. Vol. 1, No. 5, pp 922-927 (= U.S.S.R. Comp. Math. Math. Phys. Vol 1, pp 1092-1096).

FEDORENKO, R.P. (1964): The speed of convergence of one iterative process. Z. vycisl. Mat. Mat. Fiz., Vol. 4, No. 3 pp 559-564 (= U.S.S.R. Comp. Math. Math. Phys. Vol. 4, No. 3, pp 227-235).

FOERSTER, H.; STÜBEN, K.; TROTTENBERG, U (1980) : Non-standard multigrid techniques using checkered relaxation and intermediate grids. In: Elliptic Problem Solvers (M. Schultz, ed.), Academic Press New York.

FREDERICKSON, P.O. (1975): Fast approximate inversion of large sparse linear systems. Report 7-75, Lakehead University.

GIRAULT, V. and RAVIART, P.-A.(1979): Finite element approximation of the Navier-Stokes equations. Lecture Notes in Math. 749, Springer, Berlin-Heidelberg-New York.

GUSTAFSSON, I. (1978): A class of first order factorization methods. BIT, Vol. 18, pp 142-156.

HACKBUSCH, W. (1976): Ein iteratives Verfahren zur schnellen Auflösung elliptischer Randwertprobleme. Report 76-12, Mathematische Institut, Universität zu Köln.

HACKBUSCH, W. (1977): A multi-grid method applied to a boundary value problem with variable coefficients in a rectangle. Report 77-17, Mathem. Institut, Universität zu Köln.

HACKBUSCH, W. (1978): On the multi-grid method applied to difference equations. Computing, Vol. 20, pp 291-306.

HACKBUSCH, W. (1979a): On the computation of approximate eigenvalues and eigenfunctions of elliptic operators by means of a multi-grid method. SIAM J. Numer. Anal., Vol. 16, pp 201-215.

HACKBUSCH, W. (1979b): On the fast solution of nonlinear elliptic equations. Numer. Math., Vol. 32, pp 83-95.

HACKBUSCH, W. (1980a): Convergence of multi-grid iterations applied to difference equations. Math. Comp. Vol. 34, pp 425-440.

HACKBUSCH, W. (1980b): Multi-grid solutions to linear and nonlinear eigenvalue problems for integral and differential equations. Report 80-3, Math. Inst., Universität zu Köln.

HACKBUSCH, W. (1980c): Analysis and multi-grid solutions of mixed finite element and mixed difference equations. Preprint Oct. 80, Math. Inst., Ruhr-Universität Bochum.

HACKBUSCH, W. (1980d): Survey of convergence proofs for multi-grid iterations. In: Special Topics of Applied Mathematics (J. Frehse et al., eds.). North-Holland, Amsterdam.

HACKBUSCH, W. (1981a): Die schnelle Auflösung der Fredholmschen Integralgleichung zweiter Art. Beiträge zur Numerischen Mathematik, Vol. 9, pp 47-62.

HACKBUSCH, W. (1981b): On the convergence of multi-grid iterations. Beiträge zur Numerischen Mathematik, Vol. 9, pp 213-239.

HACKBUSCH, W. (1981c): Error analysis of the nonlinear multigrid method of the second kind. Aplikace Matematiky, Vol. 26, pp 18-29.

HACKBUSCH, W. (1981d): Bemerkungen zur iterierten Defektkorrektur and zu ihrer Kombination mit Mehrgitterverfahren. Rev. Roumaine Math. Pures Appl., Vol. 26, pp 1319-1329.

HACKBUSCH W. and HOFMANN, G. (1980): Results of the eigenvalue problem for the plate equation. ZAMP, Vol. 31, pp 730-739.

HACKBUSCH, W. and TROTTENBERG, U. (eds.) (1981): Multigrid Methods Proceedings, Köln Nov. 1981. Lecture Notes in Math., Springer, Berlin-Heidelberg-New York (to be published).

HEMKER, P.W. (1980a): The incomplete LU-decomposition as a relaxation method in multi-grid algorithms. In: Boundary and Interior Layers-Computational and Asymptotic Methods (J.J.H. Miller, ed.). Boole Press, Dublin.

HEMKER, P.W. (1980b): On the structure of an adaptive multilevel algorithm. BIT, Vol. 20, pp 289-301.

HEMKER, P.W. and SCHIPPERS, H. (1981): Multigrid methods for the solution of Fredholm equations of the second kind. Math. Comp., Vol. 36, pp 215-232.

KRONSJÖ, L. (1975): A note on the "nested iteration" method. BIT, Vol. 15, pp 107-110.

MALLINSON, G.D. and de VAHL DAVIS, G. (1977): Three dimensional natural convection in a box: a numerical study. J. Fluid Mech., Vol. 82, pp 1-31.

MEIS, Th. and MARCOWITZ, U. (1978): Numerische Behandlung partieller Differentialgleichungen. Springer, Berlin-Heidelberg-New York. English translation: Numerical solution of partial differential equations. Springer, New York-Heidelberg-Berlin (1981).

MOL, W.J.A. (1980): Numerical solution of the Navier-Stokes equations by means of a multigrid method and Newton-iteration. Preprint NW 92/80, Mathematisch Centrum, Amsterdam.

NICOLAIDES, R.A. (1977): On the ℓ^2 convergence of an algorithm for solving finite element equations. Math. Comp., Vol. 31, pp 892-906.

NICOLAIDES, R.A. (1979): On some theoretical and practical aspects of multigrid methods. Math. Comp. Vol. 33, pp 933-952.

OGANESJAN, L.A. and RUCHOVEC, L.A. (1979): Variacionno-raznostnye metody resenija ellipticeskich uravnenij (Variational-difference methods of solving elliptic equations). Isdatel-stvo Adademii Nauk Armjanckoj CCP, Erevan.

SCHRÖDER, J. and TROTTENBERG, U. (1973): Reduktionsverfahren für Differentialgleichungen bei Randwertaufgaben I. Num. Math., Vol. 22, pp 37-68.

SHORTLEY, G.H. and WELLER, R. (1938): Numerical solution of Laplace's equation. J. Appl. Phys. Vol. 9, pp 334-348.

STRANG, G. and FIX, G.J. (1972): An analysis of the finite element method. Prentice Hall, New York.

THOMASSET, F. (1980): Finite element methods for Navier-Stokes equations. In: Lecture Series 1980-5, Computational Fluid Dynamics. von Karman Institute, Rhode Saint Genèse.

VARGA, R.S. (1962): Matrix iterative analysis. Prentice Hall, New Jersey.

WESSELING, P. (1977): Numerical solution of the stationary Navier-Stokes equations by means of a multiple grid method and Newton iteration. Report NA-18, Delft University.

WESSELING, P. (1980a): The rate of convergence of a multiple grid method. In: Numerical Analysis (G.A. Watson, ed.). Lect. Notes in Math. 773, Springer, Berlin-Heidelberg-New York.

WESSELING, P. (1980b): Theoretical and practical aspects of a multigrid method. Report NA-37, Delft University.

WESSELING, P. and SONNEVELD, P. (1980): Numerical experiments with a multiple grid and a preconditioned Lanczos type method. In: Approximation methods for Navier-Stokes problems (R. Rautmann, ed.), Lecture Notes in Mathematics. 771, Springer, Berlin-Heidelberg-New York.

ZIENCKIEWICZ, O.C. (1977): The finite element method in engineering science. MacGraw-Hill, New York.

Higher-level Simulations of Turbulent Flows

J. H. FERZIGER

1. BACKGROUND: TURBULENT FLOW COMPUTATION METHODS

1.1 Methods of Computing Turbulent Flows: Classification

A few years ago, the author and two of his colleagues wrote a paper which attempted to classify methods of dealing with turbulent flows (Kline et al. (1978)). This paper is reviewed and extended here as a means of setting the main subject of this report in context.

There are two sub-areas that need to be dealt with in classifying methods of computing turbulent flows. These are the method by which the fluctuations are treated and the manner in which the geometry of the flow is handled. These are, of course, coupled to some extent, but it is useful to separate them for purposes of this work. We shall take up the problem of dealing with the turbulence first. According to the classification scheme in the paper cited above, there are five broad classes of methods of dealing with the turbulence; there are also subclasses of each. The five major categories are:

i) Correlations. These are the familiar correlations that give the nondimensional skin-friction coefficient as a function of the Reynolds number, Nusselt number as a function of Reynolds and Prandtl numbers, etc. They are extremely useful, but very limited. Their applicability is especially limited in high-technology applications in which the geometry plays an important role in the fluid dynamics (such as airfoils); for such problems, a new set of correlations would be needed each time the geometry of the device is changed.

ii) Integral Methods. In these methods the equations governing the fluid dynamics (which may be the equations used on level (iii) below) are integrated over at least one coordinate direction. This decreases the number of independent variables and greatly simplifies the mathematical problem to be solved. These methods allow considerable use of experimental data and physical insight and have proven quite useful. One of their principal drawbacks is that they need to be reworked when a new type of flow is to be computed.

iii) Reynolds-Averaged Equations. In this approach, one averages the Navier-Stokes equations over either time, homogeneous directions in the flow, or an ensemble of essentially equivalent flows. When averaging of any of these kinds is performed, the equations describing the mean field contain averages of products of fluctuating velocities, and there are fewer equations than unknowns--the well-known closure problem. In fact, the set of equations can never be closed by further averaging; a closure assumption or, what is the same thing, a turbulence model has to be introduced. The closure assumption must represent the unknown higher-order average quantities in terms of the lower-order quantities that are computed explicitly. This subject is undergoing a rapid expansion at the present time. It is also likely that this level should be broken into sublevels or separate levels.

iv) Large Eddy Simulation. In this approach, the equations are averaged over a small spatial region. The object is to remove the small eddies from the flow field so that an equation for the large eddies is derived. The effects of the small eddies on the large ones is then modeled. This is one of the principal subjects of this report and is discussed in considerable detail below.

v) Full Simulation. This is the numerical solution of the exact Navier-Stokes equations. The only errors made are numerical ones which, with care, can be kept as small as desired. By its nature, this approach is limited to low Reynolds numbers. This is the other principal subject of this report and will be covered in detail below.

Currently, computations at levels (iv) and (v) are limited to people with access to very large, fast computers. They are not suitable for engineering design at present and we anticipate that it will be some time before they will be (if ever). We call levels (iv) and (v) together higher-level methods of turbulence computation--hence the title of this report.

A significant point about this classification scheme is that calculations on any levels can be used to generate information that can be used on the lower levels. In applications, engineers commonly use methods at level (ii) or (iii) to produce correlations from which the design is actually done. Large eddy simulation (LES) can be used to produce information that can be used in modeling for Reynolds-averaged calculations. LES could be used in principle at the lower levels as well, but there is little need for this application. Full simulation can be used to test models for both the Reynolds-averaged equations and large eddy simulation. This will receive considerable attention in this report.

It should be noted that the nomenclature we have used for classifying methods differs from that of Schumann et al. (1980). What we have called higher-level simulations they called direct simulation, and they did not make the distinction between levels (iv) and (v). We believe the distinction important and prefer the nomenclature used in this report.

The second type of classification of methods of computing turbulent flows concerns the treatment of the geometry. This scheme contains just two categories:

a) Full Field Methods. In this approach, the same set of equations is applied everywhere in the flow field. This has the great advantage of not requiring any kind of matching in the interior of the flow and of being easier to program for computer solution. The principal drawback is that fine meshes are needed in some regions of the flow (such as near the boundaries and in shocks), and this can make the cost very high.

b) Zonal Methods. In zonal methods the flow is considered as a collection of "modules," and each module or zone is treated by a separate method. The most common example of this kind of method is the division of flows over bodies into boundary layers and potential flows which are treated by separate methods. In zonal methods, the solutions in the various zones have to be matched at their common boundaries, by an iterative process that may or may not converge well. The modules can be treated by different methods. Thus one can use an integral method for the boundary layer and the full partial differential equations for the outer flow.

The classification scheme given here differs a little from the earlier one of Kline et al. (1978). We believe that the current scheme represents an improvement in clarity. We have found it useful, and it will be one of the ways in which various methods will be compared in the 1980-81 Stanford-AFOSR Symposium on the Computation of Complex Turbulent Flows.

1.2 Classification of Turbulent Flows

An issue that is quite separate from that of how turbulent flows are computed is that of trying to classify the flows. In a field as complex as this, any classification scheme is inexact, but it is better than having no scheme at all. Thus we shall classify flows according to the phenomena that occur in them. This scheme is not new and contains three categories:

a) Homogeneous Flows. In these flows the state of the fluid is the same at every point in space; they develop in time. There is a limited number of flows of this kind; the experimental data for them have been reviewed recently by the author (Ferziger (1980)). In homogeneous flows without mean strain or shear, the turbulence decays with time; when mean strain or shear are applied, the kinetic energy of the turbulence may increase with time. The mechanism by which the turbulence length scales increase in these flows is not well understood.

b) Free Shear Flows. It is well known that free shear flows are extremely unstable. The laminar mixing layer is unstable with respect to disturbances over a wide range of wavelengths. The instability is of the Kelvin-Helmholtz type in which the perturbation grows rapidly. There is controversy about the precise mechanism of growth of the turbulent free shear layer, but it seems clear that there are large coherent regions of concentrated vorticity in all of these flows. The concentrations of vorticity cause strong large-scale motions within the flow and the vorticity tends to agglomerate further. The controversy centers on the nature of the agglomeration, cf. Roshko (1978) and Chandrsuda et al. (1977).

A subclassification of these flows is necessary. In the mixing layer (the simplest type of free shear layer), the velocity difference across the layer remains fixed as the layer develops. As a result, the layer grows linearly in space or time, indefinitely. In other flows, for example, jets and wakes, the velocity differences are reduced as the flow develops and the turbulence tends to weaken in the downstream direction.

c) Wall-Bounded Flows. The effect of a wall on a shear layer is to prevent (or at least reduce) the large-scale motions described in the previous paragraph and thus inhibit the shear layer from growing so rapidly. Thus, boundary layers and related flows grow less rapidly and have lower turbulence levels than do free shear layers. Another, weaker, mechanism of turbulence production takes over. This mechanism is less well understood than that of the free shear layer and, perhaps for that reason, seems much more complicated. It is known to involve the presence of thin regions of high- and low-speed fluid that exist close to the wall and which are long in streamwise extent (Runstadler et al. (1967), Kim (1969)) and large-scale motions of the outer part of the boundary layer, but several details remain to be filled in.

A further extension of this classification scheme was given by Bradshaw. His view is that the mean turbulent flows can be thought of as a combination of "normal" strains--the mean strains that occur in the "standard" flows--and "extra" rates of strain. There are many extra rates of strain. Some of them are: curvature, rotation, lateral divergence (in axisymmetric flows), buoyancy, blowing or suction, and wall roughness. Although these effects generally appear as small terms in the equations, they have profound effects on the structure of the turbulence and, indirectly, on the behavior of the flow as a whole. Therefore, they are very important, and we shall devote part of this report to investigating their effects on turbulent flows.

Finally, it should be noted that some complex flows may be of one type in one region and another type in another region. In particular, in flows with separation, wall boundary layers may become free shear layers and vice versa.

1.3 A Short History

There are no known analytical solutions of the Navier-Stokes equations for turbulent flows, and it is unlikely that there ever will be any. This fact, plus the obvious technological importance of turbulent flows, is the reason for the development of computational methods of predicting turbulent flows.

Prior to 1960, computers had too little capacity to do anything more than solve the ordinary differential equations of integral methods or the partial differential equations for simple, two-dimensional potential flows. Progress in this period was largely restricted to the computerization of methods that had been carried out on desk calculators up to that time.

As computers grew in sophistication, so did the problems for which people sought solutions. The 1960s saw the development of good boundary layer methods based the use of both integral methods and partial differential equations levels (ii and iii of the above scheme). The 1968 Stanford Conference (Kline (1968)) marked a milestone in this development. At that time, people were beginning to solve the Reynolds-averaged Navier-Stokes equations using simple models for relatively simple flows. Through the 1970s, the sophistication of the models grew, as did the complexity of the flows that researchers were willing to try to compute.

The first applications of what we have defined as higher-level methods were made by the meteorologists. That field has needed models for predicting the world's weather patterns for a long time. As soon as computers were large enough, meteorologists tried global weather simulations. The first three-dimensional attempt at this of which the author is aware is that of Smagorinsky (1963); this paper presented a model that will be used extensively later in this report. The grid systems used in these early calculations were necessarily very coarse, and the method used was necessarily what we have called large eddy simulation. Improvements in computers have allowed the use of finer grids, but the grids are still coarse compared to what is desired; this situation, unfortunately, will not change in the foreseeable future. Hence, subgrid-scale modeling will remain an important issue in meteorology (and oceanography) for quite some time.

The first computation of a flow of engineering interest was the simulation of channel flow by Deardorff, a meteorologist, in 1970. In this landmark paper, he laid out many of the foundations of the field. Improvements in his methods were made by Schumann (1973) and Grotzbach (1976). The latter and their group at Karlsruhe have subsequently extended the method to the computation of annular flows, the inclusion of heat transfer, and the inclusion of the effects of buoyancy.

The author's group at Stanford, which is jointly led by W. C. Reynolds, began work in higher-level simulations in 1972. Their objective was to put a sound foundation under the method of large eddy simulation by computing simple flows first. It was felt that in this way the fundamentals of the subject could slowly be put in order. The first flows chosen for study were the homogeneous turbulent flows, and quite a lot was learned about numerical methods and subgrid-scale modeling (Kwak et al. (1975) and Shaanan et al. (1975)). When the group felt that the techniques for the simulation of homogeneous flows were well developed, it was decided to go on to the study of flows which are inhomogeneous in one coordinate direction. The simplest such flows are the mixing layer and channel. The fully developed mixing layer was computed by Mansour et al. (1978), transition in the mixing layer was studied by Cain et al. (1981), and the channel flow was studied by Moin et al. (1978) and Kim and Moin (1980,1981).

Almost from the beginning it was realized that the effort in computing flows would have to be accompanied by an effort at developing better models for treating the small scales (subgrid scale models) or at least understanding the models that are in use. The method of using direct simulations for this purpose was developed by Clark et al. (1976) and extended by McMillan and Ferziger (1978), McMillan et al. (1980), and Bardina et al. (1980).

It is clear that large eddy simulation will not be a method of direct engineering applicability for some time. For that reason, the major impact the method will have is in the improvement of the understanding of the physics of turbulent flows and in helping to develop, test, evaluate, and improve models that are used in Reynolds-averaged methods. Recently, exact simulations of compressible homogeneous turbulent shear flows and homogeneous turbulent shear flow with a passive scalar were made in order to evaluate these models; cf Feiereisen et.al. (1981, and Shirani et.al. (1981).

A group under Leslie in London has been active in the field since 1975. Their early work centered on the understanding of subgrid scale models (Love and Leslie (1976) and Leslie and Quarini (1979)). Since then they have simulated the mixing of a passive scalar in homogeneous isotropic turbulence (Antonopoulos (1981)).

A number of French groups have studied subgrid scale models from a theoretical point of view and have made several contributions in this area.

Orszag and coworkers have been working since 1970 on the direct simulation of turbulent flows. Their early work centered on the prediction of homogeneous isotropic turbulence (Orszag and Patterson (1971)), and more recently they have become interested in the study of transition in wall-bounded flows (Kells and Orszag (1979), Orszag and Patera (1980)). The main interest of this group has been in the development of numerical methods (they are responsible for the widespread use of spectral methods in this field), on the study of turbulence theories, and on the prediction of transition.

Riley and Metcalf (1980) have made direct simulations of free shear flows. Their efforts have been directed at the simulation of fully developed wakes at relatively low Reynolds numbers, which may be thought of as the last stages of the decay of a turbulent wake.

Rogallo (1978, 1981) has made extensive direct simulations of all of the homogeneous turbulent flows. His results are an important resource for modelers.

1.4 Outline of This Report

In Chapter II, we shall consider the fundamentals of large eddy simulation and compare the various approaches to it.

In Chapter III we shall discuss the subgrid scale models required by large eddy simulation. We shall also study the use of large eddy simulation in the development of models for the Reynolds-averaged equations and the application of full simulation to the testing of both subgrid scale and Reynolds-averaged models.

In Chapter IV we shall discuss the numerical methods used in large eddy and full simulation. Since the numerical methods used are almost always somewhat tailored to a particular flow, we shall just touch on some of the special-purpose methods in this chapter. The latter methods will be considered in more detail in the chapters in which the flows are described.

Chapter V will be devoted to the discussion of the simulation of homogeneous flows. The flows will be categorized, and the numerical methods needed for some of the cases will be described, along with physical descriptions of the flows. We shall give the results from both full and large eddy simulations of these flows and show how they can be applied to the testing and development of models. This chapter contains a considerable amount of recent work.

Free shear flows will be considered in Chapter VI. The bulk of the chapter will be devoted to the mixing layer, which has been the principal focus of attention in this area, but we shall also look at wake simulations.

Chapter VII will be concerned with wall-bounded flows. Most of the attention will be given to channel flow, but some discussion of recent work on the boundary layer will also be given. Particular attention will be given to terms which have not been measured in the laboratory.

In Chapter VIII we shall briefly cover applications of large eddy simulation and full simulation that have not been given in the previous chapters. The most important of these applications are in meteorological and other environmental flows. However, a few applications have been made to other laboratory flows, and these will be briefly covered as well.

The concluding chapter, IX, will discuss some directions in which the work is proceeding and what can be expected from higher-level simulations in the next few years.

This report will give greater emphasis to work done in the author's group than to that of other groups. The reader is reminded that this is a consequence of greater familiarity with his own work and that of his colleagues and is in no way intended to imply that work done elsewhere is any less important.

2. FOUNDATIONS OF LARGE EDDY SIMULATION

2.1 Rationale

It is generally believed that the largest eddies dominate the physics of any turbulent flow. The differences between the large and small eddies can be summarized as follows:

a) The large eddies interact strongly with the mean flow. The small eddies are created mainly by nonlinear interactions among the large eddies.

b) Most of the transport of mass, momentum, energy, and (in flows containing more than one species) concentration is due to the large eddies. The small eddies dissipate fluctuations of these quantities but affect the mean properties only slightly.

c) The structure of large eddies is very strongly dependent on the geometry and nature of the flow. They are usually vortical, but their shapes and strengths are flow dependent. The small eddies are, on the other hand, much more universal.

d) Due to their dependence on the geometry, the large eddies are highly anisotropic. The small eddies are much more nearly isotropic and, therefore, universal.

e) The time scales of the large eddies approximate the time scales of the mean flow. For flow past a body, the large eddy scale is approximately the dimension of the body divided by the free stream velocity. The small eddies seem to be created and destroyed much more quickly.

An important consequence of these properties is that the large eddies should be harder to model than the small ones. Also, as they vary so much from flow to flow, one should not expect to find a model for the large eddies to be universal. There is hope, however, that one might be able to find a useful model of the small eddies.

This leads to the concept of large eddy simulation. In this approach, the large structures in the flow are computed explicitly and the small ones are modeled. This method should have advantages over methods in which all of the eddies are modeled.

These arguments provide the rationale for large eddy simulation. However, not all of the premises given above hold in all flows. They seem to hold in homogeneous turbulent flows and in free shear flows. In wall-bounded flows, however, the structures responsible for much of the momentum transport (and, presumably, the transport of the other properties as well) may be quite small, especially in the region close to the solid boundary. Special care is necessary in these flows; this will be discussed further in Chapter VII.

If, for now, we accept the statements made above as correct, it follows that large eddy simulation ought to have a number of advantages over Reynolds or time average methods. The most significant advantage is that much of the actual transport of mass, momentum, energy, and species is computed explicitly, and the portions of these fluxes which need to be modeled are much smaller than what is modeled in the Reynolds-averaged equations. Consequently, the overall results are less sensitive to modeling inaccuracy in large eddy simulation than in other approaches. The probability of finding a widely applicable model should be much higher.

The principal disadvantage of large eddy simulation relative to Reynolds-averaged methods is that the computations are necessarily three-dimensional and time-dependent. This means that the cost is much higher. In fact, the cost is currently high enough that, except for the simplest flows, use of the method is restricted to people with access to large amounts of time on very large computers.

The first thing that one needs to do in developing large eddy simulation is to define the large scale component of the flow field--the portion which is to be computed explicitly. There are two common approaches to doing this; they will be described and compared in the next two sections. The remainder of the chapter will present the LES equations and describe the parametr tradeoffs that must be faced in large eddy simulation.

The equations for the large scale field always contain terms which involve the small scale field, which is not computed. These terms play the same role in the large eddy equations as the Reynolds stresses play in the Reynolds-averaged equations. They are therefore called the subgrid scale (SGS) Reynolds stresses, and they must be modeled. A discussion of subgrid scale models and a comparison of them with Reynolds-averaged turbulence models is given in the next chapter.

2.2 Filtering

The first task in large eddy simulation is that of defining the large scale component of the flow field--the portion which the method will attempt to calculate. There are several ways of doing this mathematically. All are essentially equivalent to averaging the equations over a small region of space or low-pass filtering the equations in Fourier space. The starting point is the incompressible Navier-Stokes equations:

$$\frac{\partial u_i}{\partial t} + \frac{\partial}{\partial x_j} u_i u_j = -\frac{1}{\rho}\frac{\partial p}{\partial x_i} + \nu \frac{\partial^2 u_i}{\partial x_j \partial x_j} \tag{2.1}$$

which must be solved together with the continuity equation:

$$\frac{\partial u_i}{\partial x_i} = 0 \tag{2.2}$$

For homogeneous flows, we prefer to define the large scale field (also called the resolved field) by means of a convolution filter:

$$\overline{u}(\underline{r}) = \int G(|\underline{r} - \underline{r}'|)\, u(\underline{r}')\, d\underline{r}' \tag{2.3}$$

In Fourier space, this has the form:

$$\hat{\overline{u}}(\underline{k}) = \hat{G}(k)\, \hat{u}(\underline{k}) \tag{2.4}$$

Note that for this kind of filter $\hat{G}$ is a function only of the magnitude of $\underline{k}$.

A number of simple filters have been used. These are illustrated in Fig. 2.1. If the equations are simply integrated over a small control volume in space, we have the box filter; most finite-difference and finite-volume methods implicitly use this filter. Its Fourier transform is also shown in the figure. Another common choice is a sharp cutoff filter in Fourier space; this is essentially the Fourier-space version of the first filter. Both of these filters have the difficulty that their Fourier transforms have negative regions; they also are difficult to differentiate. For this reason, we prefer to use the Gaussian filter. Its Fourier transform is also Gaussian, so it is well behaved in both configuration and Fourier space, and it can be differentiated as many times as one likes in both spaces. The Gaussian is

$$G(r) = A\, e^{-6r^2/\Delta^2} \tag{2.5}$$

or, in Fourier space,

$$\hat{G}(k) = e^{-k^2\Delta^2/24} \tag{2.6}$$

The numerical factors have been chosen to make the second moment of this filter the same as that of a box filter of width Δ. $G(r)$ and $G(k)$ are Fourier inverses of each other when the variables are continuous, but not in the discrete case; in the latter case, a choice has to be made. The normalization factor, A, has been left unspecified in Eq. (2.5) in order to admit the conservation property that the integral of $G(r)$ over all space be unity whether continuous or discrete quadrature is used.

Use of this type of filter was first suggested by Leonard (1973); he showed how the concept could be generalized. It is sometimes useful to use expansions other than standard Fourier series. For example, Chebychev polynomial expansions have been used (Orszag (1978); Kim and Moin (1980)) as the basis for numerical methods. The filter can be defined in the space of the index of the expansion functions; a sharp cutoff (ignoring all components of the expansion beyond some specified N) is the simplest possibility, but it is easy to construct Gaussian-like filters as well.

When the filter (2.3) is applied to the Navier-Stokes equations (2.1) and the continuity equations (2.2), we have:

$$\frac{\partial \bar{u}_i}{\partial t} + \frac{\partial}{\partial x_j}\overline{u_i u_j} = -\frac{1}{\rho}\frac{\partial \bar{p}}{\partial x_i} + \nu \frac{\partial^2 \bar{u}_i}{\partial x_j \partial x_j} \tag{2.7}$$

and

$$\frac{\partial \bar{u}_i}{\partial x_i} = 0 \tag{2.8}$$

The difficulty comes from the nonlinear term. The approach taken by everyone in the field is to write:

$$u_i = \bar{u}_i + u_i' \tag{2.9}$$

which causes the nonlinear terms to take the form:

$$\overline{u_i u_j} = \overline{\bar{u}_i \bar{u}_j} + \overline{\bar{u}_i u_j'} + \overline{u_i' \bar{u}_j} + \overline{u_i' u_j'} \tag{2.10}$$

The first term is entirely dependent on the large scale component of the field and is computable in LES. The small scale component of the velocity field, u_i', is not computed, so the terms containing it need to be modeled. u_i' is called the subgrid scale component of the velocity field, but this is a misnomer (in this formalism), because the width of the filter (Δ) need not be related to the size of the grid on which the computations will be done. However, it has become standard nomenclature, and the set of terms involving the small scale velocity component,

$$R_{ij} = \overline{u_i' \bar{u}_j} + \overline{\bar{u}_i u_j'} + \overline{u_i' u_j'} \tag{2.11}$$

are commonly called the subgrid scale (SGS) Reynolds stresses. They must be modeled in large eddy simulations--hence the name subgrid modeling. We shall look at models for these terms in the next chapter.

The approach presented above is the one favored by the author and his colleagues at Stanford. It decouples the definition of the large scale field from the numerical solution of equations that result. We favor this method, even though it is more cumbersome than the one given in the next section, because we feel it provides more flexibility. This flexibility will be useful when we discuss methods of testing subgrid scale models in the next chapter.

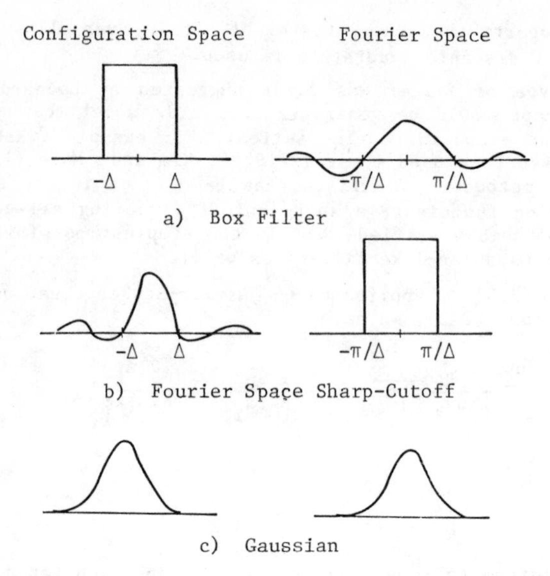

Fig. 2.1. Some filters commonly used in large-eddy simulation.

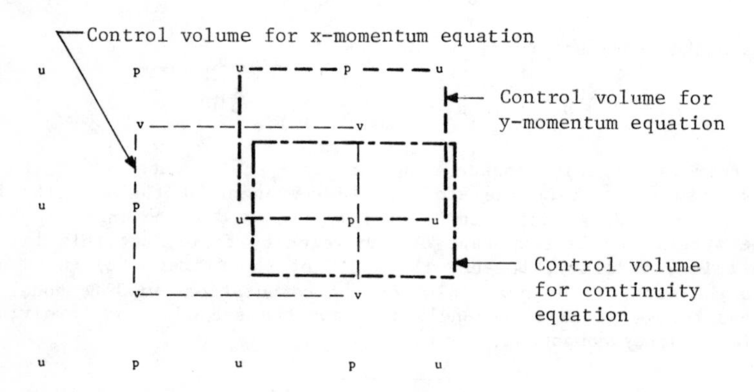

Fig. 2.2. The staggered grid and the control volumes.

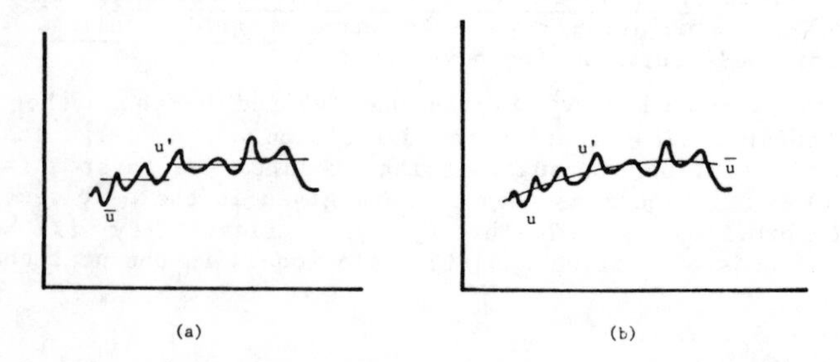

Fig. 2.3. Definition of the large-scale field by
(a) the Deardorff averaging method,
(b) filtering

2.3 The Deardorff-Schumann Approach

An alternative to the method presented in the previous section is based on the recognition that we shall be solving the equations numerically. The computerprogram will be based on a set of discretized equations. It therefore makes sense to use an approach that arrives at the discretized equations as quickly as possible. The method originally presented by Deardorff (1970) and extended by Schumann (1973) is one which accomplishes this.

The idea is to introduce the grid on which the numerical computations will be done at the outset. Deardorff and Schumann used a staggered grid, which is probably the best choice for solving the incompressible equations, but other grid systems could be used as well. The two-dimensional version of the staggered grid is shown in Fig. 2.2. One integrates each of the equations over an appropriate control volume; the control volume for the x-momentum equation is shown in Fig. 2.2. The resulting equations have the form (2.7) and (2.8), provided the operation represented by the overbar is interpreted as the volume average. Because the averaging operation is defined relative to the grid, $\overline{u_i}$ is defined only at the grid points. However, it is convenient to think of $\overline{u_i}$ as constant within the control volume. This definition of the large scale velocity differs from the one presented in the previous section. The two definitions are illustrated in Fig. 2.3.

The Deardorff definitions lead to some convenient simplifications. In particular, one can assume that:

$$\overline{\overline{u}_i\overline{u}_j} = \overline{u}_i\overline{u}_j \tag{2.12}$$

and

$$\overline{u_i'} = 0 \tag{2.13}$$

which are properties this approach shares with Reynolds-averaged modeling. The subgrid scale Reynolds stresses then reduce to:

$$R_{ij} = \overline{u_i'u_j'} \tag{2.14}$$

Models are introduced for R_{ij} and the discretized equations are similar to those commonly used on staggered grids.

In Schumann's modification of this approach, the integrals of spatial derivatives are carried out analytically. This results in equations which contain integrals over the surfaces of the control volumes. The difficulty with this approach is that four different types of averages appear: averages over the three types of faces of the grid volume and volume averages. These must be related in some way. Schumann introduced several approximations that relate the surface averages to a single volume average, but the assumptions required are difficult to evaluate and may be questionable, especially at low Reynolds numbers. It is not clear that this method has any significant advantages relative to Deardorff's.

In the Deardorff-Schumann approach, the subgrid scale velocity field is discontinuous at the edges of the control volumes, and the behavior of the subgrid scale Reynolds stress as a function of position is not very smooth. This problem and the increased flexibility in defining the filter are the primary reasons why we prefer the filtering approach to the one presented in this section.

2.4 The Large Eddy Simulation Equations

The equations of large eddy simulation are essentially (2.7) and (2.8). However, one needs to take into account Eqs. (2.10) and (2.11) as well. Also, one further modification is usually made. The subgrid scale Reynolds stress, defined by Eq. (2.11), can be decomposed into the sum of a trace-free tensor and a diagonal tensor:

$$\begin{aligned} R_{ij} &= (R_{ij} - \frac{1}{3}\delta_{ij}R_{kk}) + \frac{1}{3}\delta_{ij}R_{kk} \\ &\equiv \tau_{ij} + \frac{1}{3}\delta_{ij}R_{kk} \end{aligned} \tag{2.15}$$

Although the diagonal component of this tensor can be modeled, there is no need to do so. When the decomposition (2.12) is substituted into the filtered Navier-Stokes equations (2.7), the diagonal component produces a term which is equivalent to the gradient of a scalar. It is similar to the pressure gradient term and can be combined with it. It is therefore advantageous to define a modified pressure:

$$P = \frac{\overline{p}}{\rho} + \frac{1}{3}R_{kk} \tag{2.16}$$

The filtered Navier-Stokes equations can then be written:

$$\frac{\partial \overline{u}_i}{\partial t} + \frac{\partial}{\partial x_j}\overline{\overline{u}_i\overline{u}_j} = -\frac{\partial P}{\partial x_i} + \nu\frac{\partial^2 \overline{u}_i}{\partial x_j \partial x_j} - \frac{\partial \tau_{ij}}{\partial x_j} \tag{2.17}$$

Once a model for τ_{ij} has been introduced, these equations are to be solved numerically along with the filtered continuity equation, which is repeated here for completeness:

$$\frac{\partial \overline{u}_i}{\partial x_j} = 0 \tag{2.18}$$

2.5 Tradeoffs

In any kind of flow computation, there are tradeoffs. Higher accuracy can always be by reducing the grid size and increasing the number of mesh points. The price is paid in the form of increased computer time.

A similar tradeoff exists in large eddy simulation. Ideally, we would like all the eddies in the large scale field to behave in the manner ascribed to large eddies at the beginning of this chapter and the small eddies to behave as they are supposed to. This separation of large and small eddies is possible only at high Reynolds. At sufficiently high Reynolds numbers, the turbulent energy spectrum contains an inertial subrange in which there is essentially no turbulence production or viscous dissipation. The eddies which are larger than those in the subrange (i.e., lie at lower wavenumbers) behave like "large eddies", and those that lie at wavenumbers below the subrange are "small eddies." Since the width of the filter (Δ) is supposed to mark the boundary between the two classes of eddies, the ideal is to choose the filter width such that the corresponding wavenumber (π/Δ) lies in the subrange. If this is the case, large eddy simulation should be successful.

There are, of course, difficulties that we need to address. The principal of these are:

a) The size of the physical domain considered in the calculation needs to be sufficiently large to hold the largest eddies. We also wish the filter size to be such that all of the "small" eddies lie in or below the subrange. Finally, the computational grid size must be smaller than the filter width (this is discussed in Chapter 4). These requirements set the number of mesh points required in each coordinate direction. It is not unusual to find that the number of mesh points needed to meet these requirements is much greater than the available computer resource will allow. We are then forced to use a filter width which lies outside the subrange.

b) At low Reynolds numbers there is no subrange in the turbulence spectrum.

In either case, we are forced to use a filter width which does not lie within the inertial subrange of the turbulence spectrum. It has been argued by some that one should not do this. We believe that it is reasonable to do large eddy simulation under these circumstances. However, the model may need to be changed to account for the fact that the cutoff is not in an inertial subrange. This problem will be discussed further in the next chapter.

3. SUBGRID SCALE MODELS

3.1 The SGS Reynolds Stress

In the preceding chapter we saw that there are terms in the equations of large eddy simulation that involve the small or subgrid scale component of the velocity field, and, as this small-scale velocity field will not be computed, these terms must be modeled. This chapter will be devoted to a discussion of the models used for the so-called subgrid scale (SGS) terms.

To begin, it is well to look at the physical significance of the SGS terms. Equations (2.2) and (2.3) describe the development of the large eddies. In them, the terms containing the small scale velocity represent the interactions between the large and small eddies. On the average, kinetic energy is transferred from the large eddies to the small ones; there is energy flow in both directions, but the net flow is usually toward the small scales. Leslie and Quarini (1979) estimated that the gross transfer to the small scales is about 1.5 times the net transfer. In other words, approximately one-third of the energy transferred to the small scales is returned. We shall see later that the net energy flow may be in the reverse direction in some cases. The subgrid scale terms in Eqs. (2.2) and (2.3) must represent the effect of these transfers on the large scales. In the normal situation, the net energy transfer to the small eddies appears to be a dissipation to the large eddies--energy lost that will not reappear. Thus the model should normally be dissipative.

The terms which need to be modeled were derived in the previous chapter and can be written:

$$R_{ij} = \overline{u_i u_j} - \overline{\bar{u}_i \bar{u}_j} = \overline{\bar{u}_i u'_j} + \overline{u'_i \bar{u}_j} + \overline{u'_i u'_j} \qquad (3.1)$$

As we showed, we prefer to work with the SGS Reynolds stress defined by

$$\tau_{ij} = R_{ij} - \frac{1}{3} R_{kk} \delta_{ij} \qquad (3.2)$$

It is also worth mentioning at this point that the terms we have called the Leonard stresses:

$$\lambda_{ij} = \overline{\bar{u}_i \bar{u}_j} - \bar{u}_i \bar{u}_j \tag{3.3}$$

(which were first discussed by Leonard (1973)) may need special treatment. These terms are zero in the Deardorff-Schumann approach but not in the filtering method. Investigation has revealed that they are responsible for only a small amount of energy transfer between the large and small scales. Their major effect seems to be redistribution of energy among the various large scales.

The contents of this chapter are as follows. In the next section, equations governing the SGS Reynolds stresses will be derived and discussed. We shall also compare SGS modeling to Reynolds-averaged modeling. In Section 3, a computational method for validating SGS models will be described and some results given. This will be followed in Section 4 by a discussion of eddy viscosity models, the ones in most common use today. Section 5 will describe some of the contributions that theory has made to the state of the art in SGS modeling. Some new ideas about SGS modeling form the subject of Section 6. Higher-order modeling will be taken up in Section 7. Finally, Section 8 will discuss some effects that arise when there are extra rates of strain (in Bradshaw's sense) in the flow. We shall end the chapter with a short summary of the principal points.

3.2 The SGS Stress Equations

It is not difficult to derive a set of equations describing the dynamical behavior of the quantities R_{ij} defined by Eq. (3.1). However, the process is somewhat tedious. One takes the Navier-Stokes equations for u_i and also writes them with i replaced by j. The equation for u_i is multiplied by u_j and vice versa. Adding the two resulting equations and filtering the result yields an equation for $\overline{u_i u_j}$. By repeating the same procedure using the dynamical equation for $\bar{u}_i$, one can derive an equation for $\bar{u}_i \bar{u}_j$. Subtracting these two equations, we have the desired equation for R_{ij}:

$$\underbrace{\frac{\partial R_{ij}}{\partial t} + \bar{u}_k \frac{\partial}{\partial x_k} R_{ij}}_{\text{(Convection)}} = \underbrace{- \left[\overline{(R_{ik} + \lambda_{ik}) \frac{\partial \bar{u}_j}{\partial x_k}} + \overline{(R_{jk} + \lambda_{jk}) \frac{\partial \bar{u}_i}{\partial x_k}} \right]}_{\text{(Production)}}$$

$$+ \underbrace{\overline{\frac{p}{\rho}\left(\frac{\partial u_i}{\partial x_j} + \frac{\partial u_j}{\partial x_i}\right)} - \overline{\bar{p}\left(\frac{\partial \bar{u}_i}{\partial x_j} + \frac{\partial \bar{u}_j}{\partial x_i}\right)}}_{\text{(Redistribution)}} \tag{3.4}$$

$$\underbrace{- 2\nu \left[\overline{\frac{\partial u_i}{\partial x_k} \frac{\partial u_j}{\partial x_k}} - \overline{\frac{\partial \bar{u}_i}{\partial x_k} \frac{\partial \bar{u}_j}{\partial x_k}} \right]}_{\text{(Dissipation)}}$$

$$+ \text{ Diffusion terms}$$

There are many diffusion terms; they are not written explicitly, as we shall not need them. Here, λ_{ij} is the Leonard stress defined by Eq. (3.3). An equation for R_{kk}, the subgrid scale turbulent kinetic energy, can be derived by taking the trace of Eq. (3.4). Subtracting δ_{ij} times the resulting equation from Eq. (3.4) gives an equation for τ_{ij}.

All of the terms in Eq. (3.4) are analogous to terms in the familiar Reynolds stress equations of time-average modeling. The interpretations are also similar. However, the differences are quite important. Eqs. (3.4) contain more terms than the equations for the time-average Reynolds stresses because some items that are zero in time-average approach are not zero when filtering is used. In particular, note the appearance of the Leonard stress in the production term and, more importantly, the fact that the production term is filtered. All of the terms in Eq. (3.4) can be computed by the methods described in the next section and models for them studied, but this has not been done to date.

The most common assumption in turbulence modeling is that production and dissipation terms dominate the turbulence budget, and, as a first approximation, we can equate them and ignore the other terms. For the time-average equations, this approximation is reasonable when applied to the turbulent kinetic energy budget far from solid boundaries, but it is less valid for the component equations because the redistribution terms may be quite large. Near walls, the diffusion terms become quite important and the approximation is even more questionable. The low Reynolds numbers in this region may also affect the structure of the turbulence. Nonetheless, the "production equals dissipation argument" is frequently invoked.

For LES, the situation is somewhat different. It is important to note is that the model is assumed to represent a local spatial average of the local instantaneous small-scale turbulence. This is quite different from what is modeled in time- or ensemble-average modeling and our understanding of subgrid scale turbulence (and consequently, our ability to model it) is more limited. This is compensated for by the fact that a large eddy simulation calculation of a given flow is less sensitive to modeling errors than is a Reynolds-averaged calculation of the same flow.

In particular, because the small scales of turbulence are highly intermittent, we expect gradients of subgrid scale quantities to be relatively large. If this is the case, it is probable that the convection and diffusion terms, which are ignored in many time-averaged models, are more important in SGS modeling. On the other hand, we have recently found evidence that the pressure fluctuations and, more particularly, the pressure-strain correlations reside mainly in the large scales (this will be presented in Chapter 5), and they may be less important in SGS modeling than they are in conventional modeling. Despite these differences, most SGS models to date have relied on ideas developed for time average models.

3.3 Computational Validation of SGS Models

Two approaches are commonly used for developing and testing time-average models. One method, favored by Lumley, Reynolds and others, uses simple turbulent flows (usually homogeneous flows) to test the validity of the models and to determine the adjustable parameters. The major objection to this approach is that the structure of homogeneous flows differs considerably from the flows one really wishes to simulate, and the constants may not be valid in more complex flows. The other method, used by Spalding, Launder, and others, adjusts the parameters to fit flows similar to the ones that one wishes to calculate. This is difficult because many of the parameters must be adjusted simultaneously and this can be a difficult procedure.

It is even more difficult to develop models for the subgrid scales. Data on the small scales of turbulence are quite scarce, and direct validation of a model using experimental data is nearly impossible. Consequently, the constants have to be found by other methods. One approach is almost completed based on theory and uses the properties of the inertial subrange. Lilly (1967) and others have shown that the constant in the model can be derived on this basis. Unfortunately, it is not always possible to to assure that in a computation the cutoff between the large and small scales will lie in the subrange, so one needs to be cautious about adopting the results of this approach. Indeed, a number of authors found it necessary to modify the SGS model constant to obtain good results.

There is a second approach. With the current generation of computers it is possible to compute homogeneous turbulent flows with no approximations other than those present in any numerical simulation. At present, it is possible to do such calculations with grids as large as $64 \times 64 \times 64$ and, in a limited number of cases, $128 \times 128 \times 128$. This allows simulation at Reynolds numbers based on Taylor microscale up to approximately 40 (80 with the larger grids). The results can be regarded as realizations of physical flow fields and are an interesting and important complement to laboratory results. In particular, the computational results provide all three velocity components and the pressure at a large number of spatial points for a relatively short time span. The laboratory data typically give one or two velocity components over a longer time span at just a few spatial points.

Having a realization of a flow, we proceed in much the same manner an experimentalist would. The computed field can be filtered to give its large scale component; the small scale component is obtained by difference. We can then compute the terms that need to be modeled, and, from the large scale field, we can also compute what the model predicts these terms to be. Direct comparison between the model and the exact value is then possible. This can be done in a couple of ways.

One method is to use a scatter plot. The exact value of the SGS Reynolds stress at each mesh point is plotted against the value predicted by the model. If the model is correct, the results lie on a straight line; a totally invalid model produces a random pattern of points (usually a circle). This is a very graphic test of a model. Some scatter plots will be shown later.

The second method is to compare the model and exact results statistically. In our work we have used the correlation coefficient as a measure of the validity of a model. This is a crude test, but it seems to be sufficient for our purposes. It is important to recall that the square of the correlation coefficient is approximately the fraction of the data that the model is correctly predicting.

These are very severe tests of models--much more severe than the tests usually applied to Reynolds-averaged models. It is possible for a model which performs poorly in these tests to do well in actual simulations. However, failure of a model to do well is a signal for caution.

Use of this kind of testing for Reynolds-averaged models will be taken up in Chapter 5.

3.4 Eddy Viscosity Models

Eddy viscosity models can be "derived" from the "production equals dissipation" argument discussed earlier. This is done in a number of places and need not be repeated here. For subgrid scale turbulence, the eddy viscosity model amounts to assuming that the subgrid scale Reynolds stress is proportional to the strain in the large scale flow:

$$\tau_{ij} = 2\nu_T \bar{S}_{ij} = \nu_T \left(\frac{\partial \bar{u}_i}{\partial x_j} + \frac{\partial \bar{u}_j}{\partial x_i} \right) \tag{3.5}$$

The eddy viscosity ν_T has the dimensions of a kinematic viscosity. Most work is based on the assumption that the eddy viscosity could be represented by:

$$\nu_T = (C\Delta)^2 \, |\bar{S}| \tag{3.6}$$

where Δ is the width associated with the filter and $|\bar{S}| = (\bar{S}_{ij}\bar{S}_{ij})^{1/2}$. Recently, a number of authors have shown that this is correct only if the integral scale of the turbulence is smaller than Δ. Since LES is designed for this not to be the case, it is better to assume that:

$$\nu_T = C \, \Delta^{4/3} \, L^{2/3} \, |\bar{S}| \tag{3.7}$$

where L is the integral scale of the turbulence. Usually L is estimated from $L = q^3/\varepsilon$, where ε is the dissipation.

Eddy viscosity models have a long record of reasonable success in time-average modeling of simple shear flows, and one might expect them to do well as SGS models. In fact, they have been found to do well in some of the homogeneous flows. In particu ar, for the homogeneous flows in which there is no mean strain, one is able to predict most of the low-order statistical quantities (for example, the mean square velocity fluctuations and spectra) quite well using eddy viscosity models; that the higher-order statistics, which are sensitive to the small scales, are not well predicted should be no surprise. In the homogeneous flows with strain or shear, there is evidence (McMillan et al. (1980), Shirani et al. (1981)) that the energy transfer can be reversed and flow from the small scales to the large ones. In such cases, the model should no longer dissipate the energy of the large scales. Eddy viscosity models, which are guaranteed to dissipate energy from the large scales, cannot predict this behavior. Despite this, they may not function badly in actual simulations. The reason is that the smallest scales of the resolved field, from which the model normally extracts energy, become relatively weak in these flows, and the model may actually dissipate very little energy. Furthermore, the principal difficulty in computing these flows usually arises from the delivery of a significant amount of energy to scales larger than the computational domain. This makes the normally used periodic boundary conditions incorrect, and the results cannot be relied upon.

Eddy viscosity models are incapable of handling other classes of flows. For example, in transitional flows, we must expect that most of the energy will be in the large scales i.e., the small scales are not in equilibrium with the large scales and the "production equals dissipation" argument is incorrect. Furthermore, although Moin et al. (1978) had reasonable success in simulating channel flow with these models, later extensions by Kim and Moin (1979) and Moin and Kim (1981) clearly show the deficiencies of the model.

They found that eddy viscosity models (several were tried) were unable to maintain the energy of the turbulence. The problem is only partially due to the model, as the turbulence tends to decay even when the model is eliminated. This will be discussed in more detail in Chapter 7.

Clark et al. (1979) and McMillan and Ferziger (1979), and McMillan et al. (1980) have applied the model-testing method described above to eddy viscosity SGS models. A typical scatter plot is shown in Fig. 3.1, in which the exact subgrid scale stress is plotted against the Smagorinsky model value. It can be seen that there is a little correlation between the two data sets (the correlation coefficient is approximately .4 for the case shown), but it is even more clear that this is far from an adequate model. This result is fairly typical, although there are variations in the correlation coefficent with many of the significant parameters.

The results show that eddy viscosity models are rather poor and, in fact, they become even poorer when there is mean strain and/or shear in the flow. However, it is not easy to find more accurate models (we shall look at this below), so we may be forced to use eddy viscosity models until something better is developed. Furthermore, as McMillan and Ferziger have shown, the method can be used to predict the effect of Reynolds number on the model parameter. Their results are shown in Fig. 3.2. When these results were applied to channel flow by Moin and Kim (private communication), they did not produce the desired effects, probably for the reasons given above.

In the above, we have used the fact that the natural length scale of the SGS eddies is the width, Δ, associated with the filter. By definition, this is the scale that defines whether an eddy is large or small, and there is little reason to suspect that this is not a correct choice.

However, when the filter is anisotropic, as it should be in computing shear flows, it is not quite so clear what is the correct length scale. Almost everyone has used the cube root of the filter volume:

$$\Delta = (\Delta_1 \Delta_2 \Delta_3)^{1/3} \tag{3.8}$$

However, Bardina et al. (1980) showed that a better choice might be:

$$\Delta = (\Delta_1^2 + \Delta_2^2 + \Delta_3^2)^{1/2} \tag{3.9}$$

It is recommended that Eq. (3.9) be adopted for general use.

3.5 The Role of Theory

Theoretical insight plays a considerable role in understanding the physics of turbulence and contributes considerably to modeling it. Turbulence is, however, a problem of such complexity that the role of theory in our present state of knowledge is smaller than in most areas of physics or engineering. Progress has been frustratingly slow. A review of recent theories is given by Leslie (1973).

Most theories provide limited information about turbulence. Usually, the theories were developed for homogeneous turbulence and have proved difficult to generalize.

The theories which have attracted the most attention are Kraichnan's direct interaction approximation and others related to it. These theories are statistical in nature i.e., they attempt to make statements about averages of turbulence quantities rather than the detailed dynamics. The question of

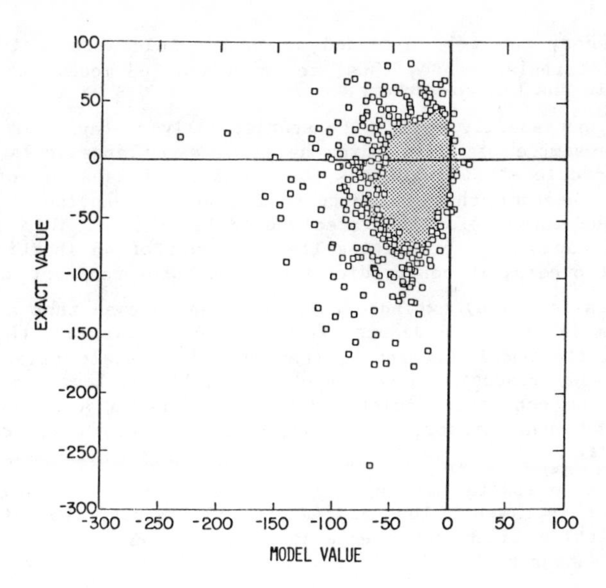

Fig. 3.1. Scatter plot of $u_i \partial\tau_{ij}/\partial x_j$ (the dissipation due to the subgrid-scale model) for the Smagorinsky model in weakly strained turbulence. The correlation coefficient is 0.4 for this case. From McMillan et al. (1980).

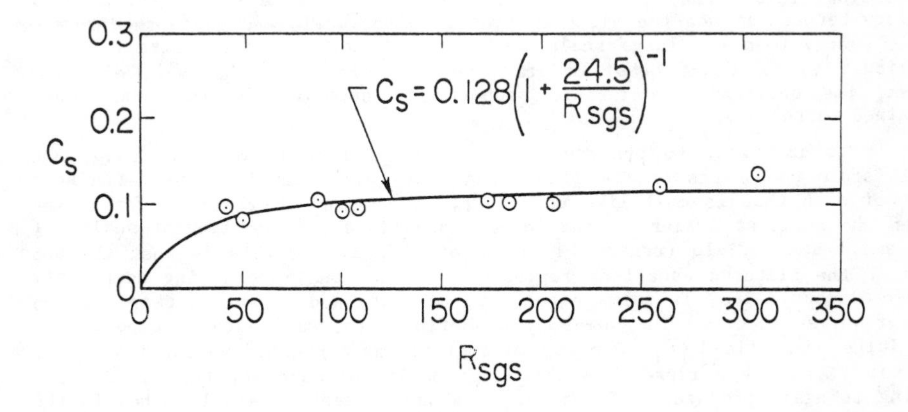

Fig. 3.2. The constant in the Smagorinsky model as a function of the subgrid-scale Reynolds number $R_{SGS} = |\overline{S}|\Delta^2/\nu$, where $|\overline{S}|$ is the r.m.s. large-scale strain. From McMillan and Ferziger (1978).

whether this theory could be extended so as to yield information about the small scales of turbulence and, thus, to provide a SGS model has been investigated by Leslie and his co-workers.

The theory necessarily deals with statistically averaged SGS turbulence. We imagine an ensemble of flows which have the same large-scale motions but different small-scale motions and ask for the average behavior of the small-scale motions. Whether this is adequate for modeling purposes is an open question, but the information generated should be helpful. This theory, like many others, is capable of predicting the existence of an inertial subrange, but, unlike most others, it can predict the Kolomogorov constant as well.

Love and Leslie (1976) extended the theory and showed that a form of the eddy viscosity model could be deduced from it. In particular, they predicted the constant in the model and showed that the large scale strain rate that appears in the eddy viscosity model ought not to be the local one but a spatial average. The constant predicted in this way is in good agreement with that obtained by other theoretical arguments and from empirical fits to experimental data.

With respect to spatial averaging of the strain rate in the eddy viscosity, the evidence is mixed. Love and Leslie (1976) found that it was important in the solution of Burgers' equation, but Mansour et al. (1978) found that it did not matter much.

A number of other issues were investigated by Leslie and Quarini (1979). In particular, they divided the SGS terms into "outscatter" and "backscatter" terms representing, respectively, the energy flows to and from the subgrid scale. They found that eddy viscosity models appear to represent the outscatter fairly well, but they could not say much about the backscatter.

Although limited, these theories are proving useful in choosing and validating models.

3.6 A Scale Similarity Model

All models, by definition, relate the SGS Reynolds stress to the large scale flow field. Eddy viscosity models view τ_{ij} as a stress and make an analogy between it and the viscous stress. These models are guaranteed to extract energy from the large scale field (i.e., they are dissipative). It is difficult to construct other models with this property. However, as noted above, the desirability of this property is questionable in sheared and strained turbulence.

It is important to observe that the interaction between the large and small scale components of the flow field takes place mainly between the segment of each that is most like the other. The major interaction is thus between the smallest scales of the large scale field and the largest scales of the small scale field (regions 1 and 2 of Fig. 3.3). This is what the SGS term in the filtered equations represents. Since the interacting components are very much alike, it seems natural to have the model reflect this. To do this requires that we find some way of defining the small scale component of the large scale field $\overline{u}_i$. One way to do this was suggested by Bardina et al. (1980). Since $\overline{u}_i$ represents the large scale component of the field, filtering it again produces a field $(\overline{\overline{u}}_i)$ whose content is still richer in the largest scales. Thus,

$$\tilde{u}_i = \overline{u}_i - \overline{\overline{u}}_i \tag{3.10}$$

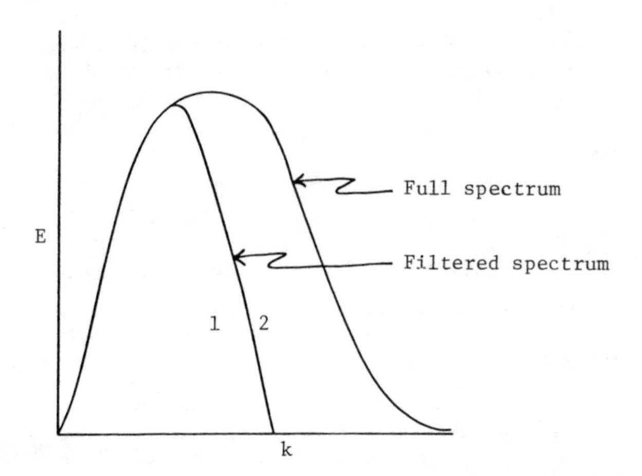

Fig. 3.3. A typical turbulence spectrum and the effect of filtering. Region 1 represents the smallest scales of the filtered or resolved component of the field. Region 2 represents the largest scales of the subgrid scale or unresolved component of the field. The scale similarity model relates these.

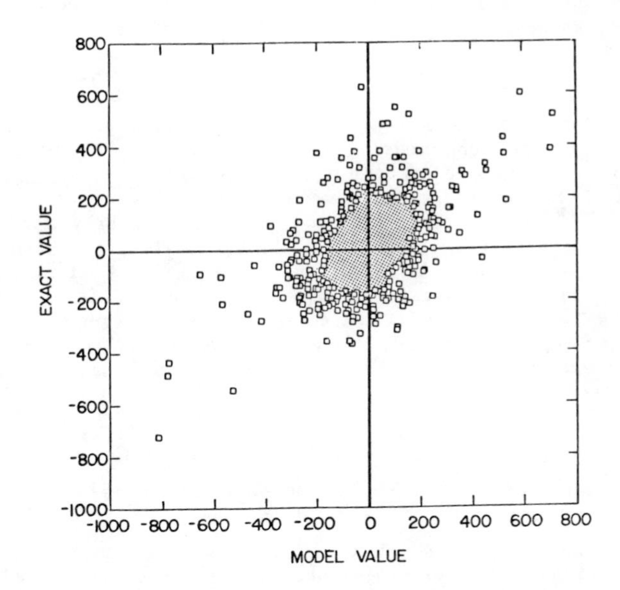

Fig. 3.4. Scatter plot of $u_i \partial\tau_{ij}/\partial x_j$ (the dissipation due to the subgrid-scale model) for the scale similarity model in strained turbulence. The correlation is improved relative to the model shown in Fig. 3.1, but the model has no average dissipation. From McMillan et al. (1980).

is a field which contains the smallest scales of the large scale component of the flow field. This suggests that a reasonable model might be

$$\tau_{ij} = c\,\tilde{u}_i\tilde{u}_j \tag{3.11}$$

or, better yet

$$\tau_{ij} = c\left(\bar{u}_i\bar{u}_j - \bar{\bar{u}}_i\bar{\bar{u}}_j\right) \tag{3.12}$$

This modification is suggested by considering the "cross-terms," e.g., $\overline{\bar{u}_i u'_j}$.

Preliminary tests have shown that this model is not dissipative, but it does correlate very well with the exact stress; a scatter plot is given in Fig. 3.4. This suggests that a combination of the two models might be better yet. The correlation is largely due to the fact that, with a Gaussian filter, the two fields in question contain much the same structures. With other filters, particularly one which is a sharp cut-off in Fourier space, the correlation is smaller. These models are currently being investigated.

3.7 Higher-Order Models

The inadequacies of algebraic eddy viscosity models in Reynolds-averaged modeling have been known for a long time. A number of more complex models have been proposed, and, since they have analogs in SGS modeling, a brief review of them is in order. We shall go into some of these models in more detail later.

Many of the improvements are based on the notion that proportionality between Reynolds stress and mean strain rate is valid, but the eddy viscosity formulation needs improvement. In these models one writes:

$$\nu_T = C_1 q \ell \tag{3.13}$$

where q and ℓ are, respectively, velocity and length scales of the turbulence. In the simplest such models, the length scale is prescribed and a partial differential equation for the turbulence kinetic energy $(q^2/2)$ is solved along with the equations for the mean flow field. These are called one-equation models; their record has not been particularly good, and most people now use still more complex models. In particular, the assumption of a prescribed length scale has been questioned, and methods of predicting the length scale have been proposed. Of these, the most widely used models are those in which an equation for the dissipation of turbulent kinetic energy (which really represents energy transferred to the small, dissipating eddies) is added to the equations used in one-equation models. The length scale is related to the dissipation ε by:

$$\varepsilon = C_2 q^3/\ell \tag{3.14}$$

and we have the so-called two-equation models. This is the most popular method of computing time-average flow fields at present.

Finally, the most recent development has been the use of the full Reynolds stress equations. In two dimensions, three PDE's are needed to define the Reynolds stress, while in three dimensions, six are required. Clearly, this is a rather expensive approach.

A way of avoiding the computational cost of full Reynolds stress methods is obtained by noting that the convective and diffusive terms can frequently be neglected. If they are, and approximations are made to the redistribution terms, the equations reduce to algebraic ones. Algebraic models have become popular in recent years. However, there is doubt as to whether the neglect of diffusion is correct near the wall.

All of these models have analogs in SGS modeling, and a number of them have been used. Let us consider them in the order in which they were introduced above.

First, consider one- and two-equation models. They have as their fundamental basis the proportionality of the SGS Reynolds stress and the large-scale stress. We saw earlier that the Smagorinsky model (an algebraic eddy viscosity model) correlates poorly with the exact SGS Reynolds stress. Clark et al. (1979) looked at the behavior of one-equation models as well as an "optimized" eddy viscosity model. In the latter, the eddy viscosity was chosen, at every point in the flow, to give the best local correlation between the SGS Reynolds stress and the large-scale strain. By definition, no eddy viscosity model can do better than this. It was found that the correlation coefficient improved somewhat relative to the Smagorinsky model (from approximately .35 to .50 in a typical case), but this still leaves the model far short of what we would like to have. The lack of correlation seems to be due to the difference between the principal axes of the two tensors. This modeling assumption needs to be changed if further improvement is to be obtained (cf. McMillan and Ferziger (1980)), and more complex models are required. Schumann (1973) also used one-equation models without finding improvement over algebraic eddy viscosity models.

Next, recall the earlier remark that convection and diffusion are likely to be more important in SGS modeling than they are in time-average modeling. This means that the approximations needed to reduce the full Reynolds stress equations to algebraic model equations are less likely to be valid in the SGS case. However, several authors have used algebraic models. The applications have been almost exclusively to meteorological and environmental flows in which stratification and buoyancy effects are important. These flows are sensitive to small variations in both properties and model, making it difficult to assess the accuracy of a model with precision. To our knowledge, no applications of these models to engineering flows have yet been made.

It is probable that, to obtain a significant improvement over the Smagorinsky eddy viscosity model, we shall need to go to full Reynolds stress models. This, of course, is not something to be looked forward to as the computing cost is likely to be more than doubled. The only use of these equations to date was in meteorological flows by Deardorff (1972, 1973a,b), who reported a computer time increase of a factor of 2.5. Furthermore, the results were not improved to the degree that he had hoped for. Although this is discouraging, Deardorff's simulation was considerably ahead of its time and had the additional difficulties associated with buoyancy, so it is hard to make definitive conclusions. Thus, we cannot conclude much about these models at present, and quite a bit of work needs to be done on them before they become useful tools of the trade.

3.8 Other Physical Effects

The author's group has done full simulations of the effects of compressibility on turbulence and the mixing of passive scalars in turbulent flows. To date, the work has concentrated on evaluating time-average models, because it was felt that this is the area in which the work has the most immediate impact.

The effects of compressibility on SGS turbulence are probably quite small. The effect on the turbulence as a whole have been found to be fairly weak, except for effects due to the propagation of acoustic pressure waves. Since the latter are large scale phenomena and the Mach number of the SGS turbulence is small, we expect that compressibility will have only a weak effect on SGS modeling.

On the other hand, SGS modeling of turbulent mixing is quite important. If we are to simulate combusting flows, it will be necessary to treat the small scales accurately, since that is where the action is in these flows. The effect of the Prandtl/Schmidt number on time-average models is moderately strong, and we expect its effect on SGS models to be even stronger. Furthermore, the specific effects due to combustion are also likely to be important on the small scales. The author intends to look at SGS modeling of mixing and combusting flows in the near future.

Another effect of considerable importance in application is buoyancy, which was mentioned earlier in connection with the meteorological simulations. Flows in which buoyancy is important and, particularly, those which are driven by buoyancy are very difficult flows to measure or simulate, and a great deal of work will need to be done in this area. Important work in this area has been done by the Karlsruhe group (Grotzbach et al. (1979)), and further work is under way in London (Leslie (1980)).

Finally, we should state that meteorologists and environmental engineers have a great interest in both mixing and buoyancy effects, and considerable effort in these areas has been made by these people. In particular, we note again the work of Deardorff cited above and that of Sommeria (1976), Schemm and Lipps (1978), and Findikakis (1981). One of the principal difficulties of these flows is that the Reynolds numbers are so large that eddies of length scale equal to the grid size are quite important. Consequently, the SGS eddies do not behave entirely like "small eddies;" they carry a significant fraction of the total energy and are therefore hard to model.

3.9 Summary of the State of SGS Modeling

From the arguments given above, we can reach the following conclusions about the current state of the art in SGS modeling.

1. Although they are inadequate in detail, eddy viscosity models can be used in simulating homogeneous turbulent flows. However, they seem to be inadequate for inhomogeneous flows, especially those in which solid boundaries are important.

2. For models in which the length scale is prescribed, the length scale of Eq. (3.9) is preferred.

3. One- and two-equation turbulence models are unlikely to provide significant improvement relative to algebraic eddy viscosity models. An exception to this might be transitional flows.

4. Full Reynolds stress models offer promise as future SGS models. However, the modeling assumptions probably need to be different from those used in time-average modeling.

5. The scale-similarity model is promising, but only when used in conjunction with other, dissipative, models.

6. Full simulations seem to be the best way available at present for testing SGS models and determining the parameters in them. Turbulence theories can also be profitably used in this regard.

7. Full simulations and large eddy simulations can both be used in time-averaged model building. This is the area in which both types of simulations will make their greatest impact in practical engineering calculations in the near future.

4. NUMERICAL METHODS

4.1 Mathematical Preliminaries

This chapter is devoted to setting out the numerical methods used in full and large eddy simulations. To some extent, numerical methods are always tailored to the problem; higher-level simulations of turbulent flows are no exception.

The partial differential equations governing a flow were given in Chapter 2. To complete the methematical setting, it is necessary to specify initial and boundary conditions. This is not easy. Higher-level simulations need details of the initial state, and experimentalists are unable to provide sufficient data about the initial state of their flows; some of the initial data therefore has to be invented. An equally serious problem is that, as the Navier-Stokes equations are nonlinear, it is not always known what boundary conditions should be specified, i.e., we may not know whether a problem is well-posed or not. There are a number of examples of people attempting to solve mathematically ill-posed problems. Another issue is that the partial differential equations have several conservation properties, and it is important that they be preserved in the numerical treatment of the problem. Finally, there are the difficulties inherent in the numerical methods themselves--accuracy, stability, and aliasing, among others. All of these need to be considered.

The equations governing incompressible flows are of mixed type; they contain elements of both parabolic and elliptic partial differential equations. This is a consequence of the momentum equations containing time derivatives, but the continuity equation not having any. As a result, one cannot advance the continuity equation in time. These equations are called incompletely parabolic by mathematicians. Means of dealing with both types of behavior are needed. The compressible equations, which are hyperbolic, are actually easier to deal with from a numerical point of view.

All of these issues will be taken up in the remainder of this chapter. Additionally, we shall need to describe the numerical approximations used in the computations. Throughout the chapter, it is important to keep in mind the kinds of flows that we are trying to simulate. They are geometrically simple turbulent flows. The fact that they are turbulent means that the high wave-number components of the velocity field are large. Large gradients of the variables can occur in any part of the flow; this has an important influence on the choice of numerical methods. On the other hand, the simplicity of the geometry helps considerably in developing accurate numerical methods.

4.2 Boundary Conditions

The simplest flows to be simulated are the homogeneous turbulent flows. By definition, these flows are statistically identical at every point in the flow. For these flows, the most convenient and most accurate boundary conditions are periodic ones. The portion of the flow within a rectangular parallelepiped is simulated, and the boundary conditions prescribe that the state of the fluid at a point adjacent to any of the boundaries is identically that on the opposite face of the parallelepiped. These conditions avoid the need for specifying the details of a highly chaotic motion on the surfaces and are the most realistic means of enforcing the idea that any point in the flow is indistinguishable from any other point.

There is one point that requires extra care. In homogeneous turbulent flows on which mean straining or shearing flow is imposed, it is convenient to solve for just the part of the flow containing the turbulent fluctuations; the mean flow is eliminated. When this is done, it is found that there are terms in the equations that do not admit the use of periodic boundary conditions. It is then necessary to do the computation in a coordinate system that deforms with the mean flow, and the ability to use periodic boundary conditions is restored. This will be taken up again in Chapter 5.

The only other flows that we shall consider in any detail in this report are inhomogeneous in one coordinate direction. Of course, this means that they are homogeneous in the other two directions, and these directions can be treated by the periodic conditions described above. There are two types of conditions we must deal with in the inhomogeneous direction; they follow from the nature of the flows we shall be simulating.

For free shear flows, we would like to prescribe the condition that the flow is at rest infinitely far from the shear layer. Dealing with an infinite region is difficult, and two methods have been used for this problem:

i) One can use a finite computational domain. At the top and bottom of the domain one specifies that the vertical derivatives of the horizontal components of the velocity are zero, and the component of the velocity normal to the boundary is set to zero. These are known as no-stress boundary conditions. Unfortunately, no-stress conditions imply the existence of image flows outside the computational domain; the images are reflections of the flow in the boundaries. To assure that the image flows do not interfere with the physical one, there must be no vorticity close to the no-stress boundary. This means that a considerable portion of the computational domain must be wasted in computing the potential part of the flow.

ii) One can use a coordinate transformation that maps the infinite domain onto a finite one. Standard numerical methods can then be used. It is important to choose a mapping that is compatible with the method used for evaluating derivatives. This issue will be dealt with in more detail later.

The second type of inhomogeneous flow that we shall consider is fully developed turbulent channel flow. Two different approaches have been taken for simulating this flow:

i) Deardorff and Schumann decided not to treat the wall directly. The reasons will be stated in detail in Chapter 7. Instead, they decided to compute only the part of the flow within and beyond the region in which the velocity profile is logarithmic. The boundary conditions must then assure that the velocity profile be logarithmic at the edge of the computational domain. In addition, it is necessary to specify something about the nature of the turbulent fluctuations at this boundary. They assume a relationship between the velocity and stress fluctuations at the boundary; this is the simplest assumption one can make, and there is no evidence for any other choice, but it has been called into question.

ii) One can compute the entire flow, including the region near the wall. The wall conditions are then the no-slip conditions that must be imposed at any solid surface. This is a much simpler boundary condition to deal with numerically. The price one pays is that all of the small structures near the wall must be computed explicitly; this leads to considerable difficulty, as we shall see in Chapter 7.

4.3 Treatment of the Spatial Derivatives: Conservation Properties

In all flow computations, the spatial derivatives are approximated in terms of the values of the dependent variables at grid points. Higher-level turbulence computations are no different from others in this respect; the methods used in these flows are also used in other types of flow simulation. Again we note that the geometric simplicity of the flows treated by higher-level simulations allows use of methods that might not be easily applied in more complex geometry.

Before giving the specific approximations to be used, it is important to discuss conservation properties. We believe that this issue is not emphasized sufficiently in the literature. The dynamical equations are essentially microscopic conservation equations. The continuity equation expresses conservation of mass. In the compressible case, the Navier-Stokes equations express momentum conservation (or what is the same thing, Newton's second law), and there is a separate energy equation to express the fact that total energy is conserved. In the incompressible case, the Navier-Stokes equations still express momentum conservation, but, in the absence of an explicit energy equation, they are also responsible for conserving the only significant energy in the flow--the kinetic energy. This leads to one of the principal difficulties in the treatment of incompressible flows.

By integration of the microscopic conservation equations over a finite volume, we obtain macroscopic conservation equations. For the incompressible form of the continuity equation we obtain the global conservation of mass relation:

$$\int_S \rho u_i dS_i = 0 \tag{4.1}$$

The Navier-Stokes equations give rise to the well-known momentum theorem:

$$\frac{d}{dt}\int_V \rho u_i dV = -\int_S \left[\rho u_i u_j + p\delta_{ij} + \mu\left(\frac{\partial u_i}{\partial x_j} + \frac{\partial u_j}{\partial x_i}\right)\right] dS_j \tag{4.2}$$

Finally, multiplying the Navier-Stokes equations by u_i and integrating over a finite volume, we obtain the equation of kinetic energy conservation:

$$\frac{d}{dt}\int_V \rho \frac{u_i u_i}{2} dV = -\int_S \left[\rho \frac{u_j u_j}{2} u_i + pu_j + \mu u_j \frac{\partial u_j}{\partial x_i}\right] dS_i - \mu \int_V \frac{\partial u_i}{\partial x_j}\frac{\partial u_i}{\partial x_j} dV \tag{4.3}$$

Each of these equations states that the conserved property changes only by flow of the property through the bounding surface; this is a consequence of the fact that there are no sources of any of these properties within the volume. If periodic boundary conditions are applied, the surface terms integrate to zero. In Eqs. (4.1), (4.2) and (4.3), S is the surface of the volume V.

The kinetic energy conservation equation (4.3) is especially interesting. The only non-surface term is the viscous dissipation term, which is usually small. It is essential to note that the kinetic energy within the control volume is not changed by the convection and pressure gradient terms and that the chain rule $(uv)' = u'v + v'u$ and the continuity equation are used in eliminating the volume integral of the pressure.

It is crucial that the numerical approximations to the equations retain these properties. For the continuity and momentum conservation, this is usually not difficult. It usually turns out that, if the equations are written in the proper form (the so-called conservative form we have used throughout), then almost any approximation will yield these conservation properties. The principal difficulty is with the kinetic energy. Normally, the verification that the numerical approximation guarantees energy conservation has to be done on a case-by-case basis. A means of avoiding this difficulty was found by Mansour et al. (1978). If the Navier-Stokes equations are written in the form:

$$\frac{\partial u_i}{\partial t} + \frac{\partial}{\partial x_j} u_i u_j - \frac{\partial}{\partial x_i} u_j u_j = -\frac{\partial}{\partial x_i}\left(\frac{p}{\rho} + u_j u_j\right) + \nu \frac{\partial^2 u_i}{\partial x_j \partial x_j} \tag{4.4}$$

rather than (2.1), the derivation of the conservation of kinetic energy equation (4.3) can be based on a symmetry property, and the use of the chain rule can be avoided. Since numerical approximations do not always have a chain rule but the symmetry property always holds, using the Navier-Stokes equations in the form (4.4) can simplify the job of finding well-behaved numerical methods.

Many workers (Deardorff (1970), Schumann (1973), Antonopoulos (1981), Shaanan et al. (1975), among others) have used the staggered-grid mesh system. The grid is shown in Fig. 2.2 for the two-dimensional case; the variables are given at the mesh points shown in the figure. The control volumes for the various equations are different and are displayed in the figure; we shall not give the finite difference equations here, as they appear in several other works. This grid system has the nice property that all of the conservation properties are obtained without difficulty, and, as we shall see in the next section, it gives no problem with the calculation of the pressure. It is the natural grid system for the incompressible equations and has been used more widely than any other. Part of the reason for the success of the staggered mesh system was explained by Shaanan et al. (1975). The approximation $\overline{\overline{u}_i\overline{u}_j} = \overline{u}_i\overline{u}_j$ which has been used by Deardorff and Schumann is valid in the staggered grid system, because the truncation errors represent the difference between these two terms (the Leonard stress) quite well. Stated otherwise, the staggered grid approximates $\overline{\overline{u}_i\overline{u}_j}$ more accurately than it does $\overline{u}_i\overline{u}_j$ and thus leads to great simplification in the finite difference equations.

If a regular grid is used, it is necessary to use a fourth-order finite difference method in order to assure that the Leonard stress is properly computed. This can be done, but the method is cumbersome (Kwak et al. (1975)).

Another popular method of computing derivatives in directions in which a flow is homogeneous is by means of Fourier transforms--the pseudospectral method. In this method one uses the discrete Fourier transform. Any function defined at a set of equally spaced mesh points $x_j = j\Delta x$, $j = 1,2,\ldots,N$, can be represented by the discrete Fourier series:

$$f(x_j) = \sum_{\ell=1}^{N} e^{ik_\ell x_j} \hat{f}(k_\ell) \qquad (4.5)$$

where $k_\ell = 2\pi\ell/N\Delta x$. This has the inverse:

$$\hat{f}(k_\ell) = \frac{1}{N} \sum_{j=1}^{N} e^{-ik_\ell x_j} f(x_j) \qquad (4.6)$$

which differs from Eq. (4.5) only by the sign of the exponent and the factor 1/N; thus, bpoth transforms can be computed in the same way. these results can be used in the following way: Given the values of the function $f(x)$ on the grid points $x_j = j\Delta x$, we can compute $f(k_\ell)$ from Eq. (4.6). When these are used in Eq. (4.5) and x_j is replaced by the continuous variable x, the result is an interpolation formula. As such, it can be differentiated with respect to x, and this provides a method of computing spatial derivatives. In fact, specializing the result to the grid points, we have:

$$\frac{\partial f}{\partial x}(x_j) = \sum_{\ell=1}^{N} ik_\ell \hat{f}(k_\ell) e^{ik_\ell x_j} \qquad (4.7)$$

The derivative df/dx can be computed by using the discrete values $f(x_j)$ to compute $f(k_\ell)$, multiplying the result by ik_ℓ, and computing the inverse transform. The result is an extremely accurate estimate of the derivative. This method is especially well adapted to the calculation of the derivatives of periodic functions, which explains its widespread application in the computation of homogeneous turbulent flows.

The practical use of the Fourier transform as a numerical tool is made possible by the existence of an extremely fast algorithm for its computation--the so-called fast Fourier transform (FFT) algorithm.

For later application it is important to note that this method could also be used to compute finite differences. It is not effective to use this as a tool for computing derivatives, but it can play an important role when we come to solving the equation for the pressure. As an example, we take the standard second-order central difference approximation:

$$\left.\frac{\delta y}{\delta x}\right)_j \simeq \frac{y_{j+1} - y_{j-1}}{2\Delta x} \qquad (4.8)$$

The derivative obtained from this formula can be put into the form of Eq. (4.7) with ik_ℓ replaced by $ik'_\ell = i(\sin k_\ell \Delta x)/\Delta x$; we call k'_ℓ the effective wavenumber. Effective wavenumbers are a good way to measure the accuracy of finite difference methods that are required to differentiate functions which contain significant high-wavenumber components, and it is not difficult to derive the effective wavenumber for various finite difference approximations. Some effective wavenumbers are plotted in Fig. 4.1.

Next, let us discuss the treatment of directions in which the flow is not homogeneous. For the free shear layer, we noted in the previous section that there are two ways of dealing with the direction normal to the flow (the shear direction). When the no-stress boundary conditions are used, one can generalize the Fourier method described above. The key idea is to expand the functions in terms of sines or cosines (using the set appropriate to the boundary

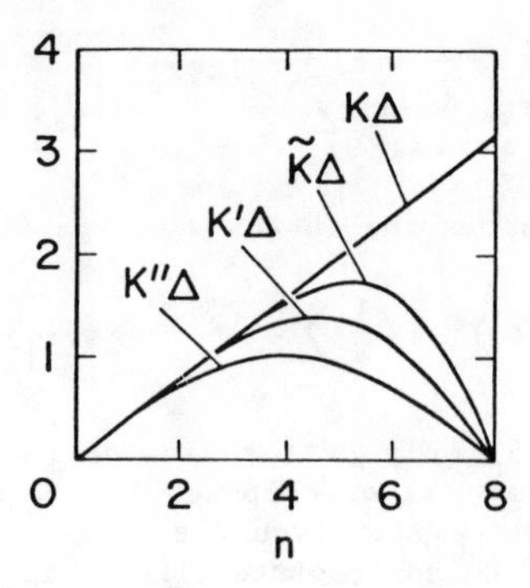

Fig. 4.1a. The effective wave member for a number of approximations. The methods used are: $k\Delta$, pseudospectral; $k''\Delta$, second-order central difference; $k'\Delta$, fourth-order central difference; $\tilde{k}\Delta$, compact method.

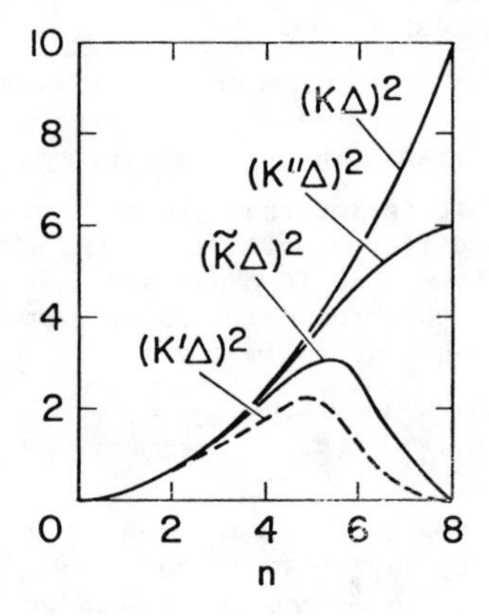

Fig. 4.1b. The square of the effective wave number in various second derivative approximations. The methods are given in Fig. 3.1. From Mansour et al. (1978).

conditions for the particular function) rather than the complex exponentials of Eq. (4.5). The numerical algorithm for computing sine and cosine transforms is equivalent to computing the exponential transform (4.5) using 2N rather than N points. Thus the cost of computing the derivatives is approximately doubled when this method is used. We noted earlier that this approach suffers from loss of accuracy due to images.

The alternative to the use of no-stress conditions is the use of a transformation which takes the physical coordinate z into a computational coordinate ζ:

$$z = h(\zeta) \tag{4.9}$$

such that $-\infty < z < \infty$ transforms to $-1 < \zeta < 1$. The derivative becomes:

$$\frac{d}{dz} = \frac{1}{h'}\frac{d}{d\zeta} \tag{4.10}$$

The trick to making this a successful method is to choose the transformation such that $1/h'$ can be expressed in terms of just a few low-order sines and/or cosines. It is then possible to obtain accuracy almost as good as that of the Fourier method for infinite regions. Details of this method are given in the report by Cain et al. (1981).

Finally we come to dealing with directions in which there are solid walls, i.e., a numerical method for treating channel flow. There are two choices that have been commonly used. The first is the use of Chebychev polynomial expansions. This is equivalent to a Fourier method on a nonuniform grid and has been used by Kells and Orszag (1980) and by Orszag and Patera (1980); see also Kim and Moin (1980).

The other method for treating the channel is to use a finite difference method on a nonuniform grid; this is equivalent to using a coordinate transformation in this direction. The choice has generally been adopted (Moin et al. (1978); Moin and Kim (1981)). We shall deal with this further in Chapter 7.

Another important issue is aliasing. Aliasing is the error introduced when two Fourier waves are multiplied; this happens implicitly when the nonlinear convective terms are computed. The waves resulting from the product of two Fourier waves contains the sum and/or difference of the original wavenumbers. These may fall outside the range of wavenumbers $(-\pi/\Delta < k < \pi/\Delta)$ which can be carried in the calculation. When this happens, the wavenumber which falls outside computational range is misinterpreted ("aliased") as one of the wavenumbers which lies inside the band. The result is a numerical error which, in mild cases, adds to the normal truncation error of a finite difference approximation and, in severe cases, can cause the calculation to become totally inaccurate or even unstable.

Aliasing can be controlled in two ways. The simplest way is to assure that the high wavenumbers are relatively unpopulated. Since these are the ones that cause the problem, eliminating them also solves the problem. In large eddy simulation, one can assure that the high wavenumbers are relatively unpopulated by using a filter which cuts off at a moderate wavenumber. In full simulations, the best way to control the problem is to keep the Reynolds number low.

The other method of controlling aliasing is to compute the portion of the field which will be aliased and explicitly eliminate it. This requires extra computation, but it allows one to include more energy in the high wavenumbers and the extra resolution gained may be worth the cost.

4.4 Time Advancement

We now consider the method of advancing the solution in time. One of the first issues that arises is that of whether to select an explicit method or an implicit one. It is important to remember that in higher-level simulations one is looking for time-accurate solutions to the equations of motion. This contrasts strongly with relaxation methods, in which the object is to find a steady state solution and to do so as quickly as possible. The point of view that we adopt is that a well-balanced, time-accurate method is one in which the errors caused by the time advancement method approximately equal those introduced by the spatial differencing method. Once spatial difference approximations and the time-advancement method have been chosen, this criterion selects the time step. The time advancement method must be stable for the time step so chosen. It is usually the case that the time step found in this way is well within the stability bounds of explicit methods, so there is no need to pay the extra cost associated with an implicit method. Thus, with a few exceptions, noted later, the time-advancement methods used in higher-level simulations are explicit. The common choices have been second-order methods such as leapfrog and Adams-Bashforth and the fourth-order Runge-Kutta method. These are standard methods of numerical analysis, so the formulas will not be given here.

For purposes of discussing time-advancement methods, it is convenient to rewrite the Navier-Stokes equations in the form:

$$\frac{\partial u_i}{\partial t} = -\frac{1}{\rho}\frac{\partial p}{\partial x_i} + H_i \tag{4.9}$$

where the viscous and convective terms have been included in H_i. There is no difficulty in time-advancing this equation by an explicit method. Most of the difficulties in solving the incompressible equations come from the lack of a time derivative in the continuity equation; the compressible equations have no such problem. One method of avoiding this difficulty is to treat the flow as if it were compressible and iteratively drive the compressibility effects to zero; the iterative nature of this process makes it inefficient, however.

A more efficient procedure is to note that application of the divergence operator and use of the continuity equation on Eq. (4.9) gives the Poisson equation for the pressure:

$$\frac{\partial^2 p}{\partial x_i \partial x_i} = \rho\frac{\partial H_i}{\partial x_i} \tag{4.10}$$

When one looks at the time-discretized version of Eq. (4.9), it is found that forcing the pressure to satisfy the Poisson equation (4.10) at time step n guarantees that continuity will be maintained at time step $n + 1$. The mixed nature of the equations is brought into clear focus. The Navier-Stokes equations (4.9) are treated as parabolic partial differential equations, but the pressure must be calculated from the Poisson equation (4.10), which is elliptic.

One further important point needs to be made here. Recall that, if the Navier-Stokes equations in the form (2.1) are used, then the derivation of the energy-conservation equation (4.3) requires use of the chain rule and the continuity equation. If we are to have numerical energy conservation, it is necessary to derive the numerical equivalent of Eq. (4.3). Assuming that the required analog to the chain rule exists, the choice made for the numerical approximation to the pressure gradient dictates the numerical approximations used in the continuity equation. Otherwise, one cannot obtain energy conservation; the usual consequence is an unstable calculation. For example, if the central difference approximation is used to estimate $\partial p/\partial x$, it must be used for the continuity equation as well. If a backward difference is used for the pressure gradient, the continuity equation must use the forward difference operator, and vice versa; this is what is done on the staggered grid.

Furthermore, one is not free to finite difference the Poisson equation (4.10) arbitrarily. The correct approximation is derived by applying the numerical divergence operator obtained in the manner described in the preceding paragraph to the finite difference version of the Navier-Stokes equations. Thus, the choice of the finite difference approximation for the pressure gradient dictates the method of differencing the Poisson equation. For example, if the central difference operator is used for $\partial p/\partial x$, it turns out that the difference operator for the Poisson equation must be the second-order central difference operator (as one might expect), but the grid spacing must be $2\Delta x$ and not Δx. We reiterate that the function of the Poisson equation is to maintain continuity in the numerical sense; it is more important to solve the correct equation than to obtain the most accurate solution to the exact partial differential equation.

The most efficient method of solving the Poisson equation is by means of the fast Fourier transform. This is the case whether one uses finite differences or the pseudospectral method, the spatial derivatives. When finite differences are used, one can solve the Poisson equation by using Fourier transforms, but one must be careful to use the effective wavenumber rather than the exact wavenumber.

The staggered grid method accomplishes all of this very efficiently. It does this so well that the need for being careful with finite difference methods is often overlooked.

There is one case in which we cannot use explicit methods. In the computation of flows with solid boundaries, it is necessary to use a very find grid in the direction normal to the wall, close to the wall. A consequence is that the time step allowed by stability is then smaller than the time step allowed by the accuracy criterion. The principal difficulty comes from the viscous term. In this case, it is necessary to treat the viscous terms containing derivatives in the normal direction implicitly. In fact, a special numerical method had to be invented for this problem; it will be sketched in Chapter 7.

4.5 Initial Conditions

The initial conditions for higher-level simulations cannot be derived directly from experimental results. The data never contain enough information to construct a complete initial field. In fact, the reported results of some experiments are quite incomplete and leave the computor so much freedom that it is always possible to find initial conditions that allow the simulation to match the experiment. From the point of view of one doing higher-level simulations, an ideal experiment reports not only the mean velocity and turbulence intensities, but information about the length scales as well. Ideally, complete spectral information should be provided.

We begin by considering the construction of a velocity field for the simulation of homogeneous isotropic turbulence; the velocity fields required by the other cases are frequently derived from this. The task is to create an initial field that has a specified energy spectrum and is divergence-free. There are several ways to do this; of these, the following is one of the easiest. There are three steps in the process:

1. Each component of the velocity at every grid point is assigned a random value. The resulting field is not divergence-free, nor does it have the desired spectrum.

2. The curl of the field is taken; the resulting field is divergence-free. The numerical operator used to take the curl must be the same as the operator used to define the divergence.

3. The Fourier transform of the velocity field is taken and in each Fourier mode is assigned an amplitude required to give the desired spectrum. The Fourier transform is inverted, and the result is the desired initial field.

This procedure is easily modified to give an initial field which is anisotropic. This can be done by biasing the random numbers used in the first step of the process.

For flows in which there is a mean velocity profile (specifically, the mixing layer and the channel), it is necessary to give the mean velocity profile in addition to the turbulence. The method of producing an initial turbulence field must also be modified. In the case of the mixing layer, we want the fluctuations to be more intense near the central plane of the flow than near the edges. Such a field can be produced in a manner similar to that described above. Insteady of allowing the field created in Step 1 to be uniformly distributed in space, we give it the desired spatial distribution. The steps for removing the divergence and producing the given spectrum are then essentially as described above.

For the channel flow it was found that the subgrid scale model destroyed too much energy and tended to make the flow become laminar if conditions of the kind described above were used. To prevent this, it was necessary to introduce large structures into the flow. These were obtained from solutions of the Orr-Sommerfeld equations. Although these are not the correct large structures for a fully developed channel flow, they are apparently similar enough to them. Randomness was added to the flow by introducing a small amount of more or less isotropic turbulence which is divergence-free and is zero at the walls.

HOMOGENEOUS TURBULENCE

5.1 Classification

A homogeneous turbulent flow is one in which each point in the flow is, in the statistical sense, equivalent to every other point. Ideally, this requires an infinite medium of fluid, every part of which experiences the same forces. In practice, close approximations to these flows are produced in wind tunnels. The mean flow is designed into the tunnel, while the turbulence is usually created by a grid (or, in a few cases, by a set of jets) and carefully controlled. The time evolution of the flow is simulated by observing its development as it moves downstream in the tunnel and invoking Taylor's hypothesis. If the gradients of mean quantities and other parameters of the flow are carefully chosen, an accurate approximation to a homogeneous flow is produced. It is not difficult to show that homogeneity requires the mean flow to be one in which the mean velocity is a linear function of all of the spatial coordinates. This severely limits the possibilities.

Nearly all turbulent flows of engineering interest are inhomogeneous; the inhomogeneity is usually the result of the shear varying through the flow. When the Reynolds stresses in these flows are modeled, five separate effects are commonly considered. They were mentioned in Chapter 3 and are repeated here:

a. Production. The creation of new Reynolds stresses via the interaction of the Reynolds stresses with the mean flow.

b. Dissipation. The destruction of turbulent energy and Reynolds stresses by the action of viscosity.

c. Redistribution. The conversion of one component of the Reynolds stress into another without change of the total turbulent energy. Much of this effect is mediated by the pressure.

d. Convection. The convection terms usually require no modeling, but their inclusion makes the local Reynolds stresses depend on the mean field in other parts of the flow.

e. Diffusion. The carrying of Reynolds stress from one part of the flow to another via the self-interactions of the turbulence.

By definition, homogeneous flows have no convection or diffusion, so we need to deal with, at most, production, dissipation, and redistribution.

Homogeneous turbulent flows can be grouped into three categories according to the phenomena contained in them. The first group contains the one flow in which the only interesting effect is dissipation. (Inertial energy transfer among the wavenumber components is, of course, an element in all flows but is not counted separately.)

Homogeneous Isotropic Turbulence. This flow, which at one time was heavily studied because it was thought that it might provide the insight into the nature of all turbulent flows, is the decay back to rest of fluid which has been set into random motion. It is still used as a means of finding turbulence model constants associated with dissipation and is usually the first flow simulated by people doing higher-level simulations.

The second group of flows contains those in which there is exchange between the various components of the Reynolds stress (redistribution) in addition to dissipation, but there is no direct production of turbulence energy. There are two such flows.

a. Homogeneous Turbulence with Rotation. The effect of rotation on isotropic turbulence is to produce anisotropy. The effect is primarily on the length scale and reduces the rate of decay of the turbulence.

b. Return to Isotropy. Turbulence which has been made anisotropic by the action of strain (see below) tends to return toward isotropy if the additional force is removed.

The final group contains the flows in which all of the phenomena that are possible in homogeneous flows actually occur. There are two major flows of this type.

a. Strained Homogeneous Turbulence. Turbulence which is initially isotropic (or nearly so) is put through a wind-tunnel section in which a fluid element is stretched in one direction and compressed in either one or two directions. The result is irrotational strain which interacts with the existing turbulence; there is considerable turbulence energy production, and the flow becomes quite anisotropic.

b. Sheared Homogeneous Turbulence. Nearly isotropic turbulence is produced in a flow which has uniform shear (a straight-line velocity profile). The effects are similar to those observed in the strained turbulence case.

The experimental data for these flows have been reviewed in a paper by the author (Ferziger (1980)).

We shall also consider flows with compressibility and mixing of a passive scalar.

All of the flows described in this chapter are ones which develop in time. It is uncertain that any of them reaches a steady state or even a self-similar state. This issue is controversial; some authors believe that a self-similar state will be reached, while others do not believe so. In any case, these flows are sensitive to the initial conditions. In turn, this means that caution is required in interpreting them and that careful documentation of the initial conditions is necessary.

All of these flows have been calculated by both full and large eddy simulation. The results show that all of the physical phenomena observed in the laboratory have been shown to be a valuable tool in evaluating turbulence models. Much of the work in the area of model validation is recent and unpublished, and we shall give a brief overview of some of the principal results. It is also worth pointing out that a complete compendium of results from full simulations of homogeneous turbulence is being assembled by Dr. R. S. Rogallo of NASA-Ames Research Center and will probably be available in the summer of 1981. His results should be an important resource for people developing turbulence models.

5.2 Isotropic Turbulence

As we have mentioned earlier, isotropic turbulence is the simplest turbulent flow. It is therefore an obvious first target for any method of simulating or modeling turbulent flows. It has long been used by the developers of Reynolds-averaged models as a basis for choosing the constant(s) associated with the dissipation. It has also been a popular choice as the first flow to be simulated by higher-level methods, and it has been used extensively as a basis for testing subgrid-scale models. We shall review this work briefly in this section.

To simulate these flows numerically, one begins with an initial condition that has the desired energy spectrum and is divergence-free. Methods of constructing such fields were described in the preceding chapter. In full simulations it is not necessary to begin the calculation with a realistic spectrum; one will develop in time. Of course, if one is trying to match an experiment, the experimental spectrum ought to be specified. In large eddy simulations of this flow, the initial spectrum is obtained by filtering the experimental spectrum.

The initial condition defines the initial Reynolds number. The Reynolds number commonly used to characterize this flow is based on the Taylor microscale λ and the turbulence intensity q. Although these may not be the optimum choices, we shall follow custom and use them. In this flow the turbulence intensity decays and the microscale increases with time, but the microscale Reynolds number decreases.

At the first few time steps, the flow field cannot be regarded as representing true turbulence. The initial field does not contain the proper higher-order statistics or correlations; only after at least some of these have developed can the field be taken as representing physical reality. We have generally taken the behavior of the skewness of the velocity derivative:

$$S = \langle (\partial u/\partial x)^3 \rangle / \langle (\partial u/\partial x)^2 \rangle^{3/2} \tag{5.1}$$

as the measure of the quality of the flow field. It is nearly zero in the initial field and quickly rises to an asymptotic value at which it tends to remain for a considerable time, except at low Reynolds numbers. The time period in which the skewness is rising is considered a "development" period. This is followed by a period in which the flow is realistically simulated. Finally, the size of the large structures grows to an appreciable fraction of the size of the computational domain, and periodic boundary conditions are no longer valid. At this point, the flow is no longer realistic, unphysical behavior is observed in the results, and the program has to be stopped.

First, let us consider full simulations of this flow. Using $64 \times 64 \times 64$ mesh points in a calculation, one is able to compute at Reynolds numbers up to $R_\lambda = 50$. This is the practical limit on most present computers; a few cases have been run on a $128 \times 128 \times 128$ grid which allows the Reynolds number doubled. These Reynolds numbers are on the low end of the experimental ones; most experiments have been run with R_λ in the range 30-400. The results of these computations match the experiment very well in terms of the decay of the turbulence intensity, the growth of the length scales, and the value of the skewness. Typical results are shown in Figs. 5.1-5.3.

The principal use to which these results have been put has been in the development and testing of subgrid scale models. Clark et al. (1979) and McMillan and Ferziger (1979,1980) used flow fields generated by full simulation of isotropic turbulence in the way suggested in Section 3.3. Some of the principal results of this work were: that the Smagorinsky model correlates very poorly with the actual SGS Reynolds stress (the actual correlation coefficient is typically .30-.40), that the width of the filter used in large eddy simulation ought to be at least twice the grid size, that changing the shape of the filter matters little, and that the model "constant" (which really ought to be called a parameter) is a function of Reynolds number that can be derived from this type of calculation. Since these results were covered in Chapter 3, we shall not repeat them here.

Full simulation has also been used to study isotropic turbulence at low Reynolds number, a purpose for which it is ideally suited. At low Reynolds numbers ($R_\lambda < 10$), it is possible to do full simulations with only $16 \times 16 \times 16$ mesh points. Interest in these flows centers on the decay rate and the skewness. The decay of isotropic turbulence can be represented by:

$$q^2 = A(t-t_o)^{-n} \tag{5.2}$$

Theory shows that the decay exponent (n) is 2.5 at very low Reynolds number, and both theory and experiment show it to be approximately 1.2 at high Reynolds number. It is therefore of interest to compute the decay exponent as a function of Reynolds number. The results are compared with experiment in Fig. 5.4.

The velocity derivative skewness defined by Eq. (5.1) is approximately .5 at high Reynolds number and can be shown to drop to zero as the Reynolds

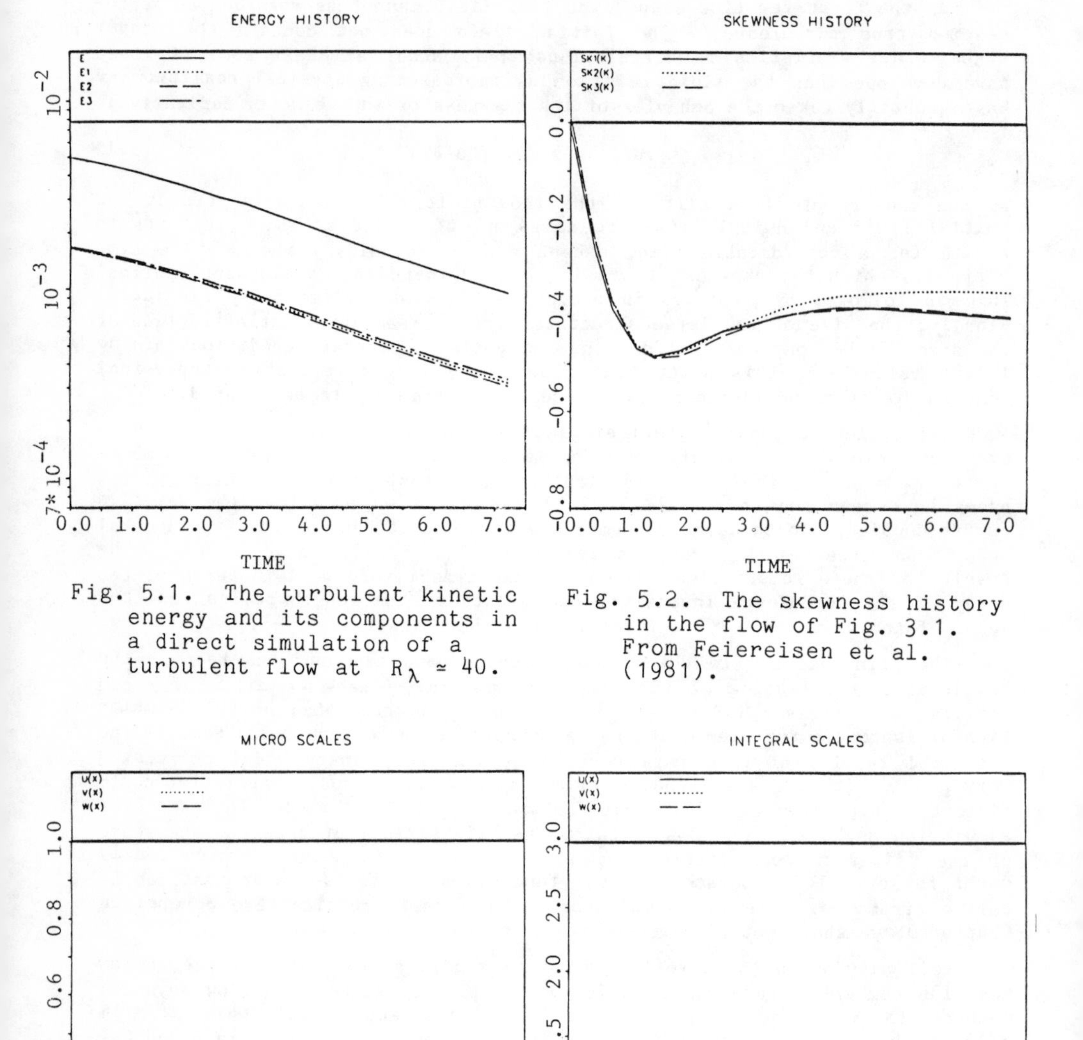

Fig. 5.1. The turbulent kinetic energy and its components in a direct simulation of a turbulent flow at $R_\lambda \simeq 40$.

Fig. 5.2. The skewness history in the flow of Fig. 3.1. From Feiereisen et al. (1981).

Fig. 5.3. The microscales and integral scales in isotropic turbulence; same flow as Figs. 3.1 and 3.2. From Feiereisen et al. (1981.

number goes to zero. The direct simulation and experimental results are shown in Fig. 5.5. Figures 5.4 and 5.5 are from a report by Shirani et al. (1981).

Now let us turn our attention to large eddy simulations of this flow. The major advantage of large eddy simulation is that, since the small eddies are modeled, the computation time is considerably reduced for a given Reynolds number. Alternatively, it is possible to go to higher Reynolds number with LES than with direct simulation.

When large eddy simulations of homogeneous isotropic turbulence were first made, the results of full simulations were not available. Consequently, the constant had to be chosen to fit the decay of the turbulent kinetic energy. It was found that the same constant can be used whether 16^3 or 32^3 mesh points were used; it was later found that the value obtained in this way agreed with those obtained from direct simulation to within 10%. It is also in good agreement with theoretical estimates (Lilly (1967)) despite the fact that these flows are at Reynolds numbers too small to support an inertial subrange. Since the constant needs to be adjusted by this amount to account for changes in numerical method (mainly changes in the spatial differencing method), this was one of the most important early successes or large eddy simulation.

It was found that it made little difference whether the primitive Navier-Stokes equations or the vorticity form of those equations were used; it made very little difference whether the model was based on the strain rate or the vorticity; and it made very little difference which filter was used. However, if pseudospectral differencing is applied to the original Smagorinsky model, the shape of the spectrum at high wavenumbers is distorted. To remedy this problem, it was found necessary to evaluate all of the derivatives that occur in the model by second-order finite-difference approximations. This is similar to the finding by Love and Leslie (1976) that the model ought to be averaged over a finite volume. A typical result obtained by large eddy simulation is shown in Fig. 5.6; other curves are similar and therefore not shown.

Finally, it was found that large eddy simulation is incapable of computing the higher-order statistical quantities such as the skewness and flatness with sufficient accuracy. These quantities are strongly affected by the small scale motions that are filtered out, and there is no way to recover the lost information; all attempts to do so failed.

Most of the results on large eddy simulation are taken from a report by Mansour et al. (1978).

5.3 Anisotropic Turbulence

Anisotropic turbulence (turbulence in which the fluctuating components are unequal so that $u_1^2 \neq u_2^2 \neq u_3^2$) usually returns to an isotropic state if not strained in any way. However, it is possible for the flow to become even more anisotropic. Thus, if the large scales are such that $\overline{u_1^2} < \overline{u_2^2}$ and the small scales have $\overline{u_1^2} > \overline{u_2^2}$ but the total field is such that $\overline{u_1^2} < \overline{u_2^2}$, it is quite likely that the turbulence will become more anisotropic with time. This is not the case in most flows, however, and the assumption that anisotropic turbulence tends to return to an isotropic state is reasonable in most flows of interest.

In the laboratory, anisotropic turbulence is usually created by straining the flow and then allowing the anisotropic turbulence to relax in the absence

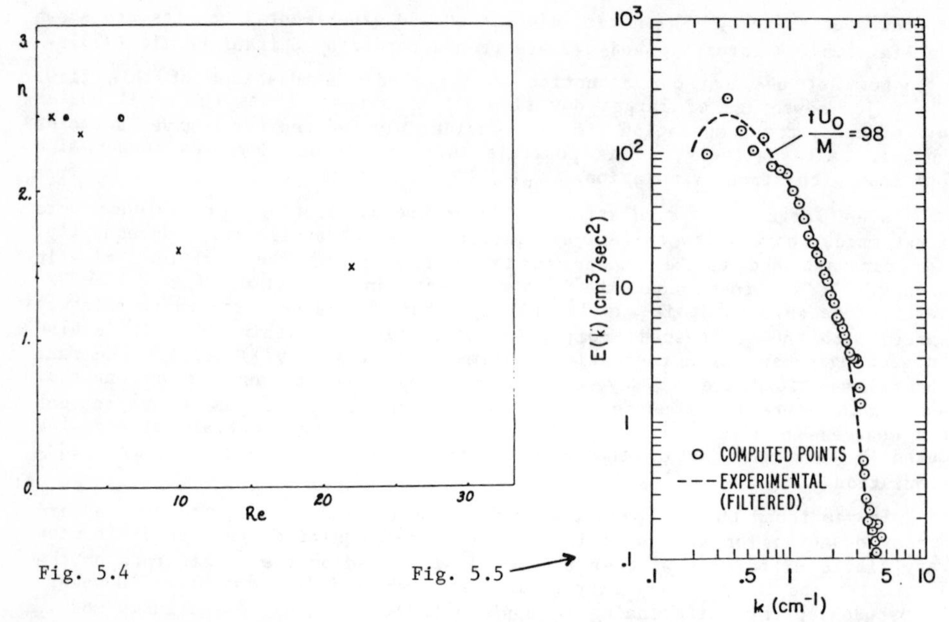

Fig. 5.4. The decay index for isotropic turbulence as a function of the Reynolds number. Crosses are computational results; other symbols are experiments. From Shirani et al. (1981).

Fig. 5.5. The filtered energy spectrum from a typical large-eddy simulation. Points are computed results; the line represents the filtered experimental data.

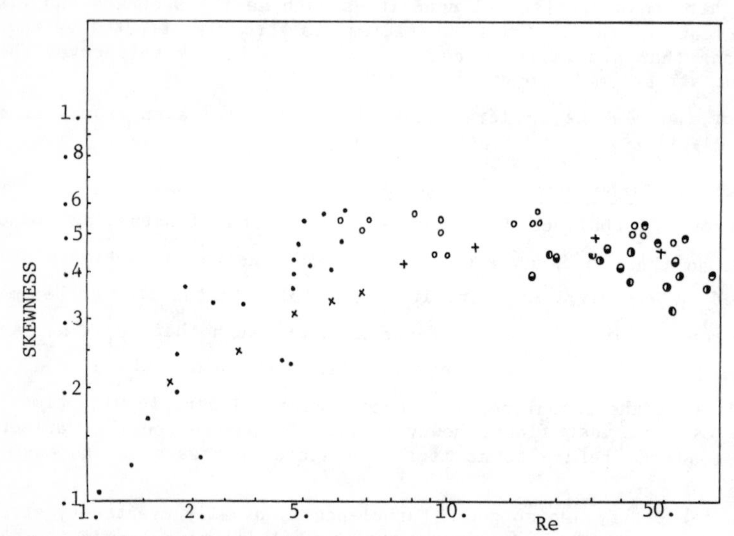

Fig. 5.6. The skewness of the velocity derivative as a function of Reynolds number for isotropic turbulence. Symbols are: x direct simulations; + large-eddy simulations; all other symbols are experimental data. From Shirani et al. (1981).

of strain. The alternative approach of using the anisotropy of turbulence created by grids has not been successful. The apparent reason is the one mentioned above--the anisotropy resides mainly in the large scales, and the flow may become more rather than less anisotropic. Creating anisotropy by straining an initially isotropic field distributes the anisotropy over the range of scales and is thus better behaved.

Simulations can emulate either of the above methods. One can simply create an initial field in which the components of the velocity fluctuations are unequal, or one can strain an initially isotropic field to produce the anisotropy. Because one has control of the anisotropy as a function of the scale size in the initial conditions in a simulation, there is no important factor favoring one method over the other. The method of creating an anisotropic initial field is preferred, as it is the simpler approach.

Full simulations of homogeneous anisotropic turbulence were made by Schumann and Herring (1976) using the method suggested above. Some of their results are shown in Fig. 5.7. We see that their flow does indeed relax toward isotropy. The tendency of the dissipation and pressure-strain terms toward their values in the isotropic flow is also evident. One should note, however, that the calculation was done on a 32^3 mesh. All of the quantities averaged in Fig. 5.7 fluctuate very strongly in both space and time, and it is likely that nearly all of the contribution to the mean values comes from a few small regions in which the fluctuations are very intense; this statement is based on some of the author's unpublished work. It is therefore likely that the uncertainty in the reported values is quite large. This is true in some of the other flows that we shall look at as well.

Schumann and Herring used their results to test two versions of Rotta's model for the return to isotropy. This model assumes that the pressure-strain term can be represented by:

$$\left\langle p' \left(\frac{\partial u_i'}{\partial x_j} + \frac{\partial u_j'}{\partial x_i} \right) \right\rangle \equiv \Phi_{ij} = -K \left(\langle u_i' u_j' \rangle - \frac{1}{3} \delta_{ij} \langle u_k' u_k' \rangle \right) \tag{5.3}$$

where one model assumes $K = C\varepsilon/q^2$, and the other assumes $K = C'q/L$, where ε is the dissipation rate and L is the integral scale. The brackets $\langle\ \rangle$ represent an average over the computational field which is assumed equivalent to an experimental time average. As can be seen from the figure, there is considerable variation in the "constant" obtained from the various runs. Clearly, this indicates that something may be wrong with the model. Schumann and Herring were not able to discern any consistent Reynolds number effects in their results.

5.4 Rotating Turbulence

The effects of rotation on turbulence are subtle and complex. In the equations of motion, the only appearance of the rotation is via the Coriolis force; the centrifugal force can be transformed away. One effect of the Coriolis force is to redistribute the kinetic energy among the components of the turbulence normal to the axis of rotation. The Coriolis force does not appear explicitly in the equation for the turbulent kinetic energy. Nevertheless, rotation has a profound effect on turbulence and, especially, on its rate of production (cf. Ferziger and Shaanan (1976)). In shear flows, rotation may in fact stabilize the flow; there is evidence that it can cause relaminarization of a turbulent boundary layer. It can also destabilize; the well-known Taylor-Görtler instability is a prime example of this.

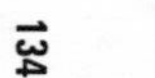

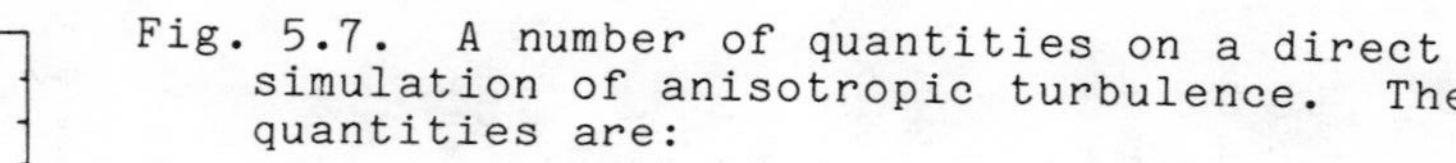

Fig. 5.7. A number of quantities on a direct simulation of anisotropic turbulence. The quantities are:

E_{ij}: Reynolds stress tensor

ε_{ij}: Dissipation tensor

Φ_{ij}: Pressure strain tensor

L_{ij}: Integral length scales

The fifth is the r.m.s. pressure gradient, and the sixth, the r.m.s. pressure fluctuations. From Schumann and Herring (1976).

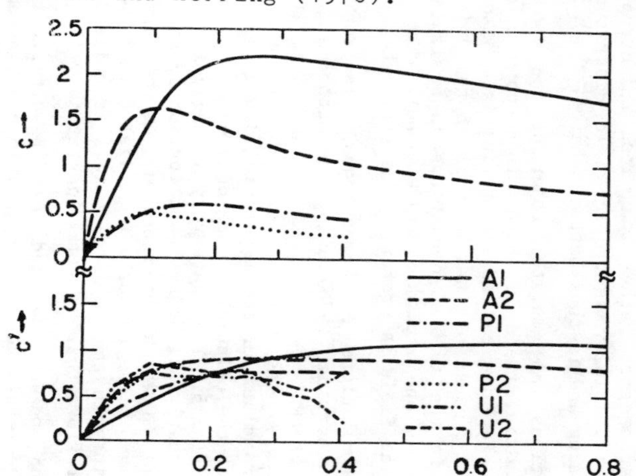

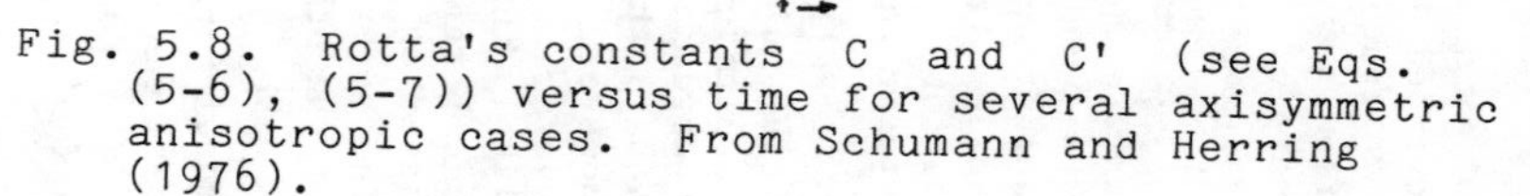

Fig. 5.8. Rotta's constants C and C' (see Eqs. (5-6), (5-7)) versus time for several axisymmetric anisotropic cases. From Schumann and Herring (1976).

The effect of rotation on isotropic turbulence is even more subtle. It seems quite likely that the principal effect is the conversion of turbulent energy into inertial waves--waves that propagate principally along the axis of rotation and which are not dissipated except near walls.

Experimentally, the study of the interaction of rotation and turbulence is very difficult. One major difficulty (about which we shall say more later) is that the fluid must be set into rotation before it passes through a grid that generates turbulence; this is a consequence of the Helmholtz theorem. Three experiments have been performed. Ibbotson and Tritton (1967) found a faster decay of the turbulence when the fluid was rotating, while Traugott (1956) found a decrease in the decay rate. The latest expeirment, and the one that is generally regarded as the best, was done by Wigeland and Nagib (1978). They found cases which went in both directions; however, the predominant effect was a decrease in the decay rate.

Since the source of the effects observed in the experiment was unknown, preliminary calculations using large eddy simulation on a 16^3 grid were made. A series of simulations using the identical initial condition with various rotation rates was made. The results, shown in Fig. 5.9, indicate that the predominant effect of the rotation may be to decrease the rate of decay of the turbulence, but there is unusual behavior, particularly at the early times. This is similar to the behavior observed by Wigeland and Nagib, but a detailed comparison is impossible.

On the basis of these results, it was surmised that rotation decreases the rate of dissipation but that this effect is masked by other effects in the early development of the flow. In order to check this hypothesis, we made full simulations of an experiment that is impossible to do in the laboratory. We allowed the turbulence to develop without rotation for a short time; this is identical to the initial stages of an isotropic turbulence experiment. When the turbulence had developed into a physically realistic field (see the preceding section for details), the rotation was "turned on." Under these conditions, it was found that increasing the rotation rate always decreased the rate of decay of the turbulence. The results are shown in Fig. 5.10.

It appears that the anomalous effects found in the experiments are caused by interactions of the rotation with the thin shear layers produced by the turbulence-producing grid, and similar effects can be produced in the simulation. These are impossible to avoid in the laboratory. In some of the experiments, interactions with the walls also play an important role.

It was also possible to search for the cause of the effect. It was found that the turbulence remains nearly isotropic, so the decrease in the rate of dissipation must be due to an increase in the length scales. Since the length scales are readily computed in these simulations, this was easily checked, and it was found that there is a large increase in the length scale in the direction of the rotation axis. A theoretical explanation for this (based on the properties of inertial waves), was given and a modification of the model was offered.

5.5 Strained Turbulence

We now come to flows in which there is turbulence production. In both the strained and sheared turbulence experiments, the turbulence decays for a short time after the start of the flow and then increases with time. The length scales of the turbulence increase more rapidly than in the unstrained decaying isotropic flow. All of this makes these flows interesting objects of study.

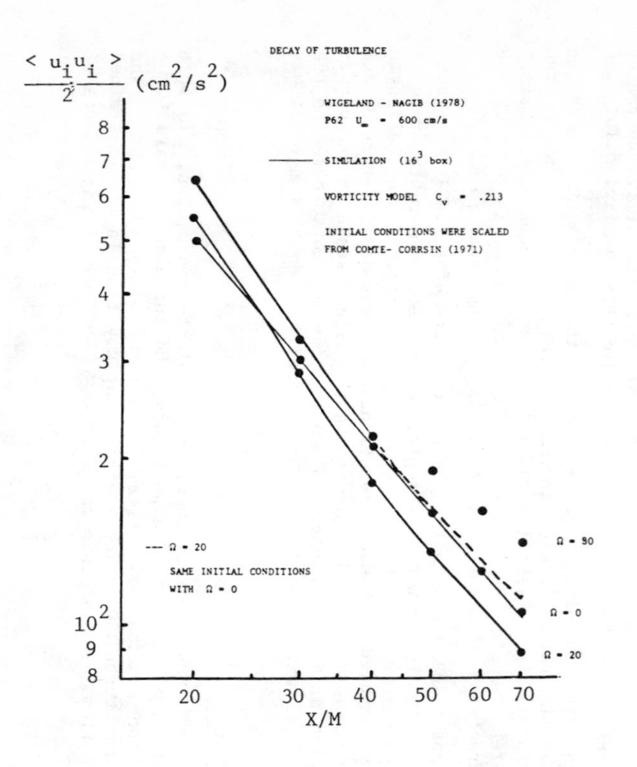

Fig. 5.9. The decay of rotating turbulence. Lines are large-eddy simulation; points are data. Rotation was present from the start of the calculation. From Bardina et al. (1981).

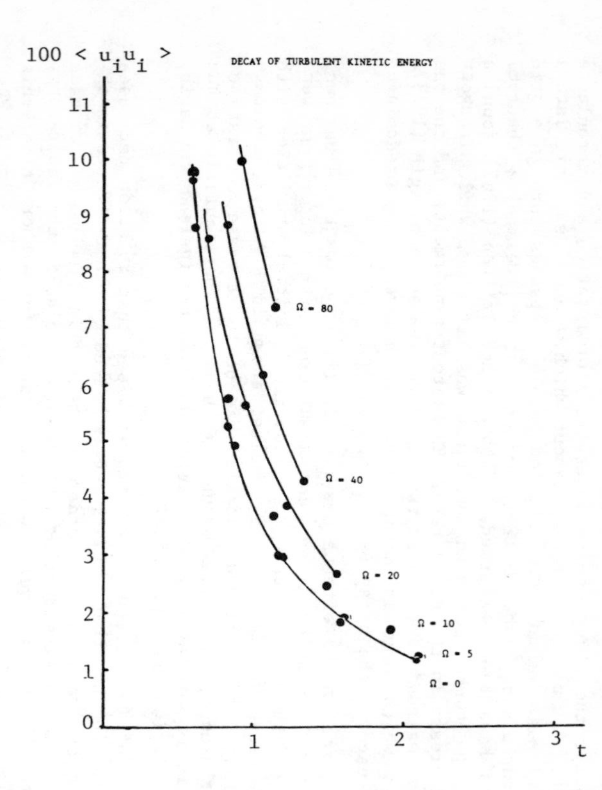

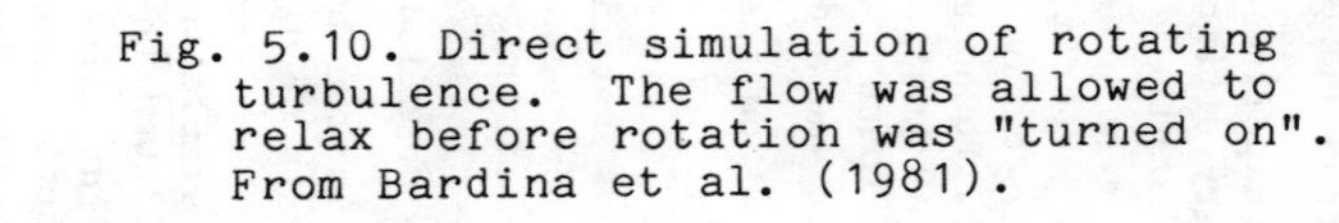

Fig. 5.10. Direct simulation of rotating turbulence. The flow was allowed to relax before rotation was "turned on". From Bardina et al. (1981).

In the laboratory, strained turbulence is created by first producing isotropic turbulence with a grid, in the same way as in the experiments described earlier. The turbulence is allowed to develop for a short time and is then made to pass through the test section. In some test sections, the cross-sectional area is kept constant but the aspect ratio in the plane normal to the flow is changed; the effect is to exert plane strain on the turbulence. In other experiments, the test section is a contraction, and the turbulence is compressed in the two directions normal to the flow and stretched in the streamwise direction; the result is axisymmetric strain.

To simulate these flows numerically, an isotropic turbulent flow field is created in the same manner as for the previous flows. The effect of the strain is turned on immediately, and the flow is allowed to develop. In order to simulate this flow correctly, it is necessary to use a straining coordinate system, one which moves with the mean flow that produces the strain. This is necessary because one of the terms due to the applied strain does not permit the application of periodic boundary conditions; the transformation removes this term. For the details of this transformation see Rogallo (1977).

Use of this transformation also introduces a difficulty. After some time, the strained coordinate system becomes quite thin in the direction which is being compressed. When this happens, the length scales in that direction become appreciable compared to the size of the computational domain in that direction. As a result, periodic boundary conditions are no longer valid, and the computation has to be stopped. This happens when the total strain $\exp(St) \simeq 2$, where $S = \partial u/\partial x$. The problem can be partially alleviated by starting with a coordinate system that is distorted in the other direction. Thus the flow contains three periods similar to those found in the flows described above. First there is a development period; this is followed by a period in which the flow is physically realistic; finally, there is a period in which the simulation is invalid, and the calculation must be stopped.

The detailed behavior of strained turbulence is dependent on the initial conditions. However, the trends are the same in all cases. As in the experiments, the turbulent kinetic energy decays until the turbulence becomes organized; then the production of turbulence increases and, somewhat later, so does the kinetic energy of the turbulence. As can be seen in Fig. 5.11, the turbulence becomes highly anisotropic. The fluctuations in the direction being compressed (the x_1-direction for the case shown in Fig. 5.11) increase most rapidly, while the fluctuations in the stretched direction (x_2) continue to decrease. The off-diagonal components of the Reynolds stress tensor are all zero in this flow.

The results of this computation could be used to test Reynolds-averaged models, but they have not been used for this purpose. The reasons are that the majority of engineering flows are shear flows, and sheared homogeneous turbulence seems more appropriate for this purpose and that the experimental data can be used as well. For this reason, Reynolds-averaged models are deferred to the following section.

McMillan and Ferziger (1980) have used strained turbulence simulations for checking subgrid scale models. They found that the Smagorinsky model becomes less accurate as the flow is strained. The correlation between the exact and model results drops from the already low value of 0.3-0.4 to nearly zero. However, the scale similarity model proposed in Chapter 3 is nearly equally valid with or without strain.

In a few cases the correlation between the exact stress and the Smagorinsky model becomes negative. On further investigation, it is found that, if

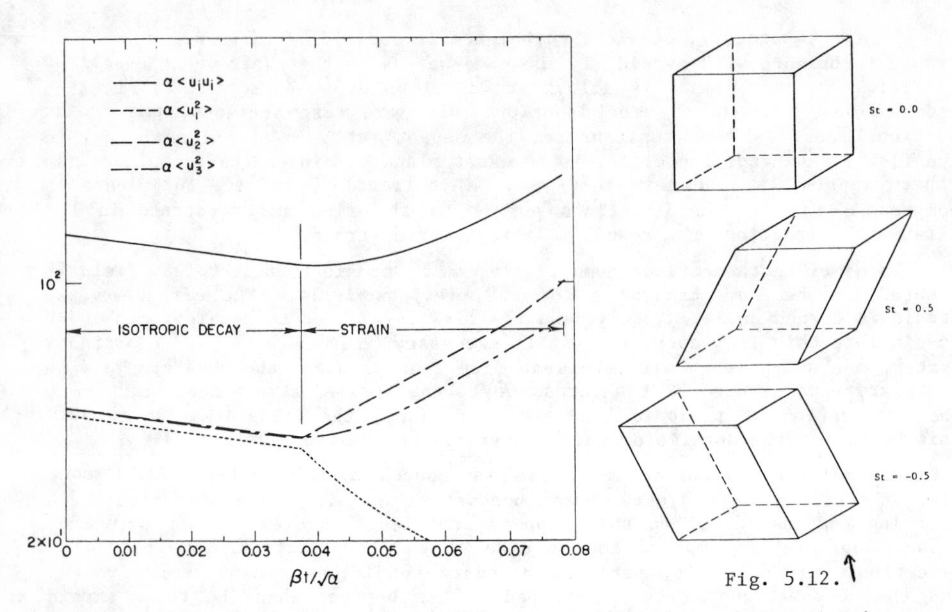

Fig. 5.11. The behavior of turbulence undergoing plane strain. From McMillan et al. (1980).

Fig. 5.12. The mesh used in sheared turbulence. Flow is started with coordinate system showed on top. When it has sheared to the position of the middle figure, the flow is interpolated onto the grid shown at the bottom. From Feiereisen et al. (1981).

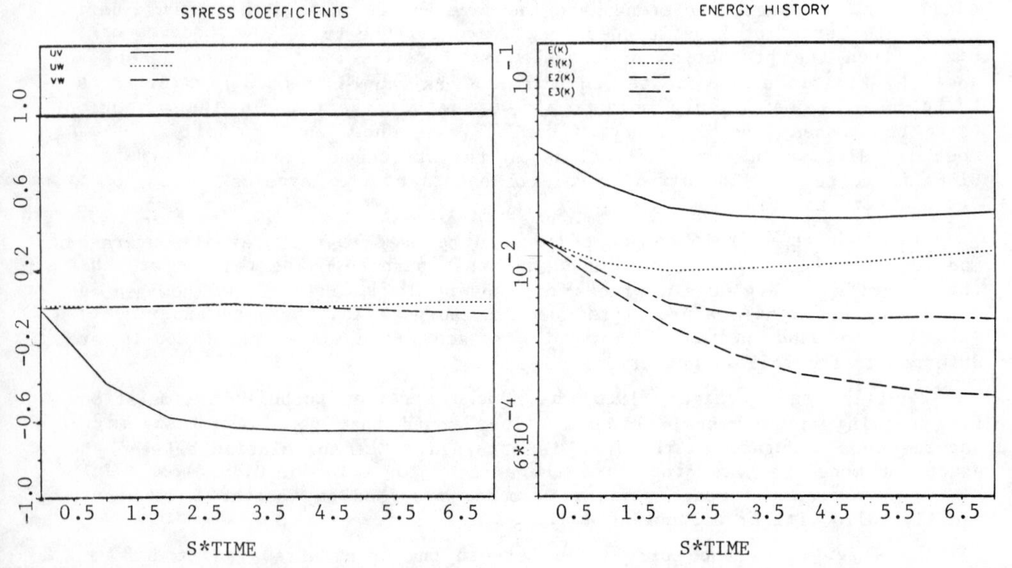

Fig. 5.13. The stress coefficiences $\langle u_i u_j \rangle / (\langle u_1^2 \rangle \langle u_j^2 \rangle^{v2}$ and energy history in sheared turbulence. From Feiereisen et al. (1981).

the strain rate is high and maintained for a long time, the energy flow is from the small scales to larger scales, i.e., from the unresolved or subgrid scales to the larger or resolved scales. This seems to be physically correct, although it has not been reported in any of the experimental results of which we are aware. it appears that the smallest scale of the turbulence may be determined by the strain rate rather than the viscosity. Direct evidence of a similar phenonemon in sheared turbulence will be presented in the following section.

5.6 Sheared Turbulence

Homogeneous turbulence interacting with mean shear behaves in a manner very similar to strained turbulence. One can regard shear as a combination of strain and rotation; the effect of the rotational component is to weaken the effect of the strain somewhat. The behavior with time is qualitatively similar to that for the strain case; after a period of decay, the turbulent kinetic energy begins to increase. The anisotropy produced is such that the streamwise component of velocity has the largest fluctuations and the normal component has the smallest fluctuations.

Homogeneous sheared turbulence is more difficult to create in the laboratory than strained turbulence. The essential reason is that, because shear has a rotational component, it cannot be suddenly introduced into the flow. It has to be created along with the turbulence. The apparatus used to produce this flow is an array of parallel channels whose flow resistances are arranged so that the velocity distribution at their exits is linear in the direction normal to the channel walls. In this way, a flow with a straight-line mean velocity profile (uniform shear) is created. With careful adjustment, the turbulence can be made to be approximately uniform across the flow. The flow is then followed down the test section, and measurements of the turbulence quantities are made at the midplane of the test section at a number of stations.

Simulation of this flow on a computer is very similar to simulation of strained flow. An initial isotropic velocity field is created in the manner described earlier. It is possible to let the flow relax before the shear is introduced, but this is not done. For this flow it is necessary to use a shearing coordinate system (one that moves with the applied linear mean flow) in order to remove the terms that forbid the use of periodic boundary conditions. The deforming coordinate system is shown in Fig. 5.12. it begins as a Cartesian system at $t = 0$ and deforms as shown until $St = 1/2$. At this point, the computational domain is on the point of becoming too narrow in the normal direction to support the use of periodic boundary conditions. This flow permits the "remeshing" of the coordinate system in the manner shown in Fig. 5.12. The shear then causes the coordinate system to become Cartesian, and the cycle is begun again. With the aid of this trick, it is possible in principle to carry on for as long as desired. In practice, the length scales in the streamwise direction eventually become too long for the computational domain, and one is forced to stop on this account. Sheared turbulence thus passes through the same three periods as strained turbulence: development, realistic representation of physics, and, finally, breakdown.

The detailed behavior of the flow may depend on the initial conditions, but the trends are essentially independent of how the calculation is started. As one can see from Fig. 5.13, the behavior of the components of the turbulence is very similar to that in the strain case. It also follows the experimental trends very well.

McMillan and Ferziger (1980) used the results of direct simulations of sheared turbulence as the basis of tests of subgrid scale models. The findings differed in no important respect from those found for strained turbulence; for this reason, we shall not give them here. However, we point out that the transfer of energy from the smallest scales to larger scales was noted in this case as well. Further evidence for this will be given below.

Let us look at the results of the simulations in somewhat more detail. Many of the results are those of Feiereisen et al. (1981), Shirani et al. (1981), and Rogallo (1981) which have not yet been published. Only partial results will be given. Three-dimensional spectra of the velocity field are shown in Fig. 5.14. We see that there is a very strong shift of the spectrum of the normal velocity component toward low wavenumbers or large scales. Care is required in dealing with the integral scales. They are the integrals of two-point correlation functions, some of which have regions in which they are negative. The negative regions can cause the integral scales to behave very erratically. The spectra probably show the length-scale behavior more accurately.

The behavior of the pressure spectrum is rather remarkable. The initial condition has a peak at a relatively high wavenumber. The pressure spectrum near the end of the physically realistic period is shown in Fig. 5.15. The spectrum is broken into two components. The decomposition is suggested by the Poisson equation for the pressure; the terms on the right-hand side of that equation can be classified according to whether or not they contain the mean velocity field. The component P_1 is a consequence of the applied mean field; it develops a k^{-5} spectrum, and the peak in the spectrum moves to the left with advancing time. The component P_2 is due to the self-interactions of the turbulence and is much more broad-band in nature. This has important consequences for pressure-strain modeling.

Finally, we show the time behavior of the terms that contribute to the spectral behavior of the turbulent kinetic energy as a function of wavenumber; these are shown in Fig. 5.16. It is seen that, as expected, the production is mainly in the large scales or low wavenumbers, and the dissipation occurs at higher wavenumbers. Finally, we note that the transfer term, which redistributes energy among the wavenumbers is negative at low wavenumbers (indicating a transfer away from the large scales) and becomes positive at higher wavenumbers. All of this is as anticipated. The surprise is that the transfer again becomes negative at the highest wavenumbers, indicating that the transfer is from both ends of the spectrum to the center. This can be taken to be a confirmation of the finding of McMillan and Ferziger discussed earlier.

Let us now look at some of the applications of these results to Reynolds-averaged modeling. Since this is the first application of this type in this report, we should first look at the possibilities. The time averages can be replaced by averages over the flow field. Although the number of mesh points is large ($64^3 = 262,144$), they cannot be regarded as statistically independent. A more realistic measure of statistical reliability is the number of large eddies captured in the computational domain. There are several ways to measure this--none of them exact--but the number of large eddies is small enough that the statistical reliability of the results is not very high. A good test of their validity is to compare results obtained from two simulations which are identical except for the set of random numbers used to initialize them.

From each realization of a shear turbulent flow, we may compute the averaged quantities as a function of time. Since the quantities vary slowly,

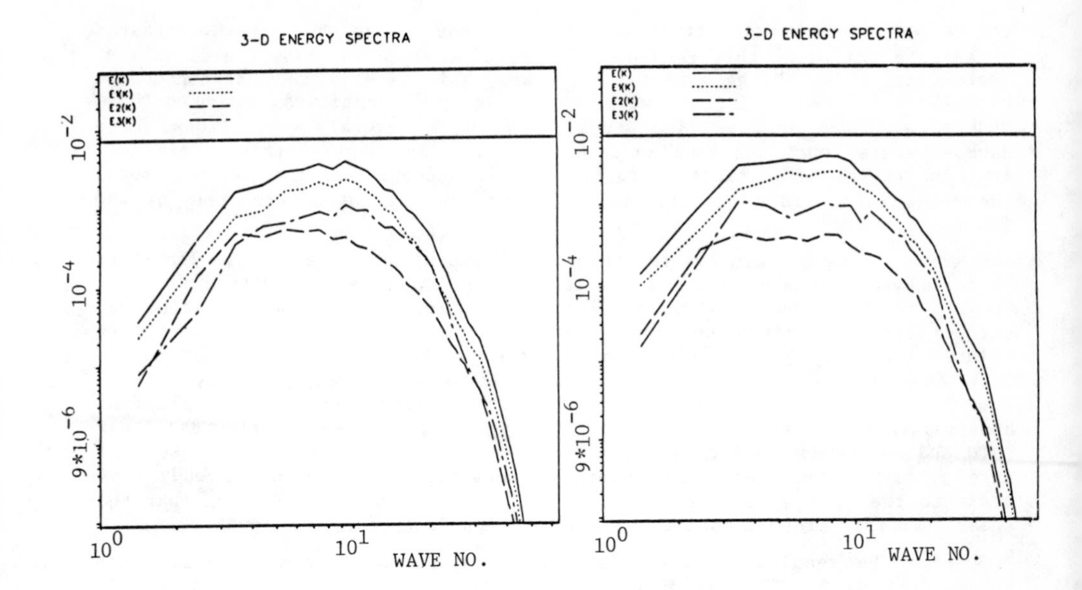

Fig. 5.14. The energy spectrum and its components at two different times in sheared turbulence. From Feiereisen et al. (1981).

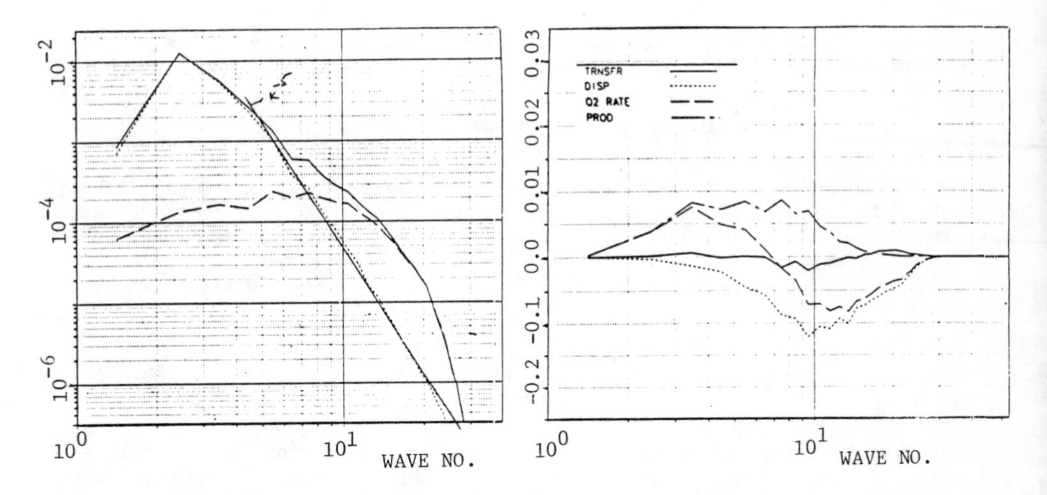

Fig. 5.15. The pressure spectrum and its components in sheared turbulence. Dashed line is "rapid" component. Dash-dot line is "Rotta" component. From Shirani et al. (1981).

Fig. 5.16. The spectra of the various terms that contribute to the change of the 3-D energy spectrum. From Shirani et al. (1981).

the values at neighboring times are not independent, and should not be treated as if they are. For this reason, we chose to analyze the flow fields only at those times at which the grid is Cartesian. This is also convenient computationally. The result is that we have the averaged quantities that need to be modeled at three or four time steps for each of several realizations. The data sets are thus much smaller than those used in subgrid scale model testing, and the kinds of tests performed need to account for this. Furthermore, one needs to consider the effects of changes in the basic parameters of the flow.

These flows contain two independent nondimensional parameters. The first is the Reynolds number. There are several length scales on which a Reynolds number can be based. The integral scale suffers from the difficulties described earlier, and we have used the microscale instead. The two should be related (possibly as a function of Reynolds number), so it does not matter much which length scale is used; however, if we try to apply the results to other flows, the choice of length scale may be very important. The second nondimensional number is the ratio SL/q, where S is the applied mean shear rate and q and L are the velocity and integral length scales. We call this parameter the shear number, and it measures the ratio of an eddy time scale to the time scale imposed on the flow. It can also be shown that the shear number is proportional to the ratio of production to dissipation.

From the results of a simulation, one can compute the Reynolds shear stress $\langle u_1 u_2 \rangle$. This is just a single quantity, and one cannot test eddy viscosity models using it alone. Eddy viscosity models could be tested by asking whether the Reynolds stress tensor, $R_{ij} = \langle u_i u_j \rangle - \frac{1}{3} q^2 \delta_{ij}$, is proportional to the rate of strain tensor

$$R_{ij} = 2\nu_T S_{ij} = \nu_T \left(\frac{\partial u_i}{\partial x_j} + \frac{\partial u_j}{\partial x_i} \right)$$

Since the model could not be tested directly, we computed the "constant" in the model defined by

$$\nu_T = - \frac{\langle u_1 u_2 \rangle}{\partial U_1 / \partial x_2} = CqL \tag{5.4}$$

and correlatee it as a function of the two nondimensional parameters given above. The result showed that C is nearly inverse to the shear number, which is equivalent to saying that $\langle u_1 u_2 \rangle / \langle q^2 \rangle$ is nearly constant; note that this result is incompatible with C's being a true constant. On further investigation, it was found that all components of the Reynolds stress anisotropy tensor:

$$b_{ij} = \frac{\langle u_i u_j \rangle}{q^2} - \frac{1}{3} \tag{5.5}$$

appear to become constant at long times in homogeneous shear flow. It is impossible to carry the calculation far enough to determine whether this is really the case or whether the b_{ij} simply change very slowly in the later stages of this flow. We are of the opinion that there is asymptotic structural similarity in this flow; this assumption has been the basis of some recent models. In many other shear flows, $\langle u_1 u_2 \rangle / \langle q^2 \rangle$ is approximately constant over a large part of the flow; for example, in the boundary layer this holds except for the region close to the wall.

Another example of model testing with these simulations is provided by the pressure-strain terms. We showed earlier that the pressure can be considered to be composed of two parts, one arising from interaction of the turbulence with the imposed mean field and the other a purely turbulent quantity. The corresponding decomposition of the pressure-strain terms is made by many modelers.

For the part of the pressure strain terms proportional to the mean strain (the "rapid" terms), one can show that, if one allows only terms which are linear in the anisotropy of the Reynolds stress, the model contains only a single constant, which for the 1,1 component can be written (Reynolds (1976)):

$$\left\langle p_1 \frac{\partial u_1}{\partial x_1} \right\rangle = 2\left(1 + \frac{A_1}{15}\right)\langle u_1 u_2 \rangle \frac{\partial U_1}{\partial x_2} \tag{5.6}$$

There are similar expressions for the other components. Given the computed values of the rapid part of the pressure-strain term, we can calculate a value of the "constant" for each of the four tensor components that are nonzero. If the model is correct, the values obtained should be the same for each tensor index and all realizations. The results showed that the "constant" is nearly independent of the Reynolds and shear numbers, but it varies by a factor of nearly seven among the various components of the tensor. These results show a deficiency in the model and suggest that an improved model should be possible, but we have so far been unable to suggest one.

The part of the pressure-strain term that results from purely turbulent interactions (the "Rotta" term) are usually modeled by:

$$\left\langle p_2 \left(\frac{\partial u_i}{\partial x_j} + \frac{\partial u_j}{\partial x_i}\right) \right\rangle = -\, C\, \varepsilon\, b_{ij} \tag{5.7}$$

This model is based on the notion that the effect of these terms is to return the flow to isotropy. It, too, is easily tested by the method used for the rapid term. It was found that the "constant" displays a great deal of variation with Reynolds number, and many of the values were below the value of 2 required for return of the turbulence to an isotropic state.

Further investigation showed that the anisotropy of the dissipation does not behave as had been expected. It is generally assumed that the dissipation is isotropic at high Reynolds numbers but may be anisotropic at low Reynolds numbers. Thus we expected to find a strong Reynolds number dependence of the anisotropy of the dissipation. In fact, we found almost no variation with microscale Reynolds number in the range from 10 to 100 (see Fig. 5.17). This does not mean that the dissipation cannot become isotropic at still higher Reynolds numbers, but it does suggest that the assumption of isotropy may be questioned.

Since the anisotropic component of the dissipation acts to reduce the isotropy of the Reynolds stress tensor, it should be included with the pressure-strain term. When the combined terms are modeled, it is found that the variation of the "constant" with Reynolds number is greatly reduced (see Fig. 5.18), and the model is fairly good. Modelers who assumed the dissipation to be isotropic have gotten reasonably good results because the anisotropy of the dissipation is implicitly included in their models.

This is a sample of some of the results obtained by Feiereisen et al. (1981) and Shirani et al. (1981). The reader is referred to those reports and

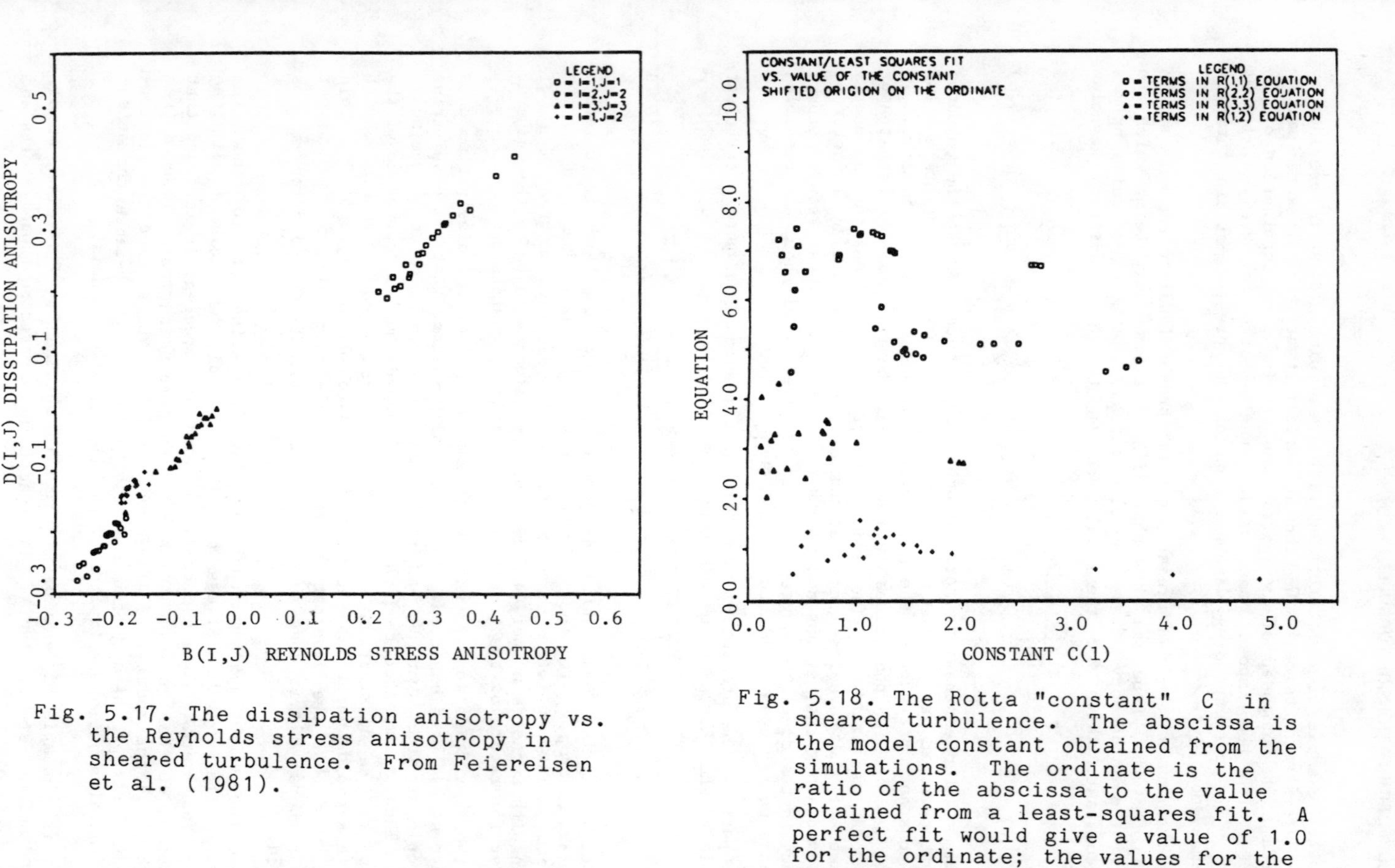

Fig. 5.17. The dissipation anisotropy vs. the Reynolds stress anisotropy in sheared turbulence. From Feiereisen et al. (1981).

Fig. 5.18. The Rotta "constant" C in sheared turbulence. The abscissa is the model constant obtained from the simulations. The ordinate is the ratio of the abscissa to the value obtained from a least-squares fit. A perfect fit would give a value of 1.0 for the ordinate; the values for the various components have been shifted for clarity. From Feiereisen et al. (1981).

forthcoming papers for more complete details.

5.7 Compressible Turbulence

It is possible to make a compressible version of the homogeneous turbulent shear flow treated in the preceding section. One need only make the velocity gradient large enough that the velocity difference across a large eddy is a significant fraction of the sound speed. It is not possible to produce this flow in the laboratory; the large velocity differences would make it impossible to maintain homogeneity. This is unfortunate, because it means that we have to believe the results of the calculation without experimental verification. We can, however, check the results at low Mach number against the incompressible experiments.

To compute this flow, the major change we need to make from the incompressible case is that the full set of compressible equations must be used. One can show that a linear velocity profile is a solution to the steady equations, and this solution can serve as the source of the shear imposed on the turbulence. In compressible computations (cf. Ballhaus (1980)), it is customary to use the continuity, momentum, and energy equations in conservation form; the dependent variable in the energy equations is usually the total energy (stagnation enthalpy). However, in the present case, this equation cannot be used without destroying the homogeneity (Feiereisen et al. (1981)), so we are forced to treat the enthalpy as one of the primary dependent variables.

The most popular numerical methods for the compressible equations are designed to relax the solution to a steady state as quickly as possible. They are not time-accurate; that is, they do not produce an accurate picture of the relaxation to steady state, and therefore they cannot be used for the purpose we have in mind. Instead, we have used a standard explicit method. The fourth-order Runge-Kutta method was chosen. The fact that all of the compressible equations contain time derivatives means that one does not need to solve a special equation for the pressure. All variables are advanced in time; the variables which are not explicitly computed from the differential equations are obtained from equations of state.

Morkovin (1963) hypothesized that compressible turbulence behaves very much like incompressible turbulence, and most models are based on this assumption. For most of the quantities in homogeneous turbulent shear flow, this hypothesis turns out to be correct. Most of the differences between the two cases are small, so we shall concentrate on the few cases in which the differences are significant.

The major difference between the incompressible and compressible flows (at least when the turbulence Mach number is not too large) is due to the appearance of acoustic waves in the latter case. The acoustic waves that are most apparent are those propagating normal to the shear, and we expect the quantities which can be affected by acoustic waves to show the most important differences from the incompressible case. The largest change is in the fluctuating velocity component normal to the shear; it is reduced relative to the incompressible case.

The most striking difference between the two flows is in the pressure and the terms associated with it. In the incompressible case, the pressure was decomposed into two parts: one arising from the mean flow that produces the shear and another that is entirely due to the turbulence. In the compressible case, there is a third term due to the presence of acoustic waves. This term turns out to be significant even at fairly low Mach numbers.

Of course, the pressure-strain terms are also affected in the same way; there are now three of them. It turns out that the third term behaves like the rapid term--the one due to the mean shear--and can therefore be combined with it. However, the "constant" is now a function of the turbulent Mach number in addition to the two dimensionless parameters of the incompressible case--the Reynolds and shear numbers. The resulting constant was fit as a function of these three parameters. The results are shown in Fig. 5.19, and the Mach number dependence is found to be significant.

Further details and results for this flow can be found in the report of Feiereisen et al. (1981).

5.8 Mixing of a Passive Scalar

By definition, a passive scalar is any quantity that can be convected by a flow and diffuse through it without affecting the velocity field. There are many applications that require knowledge of how a passive scalar behaves; any problem in which heat or mass transfer is important is in this class. Understanding the mixing of a passive scalar is also a preliminary to handling reacting flows, including combustion.

A passive scalar could be introduced into any flow treated in this chapter. In fact, only two of these have been done experimentally; these are isotropic turbulence and sheared homogeneous turbulence, so these are the cases which have been simulated. One also has to decide whether the scalar has a mean component or not. In the experiments, isotropic turbulence has been measured without a mean gradient of the scalar, and the shear flow has been performed both with and without a mean scalar gradient. To facilitate comparison with these experiments, isotropic turbulence was simulated with an isotropic scalar field, and the shear flow had a mean scalar gradient.

The equation describing the scalar concentration is:

$$\frac{\partial C}{\partial t} + \frac{\partial}{\partial x_j} u_j C = D \frac{\partial^2 C}{\partial x_j \partial x_j} \tag{5.8}$$

If there is a mean scalar field, it is subtracted from the total scalar field to obtain an equation for the fluctuating scalar field. The velocity field is also decomposed into its mean and fluctuating parts. The resulting equation for the scalar fluctuations has the same difficulty as the equation for the velocity field--the mean shear and mean scalar gradient terms do not admit the use of periodic boundary conditions. To remove this problem, the coordinate transformation made for the momentum equations has to be made here as well. It is possible to compute the velocity field prior to the computation of the scalar field, but this would require storing an enormous data set on tape and transferring it back into the machine as needed. For this reason, the velocity and scalar fields were computed simultaneously. The numerical methods used for the scalar field are identical to those used for the velocity field.

In the case of the isotropic field, the most important items to study are the decay rates of the velocity and scalar fields. The scalar field follows a decay law similar to Eq. (5.2):

$$\overline{c'^2} = B(t-t_o)^{-m} \tag{5.9}$$

where c' is the fluctuating part of the scalar field, i.e., $c = \langle C \rangle + c'$. We wish to look at the ratio m/n. The parameters on which this ratio depends are the Reynolds number and the Prandtl or Schmidt number, which is the ratio

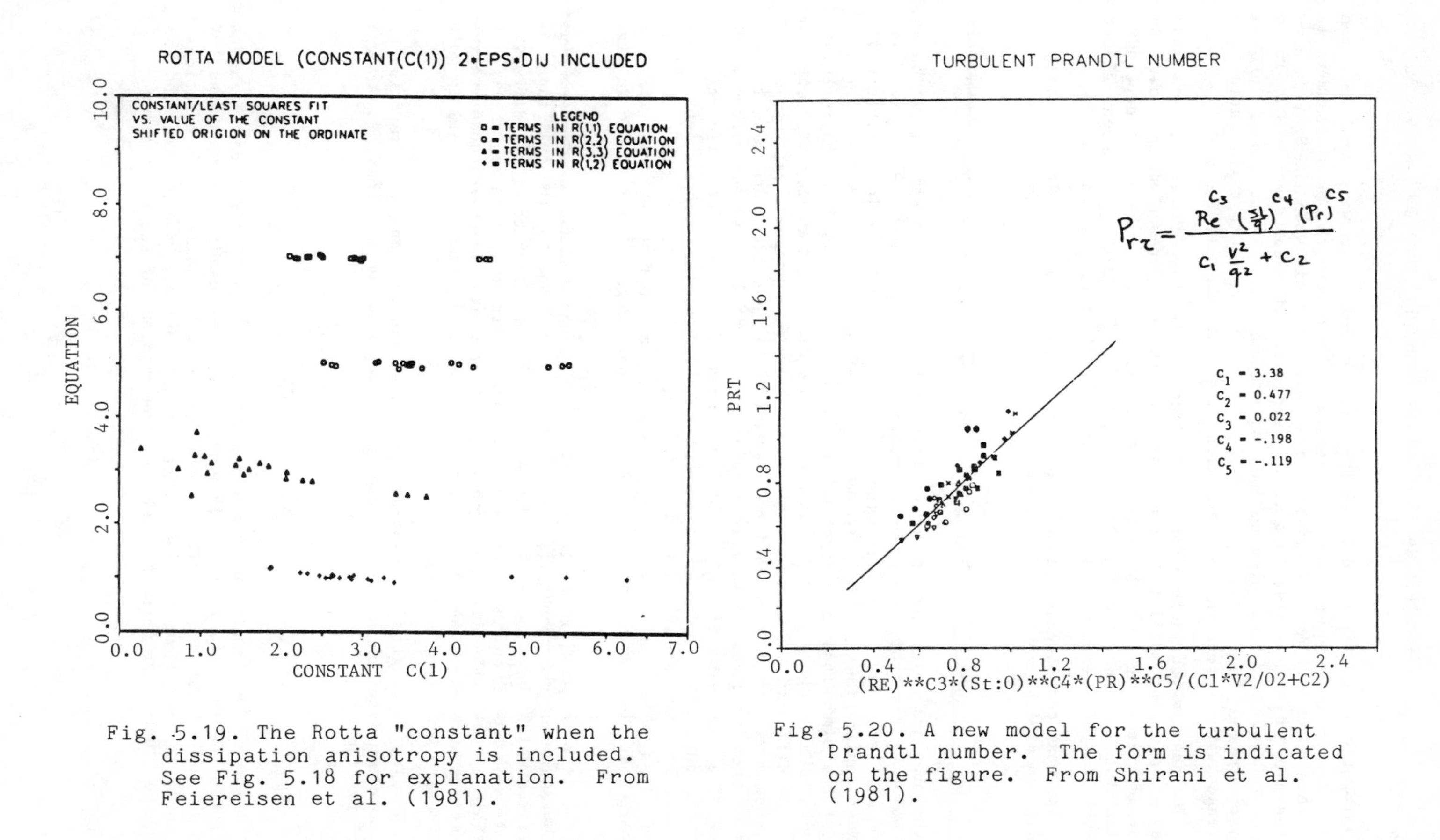

Fig. 5.19. The Rotta "constant" when the dissipation anisotropy is included. See Fig. 5.18 for explanation. From Feiereisen et al. (1981).

Fig. 5.20. A new model for the turbulent Prandtl number. The form is indicated on the figure. From Shirani et al. (1981).

of kinematic viscosity to diffusivity ($Sc = \nu/D$). It was found that the scalar decays more rapidly than the velocity field when the Schmidt number is less than unity and more slowly than the velocity field when the Schmidt number is greater than unity; this is no surprise. The dependence of the ratio m/n on Reynolds number also depends on whether the Schmidt number is less than or greater than unity. For $Sc < 1$, it is found that the ratio m/n decreases with increasing Re, and vice versa for $Sc > 1$.

The cases which include shear and a mean gradient of the scalar were analyzed in a manner similar to that used for the homogeneous shear flow. An important result is that the behavior of the scalar field becomes independent of the initial conditions after a short time. Its properties depend almost entirely on the velocity field and the mean gradient of the scalar.

The next quantity studied was the scalar flux $\langle u_i c \rangle$. This quantity is usually modeled by gradient diffusion:

$$\langle u_i c \rangle = D_{ij} \frac{\partial \langle C \rangle}{\partial x_j} \tag{5.10}$$

In the standard case, the concentration gradient is in the same direction as the velocity gradient; the nonzero gradients are $\partial U_1/\partial x_2$ and $\partial C/\partial x_2$ and there are two nonzero eddy diffusivities, D_{12} and D_{22}. The important one in most applications is D_{22}. It was computed for a number of different values of the dimensionless parameters of the flow. One can form the turbulent Prandtl/Schmidt number, Pr_T, by taking the ratio of the eddy viscosity to the eddy diffusivity. A number of models have been proposed for Pr_T, and the ones that were recommended most strongly in the literature were tested. None of them was found to be very accurate. A new model was constructed which gives D_{ij} in terms of b_{ij}, the anisotropy of the Reynolds stress tensor. Although this model models a low-order quantity in terms of a higher-order quantity, it can be made into a useful correlation by using other correlations; this model was found to be a significant improvement over the ones suggested in the literature. A test of the new correlation is shown in Fig. 5.20.

It is also possible to compute the other nonzero elements of D_{ij}. Due to the design of the computer program, this was not done for the full range of cases for which D_{22} was computed. Also, since the elements of the diffusivity tensor depend on the nondimensional parameters and, because it was not possible to match the Reynolds number used in the experiments, a quantitative comparison with experiment is not possible. However, the results are in good qualitative agreement with the data.

We also correlated the mean-square scalar fluctuations as a function of the nondimensional parameters. The principal finding was that, to a good approximation,

$$\frac{\langle c^2 \rangle}{\partial C/\partial x_2} = \frac{\langle q^2 \rangle}{\partial U_1/\partial x_2} Sc^{.1} \tag{5.11}$$

One can also construct models for the scalar field based on the ideas used for the velocity field. In particular, one can derive equations for $\langle c^2 \rangle$ and $\langle cu_2 \rangle$, which are similar to the Reynolds stress equations. The terms in them that are most difficult to model are the correlations between the fluctuating pressure field and the gradient of the fluctuating concentration $\langle p\ \partial c/\partial x_i \rangle$. They are analogous to the pressure-strain terms, and models for them can be based on models used for the latter. In particular, the pressure decomposition used in deriving pressure-strain models can be used here as well.

The model for the rapid term (the one containing the pressure derived from the mean shear) contains no adjustable constants. However, we introduced an arbitrary multiplicative constant and found good agreement between the exact and model results. The constant was found to be approximately 0.5, indicating that the arguments made in deriving the model are deficient. Another model suggested by Lumley to overcome some of the undesirable properties of the first model was tested and found to be less accurate than the first model.

The term arising from the component of the pressure that depends entirely on the turbulence was modeled by an analog to Eq. (5.7). The results show this model to be quite good--about as good as the modified Rotta model described earlier.

6. FREE SHEAR FLOWS

6.1 Overview

Free shear flows are one of the classes of flows of major technological interest. They occur in many kinds of devices, and we shall begin this chapter by briefly describing the types of free shear flows.

Free shear flows can be divided into three major categories; there are also more complex cases. The three major types are:

1. Mixing layer. This is the flow that occurs when two parallel flows of different velocity are brought together. In the laboratory this flow is created by having the fluid of different speeds on opposite sides of a dividing plate. At the end of the plate, the two streams come into contact, and the thickness of the layer in which the velocity gradient occurs grows with downstream distance.

2. Jet. A stream of high-velocity fluid issuing from an opening is called a jet. As the high-speed fluid mixes with the surrounding lower-speed fluid, the maximum velocity of the jet decreases, and the rate of growth of its thickness also decreases. The most commonly studied jets are the plane and round ones, but others, such as the rectangular jet, have been studied.

3. Wake. A wake is similar to a jet, but it is a velocity defect in an otherwise uniform stream. Like the jet, the wake has decreasing velocity gradients with downstream distance. Most wakes result from flows around bodies. The wakes form by merging of the boundary layers behind the body or from separation of the boundary layers.

We should also mention:

Complex shear layers. This is not a single type of flow, but a category containing the flows that do not fall into the above categories. Curved jets and wakes are quite common. Another important flow is one in which a laminar boundary layer separates, creating a free shear layer. The free shear layer then undergoes transition to a turbulent free shear layer which grows so rapidly that it soon reattaches to the surface. This is a common mechanism of transition from a laminar to a turbulent boundry layer.

It is also important to distinguish the early phases of free shear flows from the far-downstream flows. The early stages are sensitive to the initial conditions. Fully developed free shear layers are usually self-similar in nearly all of the measured variables and grow according to a power of the downstream distance. A majority of free shear layers occurring in applications are of the early type, but fully developed cases are also of importance.

To date, there have been large eddy and full simulations of mixing layers and full simulations of wakes. The jet has not yet been simulated (although it probably will be in the near future). Complex free shear flows have also not yet been attempted. Thus we shall devote the rest of this chapter to the mixing layer and the wake.

Nearly all laboratory free shear flows develop with downstream distance. It is much easier to simulate a layer that develops in time. One must be very careful in comparing the two cases. Consider the mixing layer. Fluid elements on the two sides of the laboratory shear layer have been in the flow for differing amounts of time. As a result, the development of the flow is not symmetric and the plane on which the mean velocity is the average of the two free stream velocities is inclined. The simulated mixing layer is, however, symmetric. The two flows may be compared if, in the laboratory flow, the velocity difference across the flow is small compared to the average velocity. This experiment requires a long apparatus, but cases exist which meet this criterion fairly well, and these are the ones to which the simulations should be compared.

6.2 Mixing Layer

As discussed above, this is the simplest of all of the free shear flows. Despite the apparent simplicity of this flow and the large number of experiments that have measured it, there is still controversy about it. Let us consider the fully developed mixing layer first and the transitional case later.

It is generally agreed that the velocity profile of the fully developed mixing layer is self-similar, and so are the components of the Reynolds stress tensor. Another point of general agreement is that the growth of the free shear layer is linear in the fully developed region. The major point of disagreement in this regime of the flow concerns the rate of growth of the layer. For the mixing layer sketched in Fig. 6.1, the growth rate parameter is conventionally defined as:

$$\frac{d\delta}{dx} = \sigma \frac{u_1 - u_1}{u_1 - u_2} \tag{6.1}$$

The measured values of σ cover the range 0.08–0.16, a much wider range than would be expected for a flow this simple. Birch (1980) recently reviewed the data and believes that there is a single correct value of this parameter, which he believes to be 0.115. However, no reason was given for the spread in the data.

There is less agreement about the early stages of the shear layer. One group, including Roshko and his coworkers and Browand and his coworkers, among others, believes that this part of the flow is essentially two-dimensional. In this view, the initial laminar shear layer rolls up into two-dimensional vortices, which then agglomerate or pair to form larger vortices of the same type (with larger spacing). This process has been observed to continue for several pairings. At this point the flow reaches the end of the apparatus.

According to this view, the important process in the growth of the mixing layer is the pairing of the vortices. However, there is evidence that streamwise vortices form in this flow. This is a kind of three-dimensionality, but it is quite regular rather than chaotic.

The other view, held by Bradshaw and others , is that the mixing layer is normally strongly three-dimensional and chaotic. According to this picture, the highly two-dimensional layers that some experimenters have observed are the result of careful arrangement of the initial conditions and design of the experimental apparatus.

Large eddy simulations of the mixing layer were made by Mansour et al. (1978). They used the vorticity equations rather than the primitive equations because the vorticity is confined to a relatively narrow region of the flow. In fact, it appears that it makes little difference which set of equations is used. The subgrid scale model had to be modified to account for this change. At the top and bottom of the computational region, no stress boundary conditions (see Section 4.5) were applied. Fourier sine and cosine transforms were used in the normal direction.

This work showed that it is possible to explain the rapid growth of the mixing layer by vortex pairing. The flow was begun with an initial condition that contained well defined two-dimensional vortices. Although there were only two vortices in the computational domain, the boundary conditions imply that they are part of an infinite array. Various perturbations to a regular vortex array were tried. It was found that small perturbations would cause the vortices to pair. Naturally, the pairing occurred more rapidly when the perturbation was larger. Surprisingly, it was found that the mean velocity profile (defined by averaging the velocity over a plane) was self-similar and that the growth of the momentum thickness of the layer was very nearly linear.

A number of three-dimensional perturbations on this basic flow were also made. First, small, random, three-dimensional disturbances were added to the initial conditions. The three-dimensionality was somewhat amplified by the pairing process, but there were only minor changes in the overall properties of the flow. Another variation was produced by the addition of streamwise vortices to the initial condition. The streamwise vortices were distorted in the pairing process, and they produced slight kinks in the large two-dimensional vortices that result from the pairing. It was conjectured that the kinks would produce larger-scale instability of the mixing layer and would then lead to considerable three-dimensionality, but this could not be demonstrated because the number of grid points was severely limited.

A simulation of the initial stages of the mixing layer was made by Cain et al. (1981). This simulation used numerical methods described in Section 4.5. The transformation of an infinite region to a finite one was used, and the modified Fourier method of taking spatial derivatives in the normal direction was used. The initial profile was a laminar mixing layer with a small random disturbance; the disturbance was strongest on the center plane of the layer.

The results turned out well. Use of the coordinate transformation and the Fourier method allowed the method to be continued until the original layer had grown by nearly a factor of ten in some cases. No effect of image layers was found, and, in most cases, the calculation was stopped only because the layer developed horizontal scales which were too large to permit application of periodic boundary conditions.

Several variations in the computational method were tried. A full simulation was made; this calculation was stopped because the turbulence in

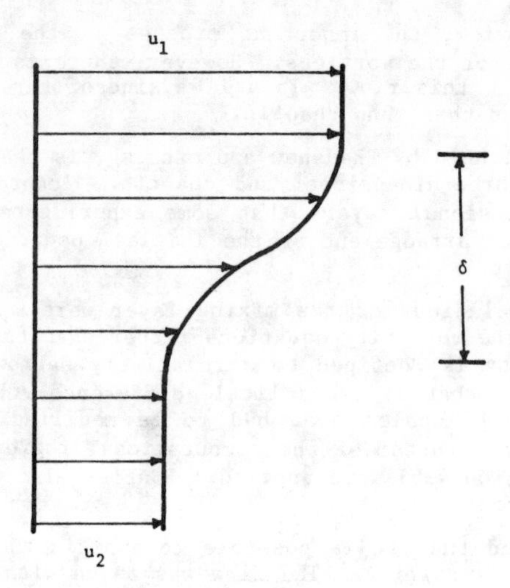

Fig. 6.1. A schematic of the mixing layer.

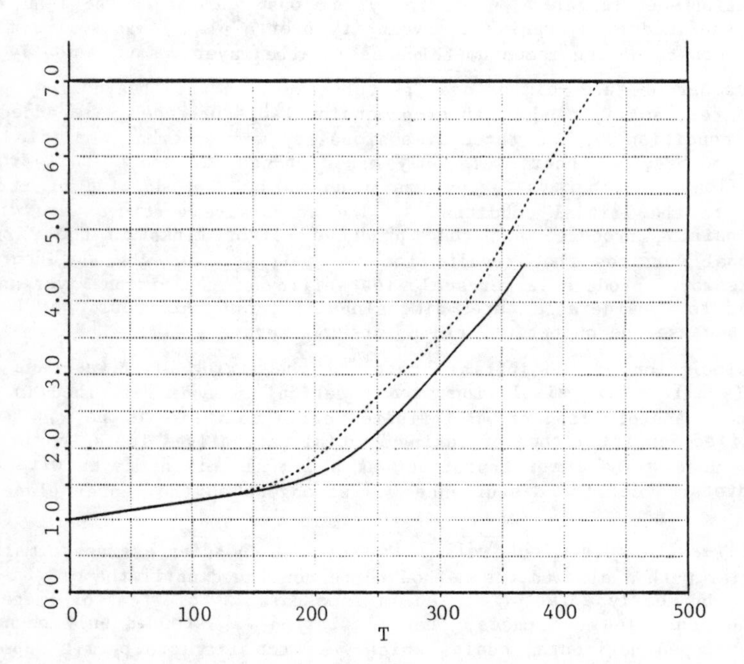

Fig. 6.2. Momentum thickness of a developing mixing layer vs. time; low initial intensity cases. From Cain et al. (1981).

the small scales became too strong. The calculation was repeated with filtering, but no subgrid scale model; the problem with the small scales disappeared, and the calculations could be carried almost twice as far in time, at which point the difficulty with the large scales appeared. A final calculation with both filtering and the subgrid scale model was made; it differed only a little from the preceding case. Thus, most of the results were obtained with filtering but no model.

Simulations were made with three levels of initial disturbance. In the low initial turbulence cases, the turbulence intensity was four orders of magnitude smaller than that of a fully developed turbulent layer; this might represent the behavior of a mixing layer produced from laminar boundary layers. The medium initial turbulence level was two orders of magnitude stronger. The high initial turbulence level cases started with turbulence intensities nearly those of the fully developed layer; these might represent a mixing layer produced from turbulent boundary layers. Cases which differed only in the set of random numbers used to generate the initial conditions were also run.

The results show that the low-turbulence cases produced a layer in which the momentum thickness grew very slowly at first but, after a latency period, grew linearly with time at a rate similar to that observed in experiments. The medium-level case gave a shorter latency period and a slightly slower rate of growth at later times. Finally, the high-turbulence level cases gave almost no latency period at all but a still slower growth rate. These results are in qualitative agreement with experimental data. They are shown in Fig. 6.2, 6.3, and 6.4.

All of the cases have mean velocity profiles that are self-similar.

The growth of the turbulence level on the center plane of the mixing layer is shown in Figs. 6.5, 6.6, and 6.7. In the low-turbulence case, the turbulent kinetic energy grows exponentially in the early stages and then levels off; the exponential growth rate is close to that of the most rapidly growing mode according to linear stability theory. The medium initial turbulence cases show similar growth, but the exponential period does not last as long. The high initial turbulence cases grow only slowly as they begin near the level for a fully developed layer. In all of the cases the kinetic energy of the turbulence overshoots the value for the fully developed layer before settling down. This has been observed in some experiments.

The profile of the turbulent kinetic energy is shown for a typical case in Fig. 6.8. The initial profile is too broad compared to the fully developed profile. This is corrected, but the profile becomes too thin before the final state is reached.

The simulations were also used as the basis for flow visualizations. A grid of "dye lines" was placed on the center plane of the flow at the initial time. The ones in the streamwise direction are essentially vortex lines in the low-intensity cases and remain so by Helmholtz's theorem. The dye lines are moved with the flow, and pictures are drawn at various times. The initial picture is shown in Fig. 6.9, and the final result is shown for two different initial fields in Figs. 6.10 and 6.11. It is clear that the layer has rolled up into vortices, but they are much more two-dimensional in one case than the other. We believe that the three-dimensional shear layer does roll up into vortical structures, but that these structures do not have spanwise uniformity except when precautions are taken to insure that the three-dimensional disturbances are weaker than the two-dimensional ones.

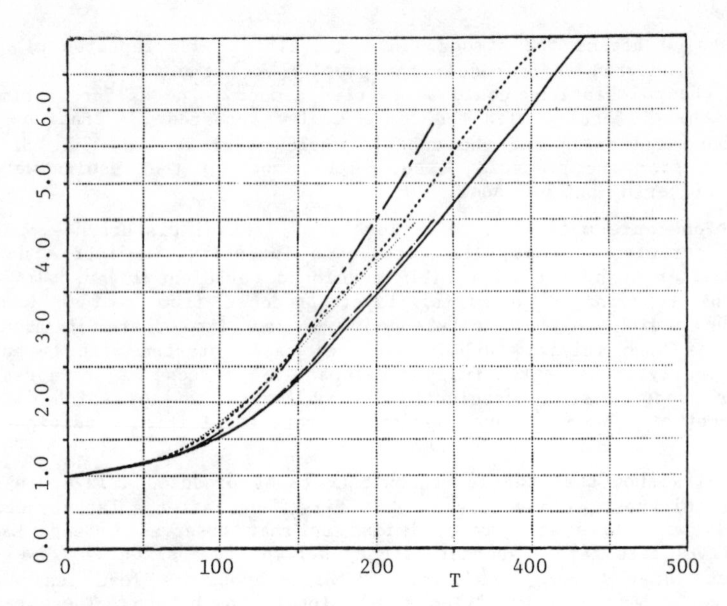

Fig. 6.3. Momentum thickness of a mixing layer vs. time; medium initial intensity cases. From Cain et al. (1981).

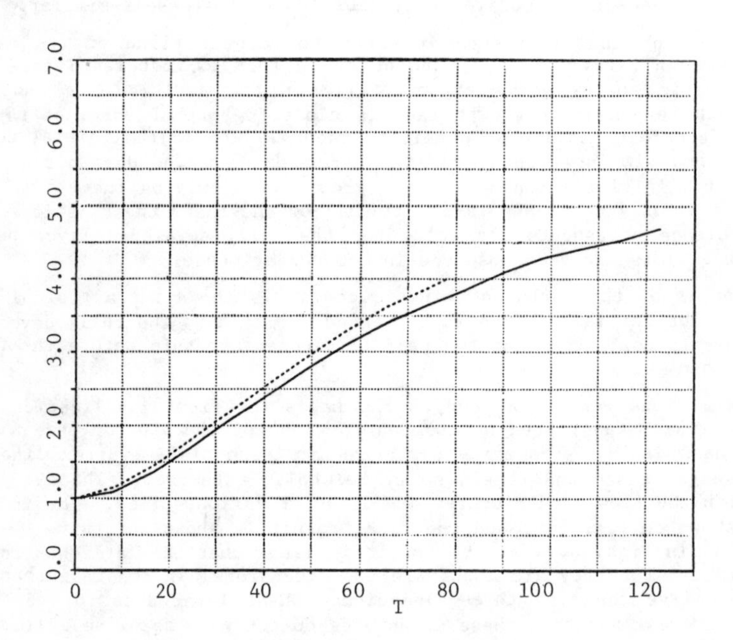

Fig. 6.4. Momentum thickness of a mixing layer vs. time; high initial intensity cases. From Cain et al. (1981).

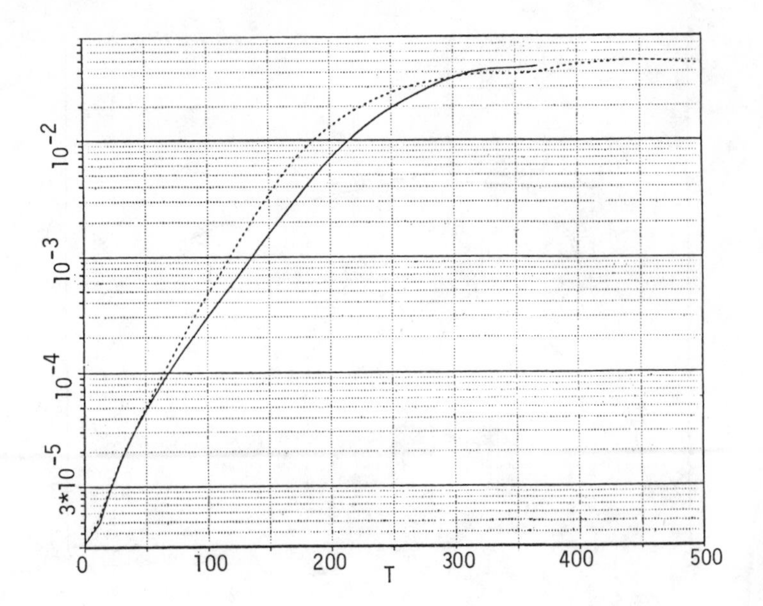

Fig. 6.5. Turbulence intensity at center of mixing layer vs. time; low intensity cases. From Cain et al. (1981).

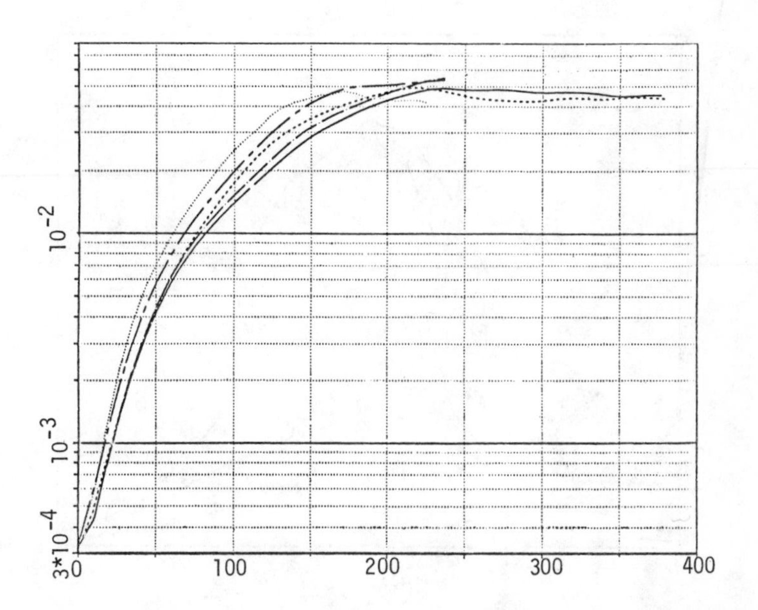

Fig. 6.6. Turbulence intensity at center of mixing layer vs. time; medium initial intensity cases. From Cain et al. (1981).

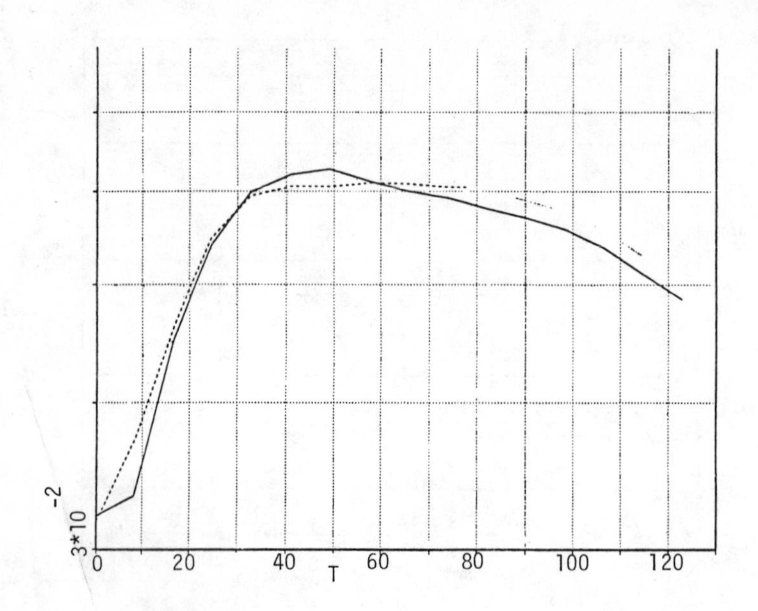

Fig. 6.7. Turbulence intensity at center of mixing layer vs. time; high initial intensity cases. From Cain et al. (1981).

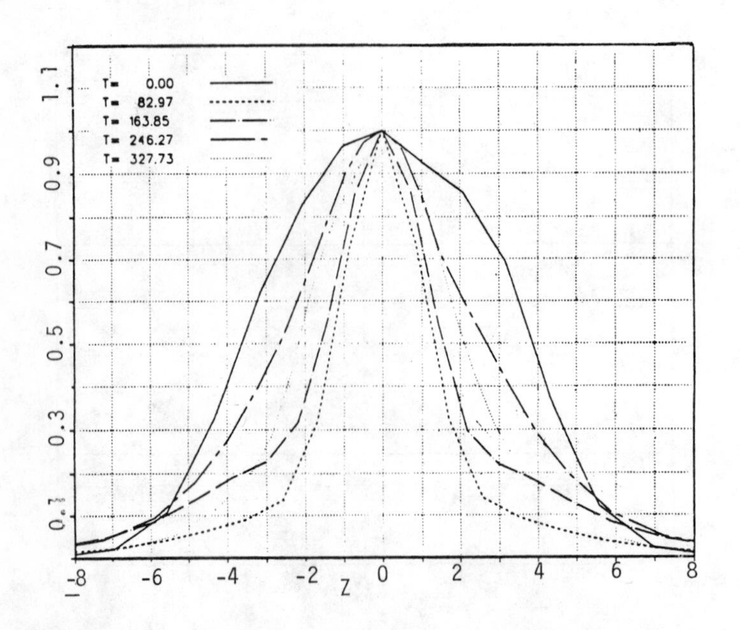

Fig. 6.8. Profiles of turbulence intensity vs. time; low initial intensity case. From Cain et al. (1981).

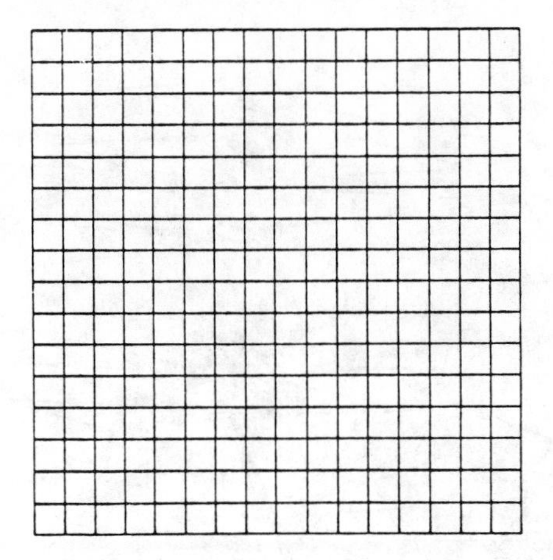

Fig. 6.9. "Dye lines" at initial time in mixing layer. From Cain et al. (1981).

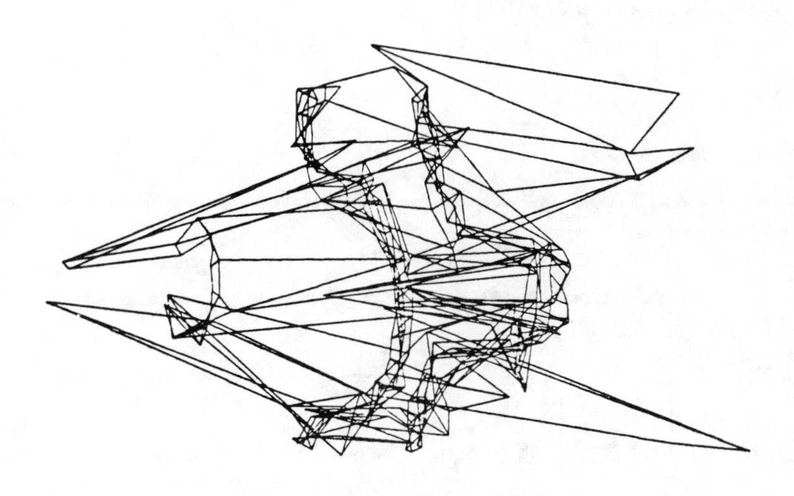

Fig. 6.10. "Dye lines" late in mixing layer development; low initial intensity case. From Cain et al. (1981).

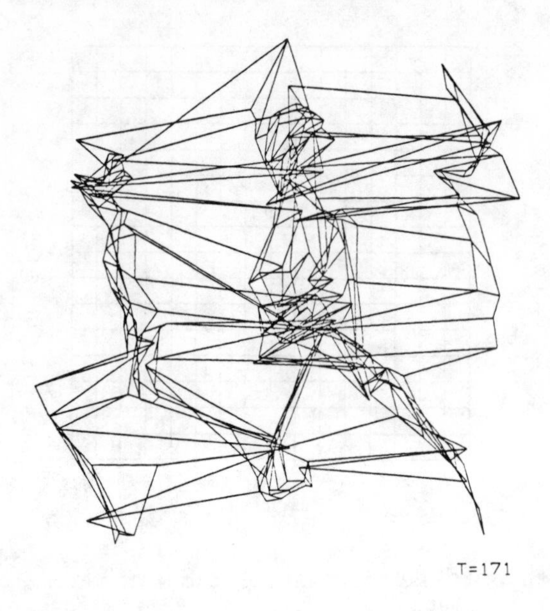

Fig. 6.11. "Dye lines" late in mixing layer development; medium intensity case (same as Fig. 6.10 except for intensity). From Cain et al. (1981).

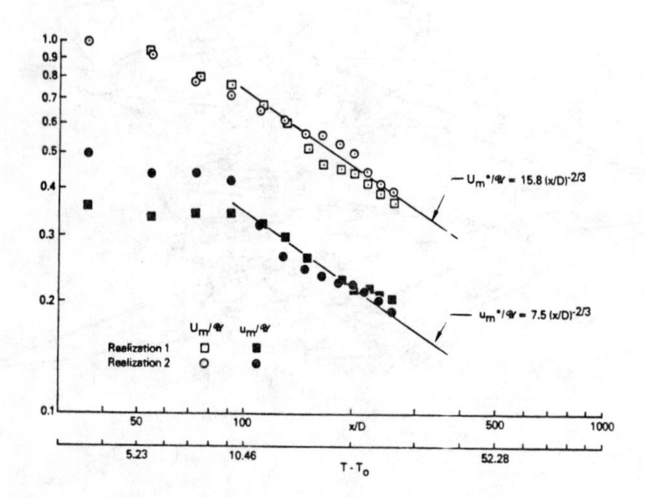

Fig. 6.12. Decay of maximum mean velocity (U_m) and maximum axial turbulent intensity (u_m) in wake. From Riley and Metcalfe (1978).

The above results were taken from the report of Cain et al. (1981).

Two-dimensional simulations of the mixing layer were made by Patnaik, Sherman, and Corcos (1976), Acton (1976), Knight (1979), Ashurst (1979), and Riley and Metcalfe (1980), among others. In these simulations, the shear layer rolls up into an array of vortices. The principal object of these studies was the determination of the effect of initial perturbations on the speed and nature of the rollup of the layer. These papers contain interesting results, but, as they are essentially outside the topic of this report, they are not covered here in detail.

Full simulations of the turbulent mixing layer were made by Riley and Metcalfe (1980a,b). These simulations are similar to the work of Cain et al., which they predated. Their calculations are performed at low Reynolds number so that no subgrid scale model is required. Their initial condition is similar to the high initial energy condition of Cain et al., but they also ran a number of cases in which a deterministic perturbation was added to the initial conditions; this perturbation was the most unstable wave of linear theory. They observed that the layer tended to roll up into vortices and found linear growth of the thickness of the layer, self-similarity of the velocity profile, and, in the case with the largest number of mesh points, constancy of the turbulent energy in the center plane of the layer. All of these observations are in agreement with experiment and the computations described above. An important contribution of this work is the demonstration that the properties of the mixing layer can be reproduced in a simulation which contains no large vortical structures in the initial conditions. They also showed that the addition of the perturbation corresponding to the most unstable wave of linear theory to the initial condition reduced the rate of growth of the layer.

6.3 Wakes

As stated in the introduction to this chapter, wakes are flows in which there is a defect in the velocity profile. As a wake develops, the velocity profile widens and the velocity gradients decrease. These factors and the fact that the rate of growth of the length scales is not as rapid in wakes as in mixing layers make wakes a little easier to simulate than mixing layers.

There are several types of wakes. The classification plays some role in determining how the flow will be simulated. A self-propelled body (one that drives itself through the fluid) leaves a wake in which the net momentum is zero; the momentum added by the propulsion just equals that due to the drag of the body. On the other hand, the wake of a towed body (or a body in a wind tunnel) has a net momentum deficit. Finally, both types of wakes can occur in plane, axisymmetric, and other geometric arrangements.

The first full simulation of a momentumless wake was made by Orszag and Pao (1974). Their work has been extended to the simulation of towed wakes by Riley and Metcalfe, in a series of papers. They concentrated mainly on the axisymmetric wake, because most of the experimental data is for this case. Despite the axisymmetry of the flow, they used a rectangular grid in their simulations; the axisymmetry is inserted via the initial conditions.

In some respects, their simulations behave very much like the simulations of the preceding section. As in all other flows, a short time period is required for the initial condition to develop into a physically realistic flow. During this period there is relatively little broadening of the wake and some decay of the turbulence. The higher-order statistics also change from their original values during this period; in particular, the velocity-derivative skewness increases. Finally, the vorticity tends to concentrate during this phase.

Figure 6.12 shows the decay of the maximum mean velocity and the maximum axial component of the turbulence. Several experiments have shown that these quantities decay as $x^{-2/3}$ with downstream distance. Since the simulated wakes are temporally developing, the analogous behavior would have these quantities decay as $t^{-2/3}$. The figure shows that the maximum mean velocity follows the expected similarity behavior quite well. The turbulence decays a little more slowly than expected. Two different realizations of this flow are shown.

Similarity arguments suggest that the radii of the wake and of the turbulence profile should increase as $t^{1/3}$; the spatially decaying wake radius increases in radius as $x^{1/3}$. Figure 6.13 shows that the simulation reproduces this behavior quite well. The decay of the integrated mean and turbulent energies are also well predicted.

The velocity profiles behave in a self-similar manner after the initial period. They agree quite well with the measured profiles except in the wings of the profile; the results are shown in Fig. 6.14. The Reynolds shear stress is also reasonably well predicted, as are some of the higher-order statistics.

To conclude this chapter, we note that full simulations seem to be able to predict free shear flows quite well. The major stumbling block to continuing the simulations further in time is the growth of the length scales with downstram distance or time. This can be partially cured by doing the simulations with larger numbers of grid points. It would be more efficient to rescale the problem after some time, but no way has yet been found to do this without invoking very serious approximations.

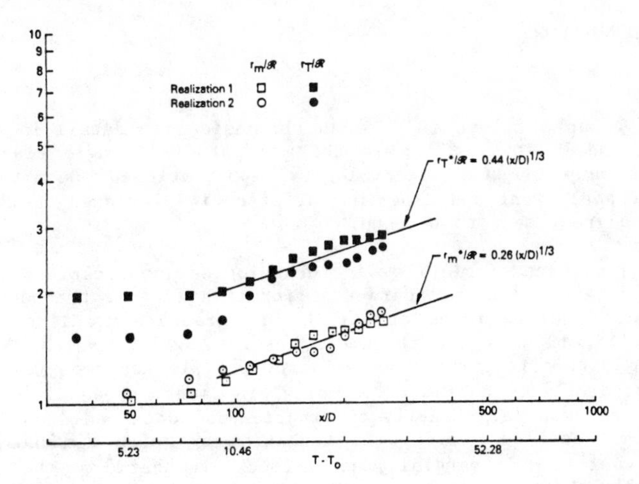

Fig. 6.13. Growth of mean wake radius (r_m) and turbulent wake radius (r_T). From Riley and Metcalfe (1978).

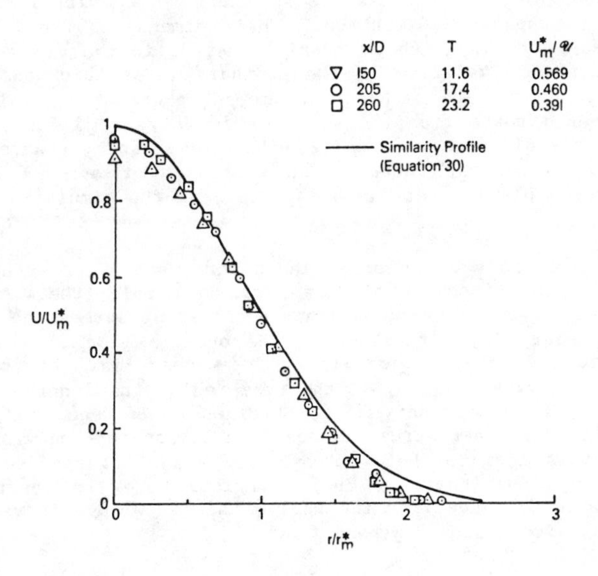

Fig. 6.14. Normalized profile of axial mean velocity in wake. From Riley and Metcalfe (1978).

7. WALL-BOUNDED FLOWS

7.1 Overview

The last group of flows that we shall consider in detail in this report is the wall-bounded flows. This is the most studied single class of flows because of its many important technological applications. Despite the enormous amount of analytical and experimental attention lavished on these flows, there remains a great deal to be done.

The most important single flow in technological applications is the turbulent boundary layer. The standard case for this flow is the boundary layer in the absence of "extra rates of strain"--no pressure gradient, curvature, rotation, suction, blowing, or roughness, etc. A great deal is known about this flow. In particular, the mean velocity profile has been well measured, and one can "predict" its behavior. (Quotes are used because all of the present prediction schemes rely heavily on experimental data and should be called "postdictive" methods.) However, the mechanism by which momentum is transferred to the wall is only partially understood. Furthermore, the information that is available about the mechanism has not been used in model construction. Thus there is still much to do. It is hoped that higher-level simulations can play a role in this, but it is clear at the outset that the task is not easy.

It is known that the mechanism of momentum transfer to the wall in the boundary layer is connected with the flow structure observed close to the wall. In the near-wall region, the flow consists of alternating "streaks" of high- and low-speed fluid; the streaks are very long in the streamwise direction and thin in the spanwise direction. Their dimensions are believed to scale with the shear stress, which is nearly constant in the vicinity of the wall; however, their size relative to the boundary layer thickness is quite Reynolds number-dependent. The mechanism of momentum transfer involves lifting of the low-speed streaks from the wall. When they are lifted, they are carried a considerable distance into the boundary layer and exchange momentum with the fluid they encounter there. The existence of streaks and their importance in the flow plays a very important role in the simulation of wall-bounded flows.

The boundary layer is made up of at least three sublayers. There is an inner layer in which the viscosity plays an important role (the viscous sublayer); here the length scales are dependent on the shear stress and are small compared to the boundary layer thickness. The outer region of the flow is essentially inviscid and behaves much like a free shear low. In fact, it is frequently called the "wake" region; in the wake region the length scales are approximately 0.1 of the boundary layer thickness. Between these two regions is one in which the shear stress is nearly constant and the viscosity is not important. In this region, the mean velocity has a logarithmic profile, and it is called the logarithmic or buffer region; here the length scales increase linearly with distance from the wall. This knowledge is very important in higher-level simulations of these flows.

The turbulent boundary layer increases in size with downstream distance. This is difficult for higher-level simulations to handle at the present time. One can consider a temporally developing boundary layer; this has been done and will be described in the last section of this chapter. Unfortunately, the velocity profile of the time-developing layer is different from that of the spatially developing layer, and the difference is significant because wall-bounded flows are quite sensitive to small changes in the velocity profile.

Most of the attention to date has been given to turbulent channel flow. It is the ideal choice for simulation, because it is the one true "equilibrium" flow of the class. It reaches a state at which none of its properties changes with downstream distance. Despite this, the physics of the near-wall flow is similar to that of the boundary layer. Thus this flow can be simulated with periodic boundary conditions without making any important assumptions that might affect the results. Of course, one must be careful that the usual criteria needed for the application of periodic boundary conditions be maintained. This flow has been simulated a number of times and will occupy the major part of this chapter.

Another very important issue in wall-bounded flows is that of transition. Laminar boundary layers are much more stable than are laminar free shear flows, but transition takes place when the Reynolds number is high enough. The ability to delay transition would enable us to reduce the drag on bodies, with obvious and important consequences. This is one of the major reasons why transition has received so much attention.

Transition in boundary layers is sensitive to relatively small changes in the velocity profile. Keeping the disturbance level small can delay transition for a long way. On the other hand, minor disturbances, such as a bit of roughness, can trigger transition.

Theory predicts that laminar channel flow is stable with respect to small disturbances at Reynolds numbers below about 5700. One can also show that it is more unstable with respect to large disturbances, but the predicted Reynolds number of transition is smaller than the Reynolds number at which transition is observed to occur. An explanation of this phenomenon will be given later in this chapter.

The next section will take up the computation of fully developed channel flow. There are two approaches to doing this, and we shall discuss them and give results obtained by both approaches. In particular, we shall describe recent results that promise to provide a great deal of interesting information about this flow.

The last section of this chapter will consider transition in wall-bounded flows. This problem has been done recently for both the channel and the time-developing boundary layer. A number of interesting results have been produced, and there is considerable hope that still more will be forthcoming in the near future.

7.2 Fully Developed Channel Flow

The dynamical behavior of fully developed channel flow is similar in many respects to that of the boundary layer. In particular, the inner layers of the two flows are quite similar. The major differences are that the channel flow requires a pressure gradient to overcome the frictional forces and that the channel flow has no region in which the flow is not completely turbulent; outside the boundary layer, the flow is potential.

Of particular importance in the simulation of the channel flow is the behavior of the length scales. What makes these flows especially hard to simulate is the fact that the spanwise length scales are much smaller near the wall than in the central portion of the flow. This means that a grid that is well adapted to capturing the streaks near the wall will be much finer than necessary near the center. On the other hand, a grid which is scaled for the central region will not be able to see the streaks at all. The variation in the length scales in the direction normal to the wall is less serious, because a variable grid size can be used in this direction.

Two approaches have been taken to simulate channel flow. In the first method, which was developed by Deardorff and extended by Schumann and co-workers, the wall is not treated explicitly. This avoids much of the difficulty with the small-scale structures that occur near the wall, and reduces the amount of computation considerably. The limit of the computational domain is placed in the logarithmic region of the flow, because this is probably the best understood part of the flow. Another argument put forward for this method is that viscous effects prohibit the existence of an inertial subrange in the inner layers, but one exists in the buffer and wake regions. The difficulty with this method is that the boundary conditions at the top and bottom of the computational domain are not well defined, and assumptions must be made. Also, this approach does not simulate much of the physics of the flow and cannot be used to study its structure and modeling.

Deardorff assumed that the derivative of the streamwise velocity in the normal directin could be written as a sum of two parts; the first guarantees the existence of a logarithmic region, and the second is responsible for the fluctuations. His expression is:

$$\frac{\partial^2 \bar{u}_1}{\partial x_2^2} = -\frac{1}{\kappa(\Delta x_2/2)^2} + \frac{\partial^2 \bar{u}_1}{\partial x_3^2} \qquad (7.1)$$

where κ is the von Karman constant (0.41) and Δx_2 is the distance of the first mesh point from the wall. This boundary condition assumes that the fluctuations of the velocity are the same in the normal and spanwise directions. The validity of this assumption is open to question; the reason that Deardorff gave for favoring it is that it produced reasonable results. Kim (private communication) has tested this boundary condition and finds it is not good at all.

Schumann's assumption is that the shear stress and the velocity are in phase at the first mesh point; according to Kim, this assumption is also inaccurate. Mathematically, his assumption is:

$$\tau_{12} = \frac{\langle \tau_w \rangle}{\langle u \rangle} u \qquad (7.2)$$

where $\langle u \rangle$ is the mean velocity at the first mesh point, $\langle \tau_w \rangle$ is the mean wall shear stress, and u is the instantaneous velocity.

For the subgrid scale model, Deardorff used the Smagorinsky model. The only modification that he found necessary was the reduction of the magnitude of the constant in the model from the value obtained from theory or the isotropic decay simulations.

Schumann modified the model. He assumed that the subgrid scale model should be composed of two parts. The first is proportional to the time-mean velocity gradient at the particular distance from the wall; the second is proportional to the deviation of the instantaneous velocity from the time-mean. He called these the inhomogeneous and locally isotropic components of the subgrid scale stress. He also used an equation for the subgrid scale turbulent kinetic energy, but found that it gave no significant improvement over an algebraic eddy viscosity model.

For the mean velocity profile, Schumann obtained very good results. The results for the components of the Reynolds stress are also quite good. Schumann also used his results for testing the Rotta model for the pressure-strain term. These results are shown in Figs. 7.1 and 7.2. It is interesting to note that the "constant" is different for the various components. However, one should be cautious about accepting these rsults, because the pressure is very sensitive to changes in the way in which the flow is computed, and we believe that large uncertainties must be assigned to these results. In fact, the results near the boundary seem to be due to the boundary conditions used. We shall have more to say about this below.

Moin et al. (1978) made the first attempt to solve the channel flow problem while treating the wall boundary conditions exactly. Doing this means that a nonuniform grid has to be used in the direction normal to the wall; the use of Chebychev polynomials is an alternative.

One of the major difficulties with this method is that the length scales of the flow become very small near the wall; the local turbulence Reynolds number also becomes very small, and it is not clear that the Smagorinsky model can be used any longer. In fact, it is possible that the overall length scales of the turbulence will be smaller than the size of the grid in this region. It is then improper to use the grid or filter size in the subgrid scale model. Instead, Moin et al. used the minimum of the Prandtl mixing length and the grid size. This modification is arbitrary but is a simple method that appears to work.

Another difficulty is that the smallness of the grid tends to make numerical methods unstable. There are two nondimensional numbers that determine the stability of a numerical method. They are the Courant number $u_2\Delta x_2/\Delta t$ and the viscous parameter $\nu\Delta t/\Delta x_2^2$. Roughly speaking, stability requires that both of these numbers be smaller than some constant of the order of unity. It turns out that the viscous condition is more stringent near the wall, and if an explicit method were used, it would be necessary to use an extremely small time step. Consequently, a method which treats the viscous terms containing derivatives with respect to the normal coordinate implicitly was devised and used. Doing this meant that the normal method of solving for the pressure via the Poisson equation had to be abandoned. We shall briefly describe the revised numerical method.

Most of the terms in the momentum equations are time-differenced using the second-order Adams-Bashforth explicit method. The exceptions are the pressure gradients and the viscous terms containing derivatives with respect to the normal coordinate, which are treated by the implicit Crank-Nicolson method. The continuity equation, which contains no time derivatives, is evaluated at the new time step. The resulting set of equations if Fourier-

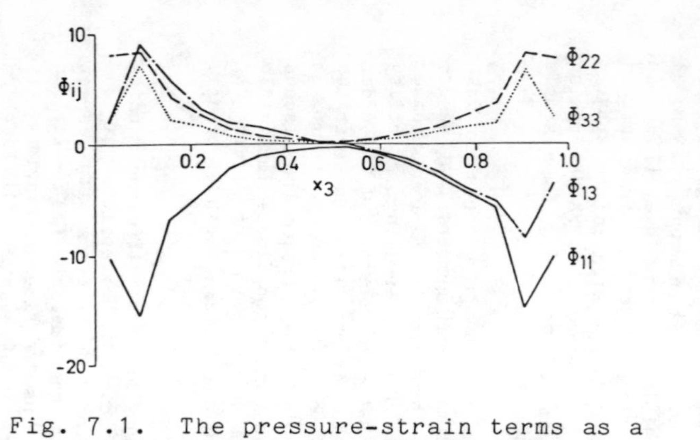

Fig. 7.1. The pressure-strain terms as a function of the normal coordinate in a channel flow. From Schumann et al. (1980).

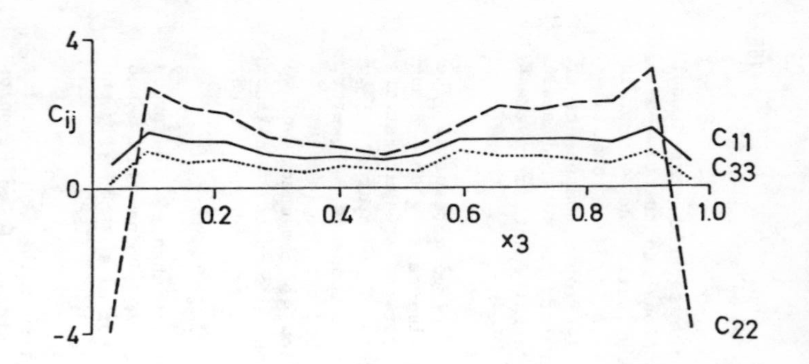

Fig. 7.2. The "constant" in the model of the pressure-strain term as a function of the normal coordinate. From Schumann et al. (1981).

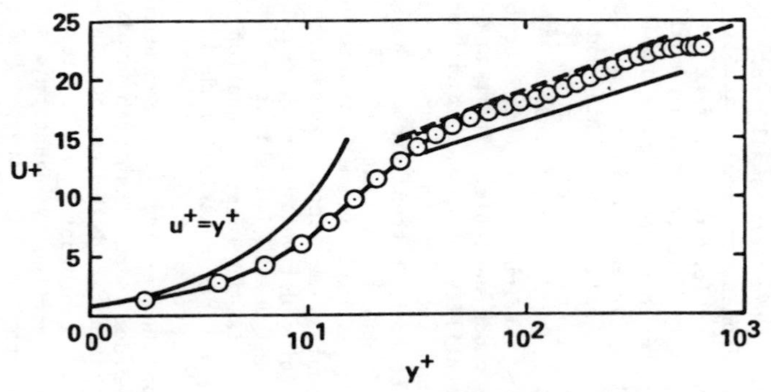

Fig. 7.3. The mean in the channel flow compared with three sets of experimental results. From Kim and Moin (1979).

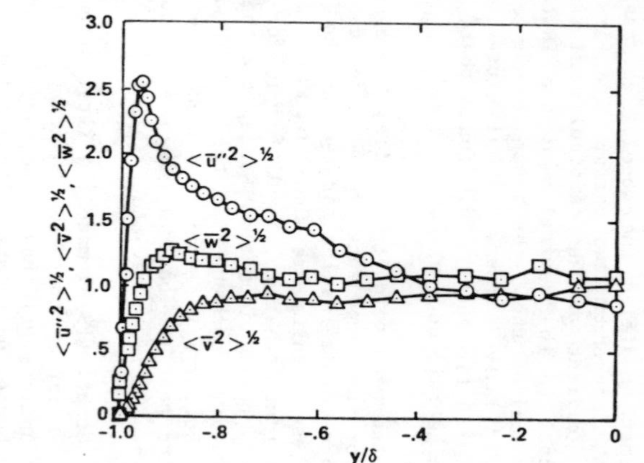

Fig. 7.4. Turbulence intensities in channel flow (resolved component only). From Kim and Moin (1979).

spanwise ones. This had also been found by Moin et al. (1978) and was unexpected. It is apparently due to fluid moving toward the wall being stopped by the wall. The vertical motions are converted into horizontal motions, and the result of this "splat" effect and the normal energy transfers is shown in the figure. Again, the quantitative results may be incorrect, but it is unlikely that the qualitative result is incorrect. More recent (and more accurate) results by Moin and Kim (1981) show a smaller, but still significant, "splat" effect.

Contours of the fluctuating velocity on a plane parallel to and close to the wall are shown in Fig. 7.6; the presence of long streaks is obvious. A similar plot for a plane close to the center of the channel is shown in Fig. 7.7; there is no evidence of streaky behavior at this plane. A number of other plots of this kind are given in their paper.

In a more recent paper, Kim and Moin (1981) have done calculations with still greater resolution and further improvements in both models and numerical methods. The results are qualitatively similar to those presented above but differ quantitatively. They have also produced a simulated flow-visualization motion picture that duplicates most of the phenomena observed in laboratory motion pictures. This application of the results should play a very important role in the future.

The splat effect is also seen in the shear-free wall layer. This is simply a turbulence near a wall which is moving at the same mean velocity as the wall. The precise nature of this flow depends on the Reynolds number. A simulation by Biringen and Reynolds (1981) captured most of the effects observed in the experiments. However, we shall not review these results in detail here.

7.3 Transition

As stated in the introduction, transition in boundary layers is a subject of great technological importance. However, transition is very sensitive to a number of factors, including the precise velocity profile, the level of the disturbance, wall roughness, etc. As a result, transition experiments are very difficult to perform reproduceably, and there is considerable scatter in the data. Naturally, simulations of these flows will be very sensitive to similar factors. Thus, a great deal of care will be necessary to simulate these flows.

Linear stability theory predicts that the laminar boundary layer profile is unstable with respect to disturbances that result in Tollmien-Schlichting waves. This instability is much less explosive than that of the free shear layer. It is generally believed that the Tollmien-Schlichting waves grow until nonlinear effects take over and complex interactions lead to the fully turbulent boundary layer. However, the late stages of transition are poorly understood.

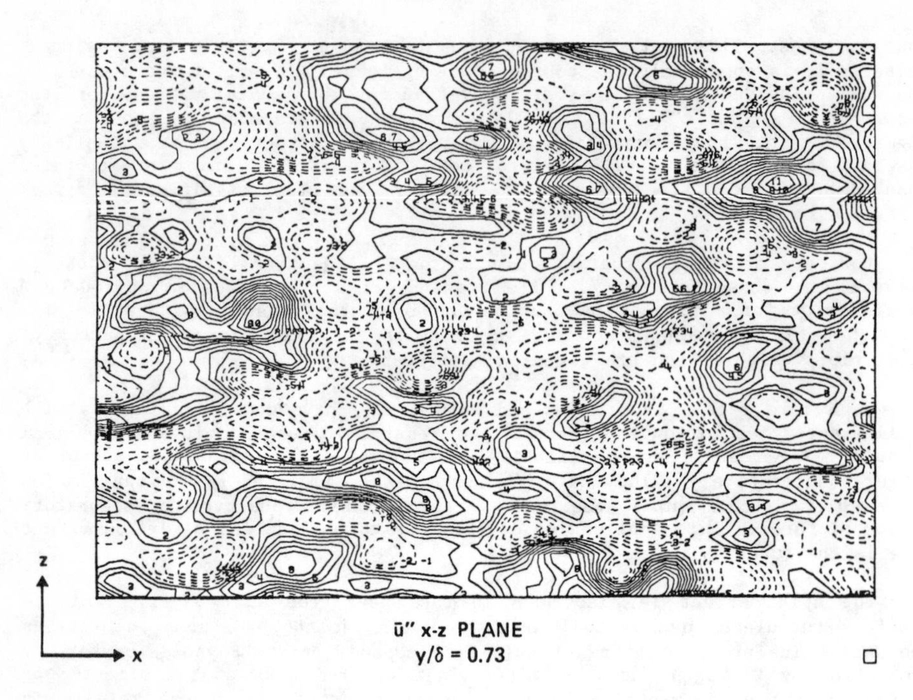

Fig. 7.7. Contours of the fluctuations of the streamwise velocity near the center of the channel. From Kim and Moin (1979).

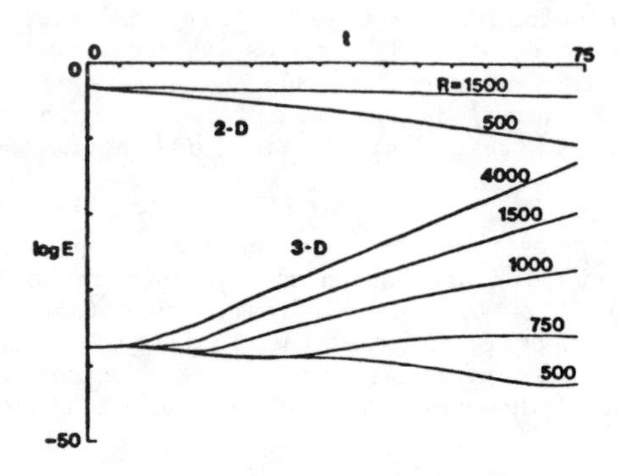

Fig. 7.8. Energy of two- and three-dimensional waves in a perturbed laminar channel flow. The 2-D waves decay while the 3-D waves grow. From Orszag and Patera (1981).

The first direct simulation of transition in wall-bounded flows was made by Kells and Orszag (1979) and Orszag and Patera (1980,1981). They chose to study channel flow at Reynolds numbers for which the flow is linearly stable. However, transition does take place at the Reynolds numbers studied. In their simulation, Orszag and Patera took a fully developed laminar channel profile (Poiseuille profile) and added finite-amplitude two-dimensional Tollmien-Schlichting waves to it; these waves are different in the channel than in the boundary layer. They found that the waves decayed slowly and that the rate of decay decreases as the Reynolds number increases; this is expected. However, they found that, when a small three-dimensional disturbance is introduced into the flow, it grows very rapidly. The growth of the three-dimensional wave is rapid enough to enter the nonlinear regime before the two-dimensional wave has decayed. At this point the simulation develops considerable energy at high wavenumbers and has to be stopped; as there is no model in the simulation, there is no way to continue. However, this simulation has provided an explanation of the instability of this flow; it is apparently due to the three-dimensional instability of stable two-dimensional waves. Orszag and Patera (1981) have done similar simulations for Couette and cylindrical tube flows.

Some of their results are shown in Fig. 7.8. The decay of the two-dimensional wave and the growth of the three-dimensional wave are quite apparent.

A simulation of the instability of the boundary layer has been made by Wray (unpublished). In order to avoid the difficulty that arises from the spatially developing boundary layer, he chose a time-developing boundary layer; physically, this corresponds to the boundary layer that develops on a suddenly started plate. Although the velocity profile of the time-developing boundary layer is different from that of the spatial layer, the calculation was started with the Blasius profile appropriate to the spatial layer. To this profile, a weak Tollmien-Schlichting wave and a weak three-dimensional random disturbance was added.

The disturbance grows very slowly at first (as expected) until it builds up to a level at which nonlinear effects become important. At this point, the rate of change of the layer becomes spectacular. The contours of the various velocity components and the vorticity develop more and more structure. Comparisons with experimental results for the parameters reveal a considerable similarity; the comparison is necessarily qualitative, but it is remarkably good.

Eventually, this simulation developed a considerable amount of energy at high wavenumbers, and it had to be stopped. There is no way to continue this simulation beyond this point without more resolution. Unfortunately, it may be that the small scales play an important role in the development of this flow, and it is not known whether the addition of a model will cure the problem.

8. OTHER APPLICATIONS

As we have stated earlier, higher-level simulation began in meteorology and oceanography. These fields have maintained an active interest in the simulation of the global circulation of the Earth's atmosphere and oceans. The methods used are similar to the ones described in this report, but there are additional difficulties. The principal of these is that thermal energy and the transport of water vapor (in the atmosphere) and salt (in the ocean) are very important in these flows, and one must deal with the effects of stratification, evaporation, and condensation. When coupled with the limitation to very coarse grids (200 km is typical in these simulations today), we see that the problems are considerably more difficult than the ones dealt with in this report. They are, however, of great importance, and considerable effort is being devoted to them. The author has only a passing knowledge of the work in these areas, and this is the reason why the subject is not covered in this report.

Methods similar to the ones given in this report have also been applied to smaller-scale environmental problems. For example, simulations have been made of local parts of the atmosphere by these methods; these are called mesoscale simulations. The author is familiar only with a few papers by Deardorff in this area; in these papers, he used a complete Reynolds stress model for the subgrid scale Reynolds stresses. Others have applied these methods to the prediction of the flow in lakes, harbors, and other small bodies of water. Of the work in this field, the author is familiar only with some of what has been done at his institution. Findikakis (1980) has recently developed a finite-element method for computing such flows.

On a still smaller scale, there have been a number of extensions of the work covered in the earlier sections of this report. Schumann and his coworkers have used the method that was described for channel flow for the simulation of flows in annuli and made other extensions. In particular, they have computed the channel and the annulus with heat transfer; in the simulations, the temperature is treated as a passive scalar. We have not dealt with this work at length in this paper for several reasons. It is covered in detail in the report of Schumann et al. (1980). Also, since the results produced by Schumann's method differ considerably from those of Kim and Moin (1979) and Moin and Kim (1981) for the channel flow, we are unsure about the accuracy of the method when applied to heat transfer. For similar reasons we have not covered their work on the effect of roughness.

Schumann, Grötzbach, and Kleiser have applied their method to natural convection flow between parallel horizontal plates. They covered a very large range of Rayleigh number and were able to predict the observed transitions from one flow regime to another. This is an excellent piece of work and was not covered because it did not fit any of the subject headings used in this paper. Grötzbach (1979) has also investigated simulated flows in vertical channels with the influence of buoyancy.

Finally, we shall mention a method that competes with the grid-based methods that are the primary subject of this report. These are methods in which the motions of vortices are followed (vortex-tracking methods). A number of interesting features of transitional and turbulent flows have been computed by this method, including flows with separation. The full capabilities of this approach and a comparison of it with the methods discussed in this report are given in a review paper by Leonard (1981). Hybrid methods which use some ideas from vortex-tracking and some from grid-based methods are also being investigated at the present time; the interested reader is referred to the paper by Couet, Leonard, and Buneman (1980).

There are undoubtedly areas that have been overlooked in this report. The author has tried his best to be complete, but in any subject area that has become this large something is likely to be missed. There is no intent to minimize any contributions that have been missed.

9. CONCLUSIONS AND FUTURE DIRECTIONS

9.1 Where Are We Now?

We hope that this report has shown that higher-level simulations of turbulent flow have reached a point in their development which allows them to play an important role in turbulent fluid mechanics. Let us now sum up where the field stands today. We start with the positive points.

a) The basic ideas of large eddy simulation seem sound. Specifically, they seem to be able to handle homogeneous turbulent flows and free shear flows quite well. For wall-bounded flows, the importance of small structures near the wall is a problem, and these flows are difficult to deal with, but good progress has been made.

b) Direct simulation of many interesting flows are now feasible. We are limited to low Reynolds numbers, but this restriction may not be important in some flows, as the large scales may be nearly Reynolds number independent. Alternatively, one can regard the viscosity as a simple subgrid scale model and pretend that a higher Reynolds number flow is being simulated. Both of these approaches have been taken. Orszag has used the concept of "Reynolds number similarity" with considerable success. Rubesin (1979) regarded direct simulations as large eddy simulations, also with considerable success.

c) Higher-level simulations have come to the point at which they are able to provide information on quantities that are difficult to measure in the laboratory. In this role, they are able to evaluate turbulence models in a way that is very difficult to do by any other method.

d) Higher-level simulations are able, in some cases, to simulate flows that are very difficult or impossible to perform in the laboratory. Some examples are flows with rotation and/or compressibility.

e) It is now possible to do flow visualizations using full and/or large eddy simulations. These visualizations reproduce much of what is seen in the laboratory. They also offer flexibility that is difficult to match in the laboratory. They can be used to look in detail at specific regions, and can even be used backwards in time.

Now let us consider some of the difficulties.

a) The most obvious problem is that these methods require large amounts of computer time. Although some of the simpler flows can be done in a few minutes on large machines, running times of the order of hours are not unusual for the more difficult flows. Use of these methods must be restricted to individuals with access to the machines that can do these simulations. Some means of assuring that the problems of greatest interest are done is necessary.

b) Although some flows are amenable to full simulation, Reynolds number similarity does not hold for all flows, so it is not possible to treat low Reynolds number flows as models of a high Reynolds number flow in all cases. Better subgrid scale models will be necessary if high Reynolds number flows are to be simulated, but it may be very difficult to find models with wide applicability.

On balance, the contribution of higher level simulations seems to be more than worth the cost, and the approach is just beginning to reach its potential. With new generations of computers, it should be possible to do much more with these methods.

9.2 Where Are We Going?

It is clear that a great deal remains to be done in turbulence simulation. There are many directions which can be taken in the future, and, with more groups beginning to do these types of simulations, we expect the area to broaden rapidly. Of course, it is difficult to predict the future with any precision, but it is always interesting to try. Let us look at what can be expected in the next few years.

a) One obvious direction in which higher-level simulations will be extended is toward the simulation of a larger number and greater variety of flows. There are many possibilities, so the following list cannot be all-inclusive.

i) The flows which have already been simulated can be done with additional effects. Thus, to any of the flows treated in this report we can add rotation, curvature, heat transfer, passive scalars, and/or pressure gradients alone or in combination. In the wall-bounded flows we can also add wall roughness and blowing or suction. Many of these effects are quite important in engineering flows and should be considered at an early date.

ii) To date, no method has been found for dealing with inflow or outflow boundaries. The outflow boundary can probably be handled by the usual method of requiring the streamwise derivatives to be zero at the outlet. The inflow condition is much more difficult, because it is necessary to prescribe a realistic representation of the turbulence in order not to require too much of the computation; to do this would waste a very large part of the computational resource. Being able to handle inflow and outflow boundaries is central to the computation of many flows of interest.

iii) There are some fairly simple flows which have not been done. Included among these are the jet and the wall jet.

b) Simulation of wall-bounded flows is much simpler if the kinds of boundary conditions used by Deardorff and Schumann can be applied. Chapman (1980) estimated that the savings to be realized in this way could make the difference between practical use of the higher-level simulations and their continuing to be confined to research. Accurate boundary conditions of that type need to be searched for.

c) Use of higher-level simulations in conjunction with flow visualization and statistical methods should become a very powerful tool for investigating the structure of turbulent flows. It is possible that such an approach may be able to tie the structure of turbulent flows to the modeling. This is highly speculative, but, if it can be done, it could be an important step forward. We may become "computational experimentalists."

d) The interaction of higher-level simulations and conventional models should become stronger. We can envision a time when people developing new models will routinely validate them using higher-level simulations. Certainly, we can expect higher-level simulations to be helpful in determining the constants in the models. It is worthwhile to set up a facility which is available for this purpose.

e) We expect that there will be considerable work on the improvement of subgrid scale models, but the direction this work will take is not obvious.

f) Higher-level simulations will be extended to include a number of phenomena that are not currently treated. Sound generation should be relatively easy, as it seems to depend mainly on the large scales. Combustion should be very challenging, because the chemical reaction depends on intimate mixing at the small scales.

g) Something has to be left to the reader's imagination.

Acknowledgments

A work of this kind can never be done by one person. The author is indebted to a great many people for making this work possible. I shall try to thank everybody, but it is possible that someone may be left out; if this is the case, I take this opportunity to apologize.

Let me begin with my colleagues. Prof. W. C. Reynolds has been the coleader of the effort in higher-level simulations at Stanford since the beginning, when I knew essentially nothing about turbulence; he has taught me a very large part of what I now know. Prof. S. J. Kline has also taught me a great deal about turbulent flows and has been a very important friend when I have needed one. Profs. J. P. Johnston and R. J. Moffat have also played important roles in this work; from them I have learned a lot about the art of experimenting in turbulence. Dr. S. Schechter of SRI and Profs. J. Oliger and G. H. Golub of the Computer Science Department have taught me a great deal about numerical methods. Dr. S. Biringen was a one-year visitor who contributed an important piece to this work. Prof. D. C. Leslie and his group were outstanding hosts during a sabbatical stay at Queen Mary College, London; I learned much about turbulence theory from them.

Next I must thank my students, who have done most of the actual work and all of the really hard work. Drs. D. Kwak and S. Shaanan provided the early basis of the method; Dr. R. A. Clark developed the method of computational model validation; Dr. N. N. Mansour pioneered the free shear layer work; Dr. P. Moin did the same on the channel flow. More recently, Drs. W. J. Feiereisen, A. B. Cain, and E. Shirani and Mr. J. Bardina have made important contributions; their work figures very prominently in this report, and I am indebted to them for allowing me to use some of their unpublished results.

A number of colleagues at NASA-Ames Research Center have played important roles in this work. In particular, I want to thank Drs. D. R. Chapman, M. Rubesin, V. Peterson, and H. Mark for the constant attention they gave this work. Drs. A. Leonard, R. S. Rogallo, P. Moin, J. Kim, and A. Wray, among others, have provided many useful ideas and many of the results contained in this report.

I also wish to thank Drs. J. N. Nielsen and O. J. McMillan of Nielsen Engineering and Research for allowing us to extend the area of work and for important contributions.

Lastly, and perhaps most importantly, I must thank the organizations who provided the support for this work. NASA-Ames Research Center has supported this work financially since its inception; their generosity in allowing us to use their computers has been even more important than their direct assistance. The National Science Foundation has supported the work on the mixing of a passive scalar over the last two years and has allowed us to broaden the base of our activity. The Office of Naval Research supported work at Nielsen Engineering and Research on model validation.

Mrs. Ruth Korb typed this manuscript in record time under a great deal of pressure and without complaint. Her work over many years has never been sufficiently appreciated.

This work was originally written as a set of notes for lectures given at the Von Kármán Institute of Fluid Dynamics in Brussels in March, 1981. I wish to thank the Institute and Professors W. Kollmann and J. Essers in particular for the invitation that compelled me to tackle a job that might not have been done otherwise, and for their kind hospitality at the Institute.

References

Acton, E., "The Modeling of Large Eddies in a Two-Dimensional Shear Layer," J. Fluid Mech., 76, 561-592 (1976).

Antonopoulos-Domis, M., "Large Eddy Simulation of the Decay of a Passive Scalar in Isotropic Turbulence," J. Fluid Mech. , 104, 55 (1981).

Ashurst, W. T., "Calculation of Plane Sudden Expansion Flow via Vortex Dynamics," Proc. 2nd. Conf. on Turbulent Shear Flows, London, 1979.

Ballhaus, W., "Prediction of Transonic Flows," in Numerical Fluid Dynamics, W. Kollman, ed., Hemisphere, Washington, 1979.

Bardina, J., J. H. Ferziger, and W. C. Reynolds, "Improved Subgrid Scale Models for Large Eddy Simultion," AIAA paper 80-1357, 1980.

Birch, S., "Free Shear Layers," in Proc. 1980 AFOSR-Stanford Conference on Complex Turbulent Flows (S. J. Kline, B. J. Cantwell, and G. M. Lilley, eds.), Dept. of Mech. Engrg. Stanford University, 1981.

Browand, F. K., and T. R. Troutt, "Vortices in the Turbulent Mixing Layer: The Asymptotic Limit," Presented at 32nd Meeting of APS, Division of Fluid Dynamics, Univ. of Notre Dame, Nov. 18-20, 1979.

Brown, G. L., and A. Roshko, "On Density Effects and Large Structures in Turbulent Mixing Layers," J. Fluid Mech., 64, 775 (1974).

Cain, A. B., W. C. Reynolds, and J. H. Ferziger, "Simulation of the Transition and Early Turbulence Regions of a Mixing Layer," Report TF-14, Dept. of Mech. Engrg., Stanford Univ., Stanford, CA., 1981.

Chandrsuda, C., and P. Bradshaw, "An Assessment of the Evidence for Orderly Structure in Turbulent Mixing Layers," Report 75-03, Aeronautics Dept., Imperial Coll., London (1975).

Chapman, D. R., "Computational Aerodynamics Development and Outlook," AIAA J., 17, 1293 (1979).

Chapman, D. R., and G. D. Kuhn, "Simulation of the Wall Region of a Turbulent Boundary Layer," AIAA paper 81-1024, 1981.

Clark, R. A., J. H. Ferziger, and W. C. Reynolds, "Evaluation of Subgrid Scale Turbulence Models Using a Fully Simulated Turbulent Flow," J. Fluid Mech. 91, 92 (1979).

Coles, D. E., and E. Hirst, eds., Computation of Turbulent Boundary Layers--1968 AFOSR-IFP-Stanford Conference, Vol. II, Dept. of Mech. Engrg., Stanford Univ., 1968.

Cooley, J. W., and J. W. Tukey, "An Algorithm for the Machine Calculation of Complex Fourier Series," Math Comput., 19, 297-301 (1965).

Couet, B., O. Buneman, and A. Leonard, "Three-Dimensional Simulation of the Free Shear Layer Using the Vortex in Cell Method," Proc. 2nd Conf. on Turbulent Shear Flows, London, 1979.

Deardorff, J. W., and G. E. Willis, "The Effect of Two-Dimensionality on the Suppression of Thermal Turbulence," J. Fluid Mech., 23, 337-353 (1965).

Deardorff, J. W., and G. E. Willis, "Investigation of Turbulent Thermal Convection between Horizontal Plates," J. Fluid Mech., 28, 675-704 (1967).

Deardorff, J. W., and G. E. Willis, "Investigation of Thermal Convection between Horizontal Plates," J. Fluid Mech., 28, 675 (1968).

Deardorff, J. W., "A Numerical Study of Three-Dimensional Turbulent Channel Flow at Large Reynolds Number," J. Fluid Mech., 41, 452-480 (1970).

Deardorff, J. W., "On the Magnitude of the Subgrid Scale Eddy Coefficient," J. Comp. Phys., 7, 120-133 (1971).

Deardorff, J. W., "Three-Dimensional Numerical Study of Turbulence in an Entrained Mixed Layer," Boundary Layer Meteorology, 1, 199-226 (1974).

Deardorff, J. W., "Three-Dimensional Numerical Modeling of the Planetary Boundary Layer," Workshop in Micrometeorology, A.M.S., 271-311 (1973).

Deardorff, J. W., "Simulation of Turbulent Channel Flow," J. Fluids Engrg., 95, 429 (1973).

Deardorff, J. W., "Three-Dimensional Numerical Study of Turbulence in an Entraining Mixed Layer," Boundary Layer Meteorology, 1, 199 (1974).

Feiereisen, W. J., W. C. Reynolds and J. H. Ferziger, "Simulation of Compressible Turbulence," Report TF-13, Dept. of Mech. Engrg., Stanford University (1981).

Ferziger, J. H., "Large Eddy Numerical Simulation of Turbulent Flow," AIAA Journal, 15, 1261 (1977).

Ferziger, J. H., and D. C. Leslie, "Large Eddy Simulation--A Predictive Approach to Turbulent Flow Computation," AIAA paper 79-1441, 1979.

Ferziger, J. H., "Homogeneous Turbulent Flows: A Review and Evaluation," prepared for the 1980-81 Stanford-AFOSR-HTTM Conf. on Complex Turbulent Flows, Dept. of Mech. Engrg., Stanford Univ. (1980).

Findikakis, A., "Large Eddy Simulations of Stratified Flows," Dissertation, Dept. of Civil Engrg., Stanford University, Stanford, CA., 1981.

Fox, D. G., and J. W. Deardorff, "Computer Methods for Simulation of Multidimensional, Nonlinear, Subsonic, Incompressible Flow," J. Heat Transfer, 45, 337-346 (1972).

Fox, D. G., and D. K. Lilly, "Numerical Simulation of Turbulent Flows," Rev. of Geophys. and Space Phys., 10, 51-72 (1972).

Fox, D. G., and S. A. Orszag, "Pseudospectral Approximation to Two-Dimensional Turbulence," J. Comp. Phys., 11, 612-619 (1973).

Gottlieb, D., and S. A. Orszag, "Numerical Analysis of Spectral Methods: Theory and Applications," NSF-CBMS Monograph 16, SIAM, Philadelphia (1977).

Grötzbach, G., G. Lörcher, and U. Schumann, "Anwendung und experimentelle Absicherung der direkten numerischen Simulation turbulenter Strömungen," Reaktortagung, Nürnberg, 8.43-11.4, 145-148 (1975).

Grötzbach, G., "Direkte numerische Simulation turbulenter Geschwindigkeits-, Druck-, und Temperaturfelder bei Kanalströmungen, Dissertation, Univ. of Karlsruhe (1977).

Grötzbach, G., "Numerische Experimente zur Untersuchung des Wärmtetransports in turbulenter Flüssigmetallströmung," Reaktortagung, Mannheim, 29.3-1.4, 7-10 (1977).

Grötzbach, G., "Numerische Untersuchung der Quervermischung bei auftreibsbeinflusster turbulenter Konvektion in einem vertikalen Kanal," KFK 2648 (1978).

Grötzbach, G., Direct Numerical Simulation of Secondary Currents in Turbulent Channel Flows," Lect. Notes in Phys., 76, 308-319, Springer Verlag (1978).

Grötzbach, G., Convective Velocities of Wall Pressure Fluctuations in a Turbulent Channel Flow Deduced from a Computer-Generated Movie," Lect. Notes in Phys., 76, 320-324, Springer Verlag (1978).

Grötzbach, G., Numerical Investigation of the Influence of Secondary Flows on Characteristic Turbulence Data, KFK 2553 Kernforschungszentrum Karlsruhe (1978).

Grötzbach, G., "Direct Numerical Simulation of Laminar and Turbulent Benard Convection," Report, Kernforschungszentrum Karlsruhe (1979).

Grötzbach, G., and U. Schumann, "Direct Numerical Simulation of Turbulent Velocity-, Pressure-, and Temperature-Fields in Channel Flows," Proc. of the Symp. on Turbulent Shear Flows, Penn. State Univ., Apr. 18-20, 1979.

Grötzbach, G., "Numerical Investigation of Radial Mixing Capabilities in Strongly Buoyancy-Influenced Vertical Turbulent Channel Flows," Nucl. Engr. and Design, 54, 49 (1979).

Grötzbach, G., "Numerical Simulation of Turbulent Liquid Metal Flows in Plane Channels and Annuli," Report KFK 2968, Kernforschungszentrum, Karlsruhe (1980).

Ibbetson, A., and D. J. Tritton, "Experiments on Turbulence in a Rotating Fluid," J. Fluid Mech., 68, 639 (1975).

Kim, H. T., S. J. Kline, and W. C. Reynolds, "The Production of Turbulence Near a Smooth Wall in a Turbulent Boundary Layer," J. Fluid Mech., 50, 133 (1971).

Kim, J., and P. Moin, "Large Eddy Simulation of Turbulent Channel Flow," AGARD Symposiium on Turbulent Boundary Layers, The Hague, 1979.

Kleiser, L., and U. Schumann, "Simulation of 3-D Incompressible Flows between Parallel Plates with a Fourier-Chebyshev Spectral Method," EUROVISC Working Party on Transition in Boundary Layers, Univ. Stuttgart, Oct. 2-3, 1978.

Kline, S. J., M. V. Morkovin, G. Sovran, and D. J. Cockrell, eds., Computation of Turbulent Boundary Layers--1968 AFOSR-IFP-Stanford Conference, Vol. I. Dept. of Mech. Engrg., Stanford University, 1968.

Kline, S. J., J. H. Ferziger, and J. P. Johnston, "Calculation of Turbulent Shear Flows: Status and Ten-Year Outlook," ASME J. Fluids Engrg., 100, 3 (1978).

Kline, S. J., and B. J. Cantwell, eds., Proc. of the 1980-1981 Stanford-AFOSR-HTTM Conf. on Complex Turbulent Flows, Dept. of Mech. Engrg., Stanford Univ. (1981).

Knight, D. D., "Numerical Investigation of Large Scale Structures in the Turbulent Mixing Layer," Proc. 6th Biennial Symp. on Turbulence, U. of Missouri, Rolla, Oct. 8-10, 1979.

Launder, B. E., G. J. Reece, and W. Rodi, "Progress in the Development of a Reynolds Stress Turbulence Closure," J. Fluid Mech., 68, 537 (1975).

Leonard, A., "Energy Cascade in Large Eddy Simulations of Turbulent Fluid Flows," Adv. in Geophysics, 18A, 237 (1974).

Leonard, A., "Vortex Methods for Flow Simulation, J. Comp. Phys. (to be published).

Leslie, D. C., "Developments in the Theory of Turbulence," Oxford U.P., 1973.

Leslie, D. C., and G. L. Quarini, "The Application of Turbulence Theory to the Formulation of Subgrid Modeling Procedures," J. Fluid Mech., 91, 65 (1979).

Lilly, D. K., "On the Computational Stability of Numerical Solutions of Time-Dependent, Nonlinear, Geophysical Fluid Dynamic Problems," Mon. Wea. Rev., 93, 11 (1965).

Lilly, D. K., "The Representation of Small Scale Turbulence in Numerical Simulation Experiments," Proc. IBM Sci. Comp. Symp. on Env. Sci., IBM-Form No. 320-1951, 195-210 (1967).

Love, M. D., and D. C. Leslie, "Studies of Subgrid Modeling with Classical Closures and Burgers' Equation," Proc. Symp. on Turbulent Shear Flows," Penn. State Univ., 1976.

Mansour, N. N., P. Moin, W. C. Reynolds, and J. H. Ferziger, "Improved Methods for Large Eddy Simulations of Turbulence," Proc. 1st Int'l. Symp. on Turbulent Shear Flows, Penn. State Univ., 1977.

McMillan, O. J., and J. H. Ferziger, "Direct Testing of Subgrid Scale Models," AIAA Journal, 17, 1340 (1979).

McMillan, O. J., J. H. Ferziger, and R. S. Rogallo, "Tests of New Subgrid Scale Models in Strained Turbulence," AIAA paper 80-1339, 1980.

Metcalfe, R. W., and J. J. Riley, "Direct Numerical Simulations of Turbulent Shear Flows," Proc. 7th Int'l. Conf. on Numerical Methods in Fluid Dynamics, Stanford University, 1980.

Moin, P., and J. Kim, "Improved Simulation of Turbulent Channel Flow," submitted to J. Fluid Mech., 1981.

Moin, P., W. C. Reynolds, and J. H. Ferziger, "Large Eddy Simulation of an Incompressible Turbulent Channel Flow," Report TF-12, Dept. of Mech. Engrg., Stanford University, Stanford, CA., 1978.

Morkovin, H. V., "Effects of Compressibility on Turbulence," in Mechanique de la Turbulence," CNRS, Paris, 1962.

Orszag, S. A., "Numerical Methods for the Simulation of Turbulence," Phys. Fluid, Suppl. II, 250-257 (1969).

Orszag, S. A., "Numerical Simulation of Incompressible Flow within Simple Boundaries. I. Galerkin (Spectral) Representations," Stud. Appl. Math. 50, 293-327 (1971).

Orszag, S. A., "Numerical Simulation of Incompressible Flows within Simple Boundaries," J. Fluid Mech., 49, 75-112 (1971).

Orszag, S. A., "Galerkin Approximations to Flows within Slabs, Spheres, and Cylinders," Phys. Rev. Let. 26, 1100-1103 (1971).

Orszag, S. A., and G. S. Patterson, Jr., "Numerical Simulation of Turbulence," in Statistical Models and Turbulence, 127-147, Springer-Verlag, Berlin (1972).

Orszag, S. A., and M. Israeli, "Numerical Flow Simulation by Spectral Methods," in Numerical Models of Ocean Circulation, 284-300, Washington (1973).

Orszag, S. A., "Numerical Simulation of Turbulent Flows," Flow Research Rept. No. 52, M.I.T., Cambridge (1974).

Orszag, S. A., and Y. H. Pao, "Numerical Computation of Turbulent Shear Flow," Adv. Geophys., 18A, 225-236 (1974).

Orszag, S. A., and M. Israeli, "Numerical Simulation of Viscous Incompressible Flows," Ann. Rev. of Fluid Mech., 6, 281-318 (1974).

Orszag, S. A., "Comparison of Pseudospectral and Spectral Approximation," Stud. in Appl. Math., 51, 253-259 (1975).

Orszag, S. A., "Turbulence and Transition: A Progress Report," Lect. Notes in Phys., 59, 32-51 (1976).

Orszag, S. A., and L. C. Kells, "Transition to Turbulence in Plane Poiseuille and Plane Couette Flow" (to be published).

Patnaik, P. C., F. S. Sherman, and G. M. Corcos, "A Numerical Simulation of Kelvin-Helmholtz Waves of Finite Amplitude," J. Fluid Mech., 73, 215-240 (1976).

Patterson, G. S., Jr., and S. A. Orszag, "Spectral Calculations of Isotropic Turbulence: Efficient Removal of Aliasing Interactions," Phys. Fluids, 14, 1538-1541 (1971).

Reynolds, W. C., "Computation of Turbulent Flows," Ann. Rev. Fluid Mech., 8, 183-208 (1976).

Riley, J. J., and G. S. Patterson, "Diffusion Experiments with Numerically Integrated Isotropic Turbulence," Phys. Fluids, 17, 292-297 (1974).

Riley, J. J., and Metcalfe, R. W., "The Direct Numerical Simulations of the Turbulent Wakes of Axisymmetric Bodies--An Interim Report," Report No. 135, Flow Research Corp., Kent, Wash., 1978. Also available as NASA CR-152282.

Riley, J. J., and R. W. Metcalfe, "Direct Numerical Simulations of the Turbulent Wake of an Axisymmetric Body," in Turbulent Shear Flows. II (L. Bradbury et al., eds.), Springer, 1980.

Riley, J. J., and R. W. Metcalfe, "Direct Numerical Simulation of a Perturbed Turbulent Mixing Layer," AIAA paper 80-0274, 1980.

Rogallo, R. S., "An ILLIAC Program for the Numerical Simulation of Homogeneous Incompressible Turbulence," NASA TM-73, 1978.

Rose, H. A., "Eddy Diffusivity, Eddy Noise, and Subgrid Scale Modeling," J. Fluid Mech., 81, 719-734 (1977).

Rose, H. A., and P. L. Sulem, "Fully Developed Turbulence and Statistical Mechanics," J. de Physique, 39, 441-484 (1978).

Roshko, A., "Structure of Turbulent Shear Flows: A New Look," AIAA Journal, 14, 1249-1357 (1976).

Rubesin, M. W., "Subgrid- or Reynolds Stress Modeling for Three-Dimensional Turbulence Computations," Proc. AGARD Meeting on Turbulent Flows, The Hague, 1979.

Runstadler, P. W., S. J. Kline, and W. C. Reynolds, "An Investigation of the Flow Structure of the Turbulent Boundary Layer," Report MD-8, Dept. of Mech. Engrg., Stanford Univ. (1963).

Schemm, C. E., and F. B. Lipps, J. Atmos. Sci., 33, 1021 (1976).

Schumann, U., "Ein Verfahren zur direkten numerischen Simulation turbulenter Stromungen in Platten- und Ringspaltkanalen und uber seine Anwendung zur Untersuchung von Turbulenzmodellen," dissertation, Univ. of Karlsruhe, 1973.

Schumann, U., "Results of a Numerical Simulation of Turbulent Channel FLows," in Int'l. Meeting on Reactor Heat Transfer (M. Dalle-Donne, ed.), 230-251 (1973).

Schumann, U., "Subgrid Scale Modeling for Finite Difference Simulations of Turbulent Flows in Plane Channels and Annuli," J. Comp. Phys., 18, 376 (1975).

Schumann, U., "Experiences with the Spectral Method for Three-Dimensional Turbulence Simulations," GAMM-Conf. on Numerical Methods in Fluid Mech., DFVLR, 209-216, Koln, Oct. 8-10, 1975.

Schumann, U., and J. R. Herring, "Axisymmetric Homogeneous Turbulence: A Comparison of Direct Spectral Simulations with the Direct Interaction Approximation," J. Fluid Mech., 76, 755-782 (1976).

Schumann, U., and G. S. Patterson, "Numerical Study of Pressure and Velocity Fluctuations in Nearly Isotropic Turbulence," J. Fluid Mech., 88, 685-709 (1978).

Schumann, U., and G. S. Patterson, "Numerical Study of the Return of Axisymmetric Turbulence to Isotropy," J. Fluid Mech., 88, 711-735 (1978).

Schumann, U., G. Grötzbach, and L. Kleiser, "Direct Numerical Simulation of Turbulence," in Prediction Methods for Turbulent Flows (W. Kollmann, ed.), Hemisphere, New York (1980).

Sicilian, J. M., and A. Leonard, "The Use of Fourier Expansions in Turbulent Flow Simulations," Report No. TF-2, Thermosciences Div., Stanford Univ., (1974).

Siggia, E. D., and G. S. Patterson, Jr., "Intermittency Effects in a Numerical Simulation of Stationary Three-Dimensional Turbulence," J. Fluid Mech., 86, 567-592 (1978).

Smagorinsky, J., "General Circulation Experiments with the Primitive Equations. I. The Basic Experiment," Mon. Wea. Rev., 91, 99 (1963).

Sommeria, G., J. Atmos. Sci., 33, 216 (1976).

Traugott, S. C., "Influence of Solid-Body Rotation on Screen-Produced Turbulence," Quart. J. Mech. Appl. Math., 104 (1954).

Tennekes, H., and J. L. Lumley, A First Course in Turbulence, MIT Press, 1972.

Wigeland, R., and H. M. Nagib, "Grid-Generated Turbulence With and Without Rotation," Fluids and Heat Transfer Report R78-1, Ill. Inst. of Tech., 1978.

Winant, C. D., and F. K. Browand, "Vortex Pairing: The Mechanism of Mixing Layer Growth at Moderate Reynolds Number," J. Fluid Mech., 63, 237-255 (1974).

Wray, A. A., M. Y. Hussaini, and D. Degani, "Numerical Simulation of Transition to Turbulence," 2nd. GAMM-Conf. on Numerical Methods in Fluid Mechanics (E. H. Hirschel & W. Geller, eds.), DFVLR, Koln, 247-254 (1977).

Numerical Methods for Two- and Three-dimensional Recirculating Flows

R. I. ISSA

ABSTRACT

The lectures present a brief review of the main alternative practices encountered in the application of finite-difference/volume methods to the calculation of two- and three-dimensional, laminar or turbulent, recirculating flows, with emphasis being put on the choice of method of solution of the difference equations. The discussion is presented within the context of generality and applicability to engineering problems, and the assessment of the various options is made from this viewpoint.

1. INTRODUCTION

1.1 Objectives

An exhaustive survey, or a comprehensive theoretical analysis of numerical methods suitable for two- and three-dimensional flows, as the title of this article might suggest it is, would be far beyond the scope of the present contribution. Given such a limited scope, it is necessary to confine attention to only some of the major aspects of the subject, and to accord the others a rather brief treatment. Presently, consideration is restricted to methods of the finite-difference/volume variety with emphasis being put on the overall structure of algorithms rather than on particular difference schemes. Also, the lectures are intended to serve as an overview of the available options, and the criteria by which these are assessed, as opposed to a treatise of formal mathematical analyses of the various techniques.

The subject matter is discussed in the context of practicability and generality of application to engineering problems; therefore, both laminar and turbulent flows with variable properties are necessarily accommodated in the discussion. Also, most engineering applications involve flows in the low subsonic regime, the outlook of the article is hence biased in favour of this regime.

The main topics covered are: type of spatial discretization; choice of grid arrangement and its consequences; the options for temporal discretization and linearisation of the equations; choice of pressure versus density as a working variable; methods for the solution of the linearised equations; and finally but most significantly means of dealing with the pressure-velocity coupling of the equations.

1.2 General Considerations

In the past, most computational methods for fluid flow were developed primarily for problems arising in the areas of defence and space exploration where computing resources were limited only by computer capacity. Many of the present day applications are, however, aimed at industrial problems where such resources are limited and where economic considerations play an important role in decision-making. Development of contemporary numerical methods must therefore take economic competitiveness into account as an influential factor.

Examples of industrial applications of computational methods are: the prediction of in-cylinder processes and the flow in valve passages of reciprocating internal combustion engines (see, e.g. refs. [5] to [7]); the three-dimensional turbulent flow in passages of nuclear reactor fuel assemblies (see, e.g. refs. [8] and [9]); and the two-phase, three-dimensional flow in chemical-process, agitator vessels (see, e.g. ref. [10]). These are but a few examples of the flow cases encountered in engineering; they do, however, point out the wide range of problems to be dealt with. The flow may be in a steady state or may be time-varying, laminar or turbulent, compressible or incompressible, and can be multiphase, with or without combustion; moreover, the flow geometry is often complex.

In attending to such diverse problems, one need not resort to a single unified method for dealing with all; indeed, this is not the case in reality. However, the viewpoint put forward here is that it is in fact advantageous to develop a general unified approach, at least as far as methodology is concerned. For, generality of approach yields versatile and flexible methods capable of extension and development to totally new or partially different situations (e.g. from 2 to 3 dimensional configurations). This is a desirable feature from both the viewpoints of judicious use of skilled labour resources and speed of implementation.

An example of methods that are of restricted generality is one which uses the vorticity and stream-function as main dependent variables (e.g. ref. [1]). In such methods, the momentum and continuity equations for incompressible flow are manipulated so that the pressure is eliminated and the new aforementioned variables are introduced as dependent variables in the resultant equations in place of the velocities. The approach is limited to incompressible flow and its extension from two to three dimensions necessitates the introduction of a vector potential in place of the stream-function, hence increasing the number of equations and variables to be solved for in the problem. Techniques such as this, as well as those applicable to only specific problems are excluded from the present survey on the grounds of lack of generality.

As well as being versatile, computational procedures are also required to be efficient and reliable. The attributes of efficiency, versatility and reliability are not readily separable from each other. It is obvious that to be competitive a scheme must be efficient. However, efficiency is sometimes gained at the expense of flexibility and/or reliability. Thus, for example, substantial savings can be achieved by using an efficient but complex technique for computing processes such as combustion and turbulent transport which involve the solution of numerous equations. Should introduction of yet more equations or possibly new boundary conditions into the scheme be desired, expert knowledge or intensive labour may be called for, both of which can be deterimental to overall economy. Another point to note that stability and convergence problems are often encountered in fluid flow computations. The effort involved in remedying the situation can be appreciable and may well wipe out all the benefits of an otherwise economical technique. From a practical point of view, all these factors must, therefore, be weighed against each other when a choice is made; these lectures reflect such considerations.

1.3 Turbulence Effects

As practical applications are of primary concern here, the problem of simulating the effects of turbulent transport phenomenon arising in most of these applications must be attended to. Big advances in this field have been made lately by use of direct and large-eddy simulations. These approaches involve the solution of the full three-dimensional, transient Navier-Stokes equations, either over the whole range of eddy sizes (as in the case of the former), or over a range of eddy sizes larger than those comparable to the grid dimensions (as in the case of the latter). The computing times and storage requirements of these calculations are enormous, even for the simplest of turbulent flow cases such as fully developed Couette flow. Hence, these techniques cannot be regarded as a practical means of computing turbulent flows; for the time being, given present day computer capacity, they remain, though valuable as they are, as research tools for investigating the phenomenon of turbulence.

Turbulence modelling on the other hand, has fast become a plausible means of accounting for the effects of turbulence in practical calculations. This is because the additional effort required involves the solution of only a few more transport equations having a similar form to that of the momentum equations. The latter equations themselves are either time- or ensemble-averaged (see, e.g. refs. [3] and [4]), and likewise the dependent variables appearing in them are taken to be time or ensemble averaged quantities. Thus the need for tracing the history of individual eddies is obviated: indeed, it is this feature which bestows relative simplicity to turbulence modelling. However, it is also the very same feature which precludes the possibility of exact simulation of the effects of turbulence, since averaging of the equations inevitably robs them of their precision to reproduce these effects. Nevertheless, turbulence models have proved to be invaluable in yielding reasonably accurate results for the majority of problems, and in some cases, even highly accurate ones.

There are many well established turbulence models to choose from (see, e.g. ref. [2]). The most popular one appears to be, at present, the two-equation model in which the turbulence kinetic energy and its rate of dissipation are used to characterise turbulence. In such a model, the turbulent fluxes and stresses are represented by diffusion-like terms containing a 'turbulent diffusion' coefficient which is related to the two turbulence parameters. More sophisticated models involve transport equations for the turbulent stresses and fluxes themselves, in which case, these fluxes and stresses appear explicitly in the corresponding equations for the mean averaged quantities to which they relate.

The treatment presented herein implicitly accounts for the inclusion of a turbulence model and does not discriminate in favour of any particular one of the aforementioned models.

2. EQUATIONS AND FINITE-DIFFERENCE FORMULATION

2.1 Governing Equations

As mentioned above, a turbulence model is assumed to be incorporated into the analysis, in which case all the governing equations presented herein are taken to be in ensemble-averaged form. All variables appearing in these equations are likewise meant to stand for ensemble-averaged quantities.

The transport equations governing the flow field, in Cartesian tensor notation (which is used for convenience), are:

$$\frac{\partial \rho}{\partial t} + \frac{\partial}{\partial x_i}(\rho u_i) = 0 \tag{1}$$

and

$$\frac{\partial}{\partial t}(\rho u_i) + \frac{\partial}{\partial x_j}(\rho u_j u_i) = -\frac{\partial p}{\partial x_i} + \frac{\partial}{\partial x_j}\sigma_{ij} + S_u \tag{2}$$

where σ_{ij} is the stress tensor which is taken to include both laminar and turbulent stresses, and S_u contains the body forces.

Depending on the nature of the turbulence model used, the constitutive relation for σ_{ij} can be given either by:

$$\sigma_{ij} = [\mu(\frac{\partial u_i}{\partial x_j} + \frac{\partial u_j}{\partial x_i}) - \frac{2}{3}\mu\delta_{ij}\frac{\partial u_k}{\partial x_k}] \tag{3}$$

which applies for models involving the concept of eddy-diffusivity where μ is an 'effective' diffusion coefficient incorporating the effects of turbulence, or by:

$$\sigma_{ij} = [\mu(\frac{\partial u_i}{\partial x_j} + \frac{\partial u_j}{\partial x_i}) - \frac{2}{3}\mu\delta_{ij}\frac{\partial u_k}{\partial x_k}] - \rho\overline{u_i'u_j'} \tag{4}$$

which applies for models where the turbulent stresses $-\rho\overline{u_i'u_j'}$ are calculated from their own transport equations and μ stands for the laminar viscosity. In either case, the momentum equation (2) can be cast in the form:

$$\frac{\partial}{\partial t}(\rho u_i) + \frac{\partial}{\partial x_j}(\rho u_j u_i) = -\frac{\partial p}{\partial x_i} + \frac{\partial}{\partial x_j}[\mu(\frac{\partial u_i}{\partial x_j} + \frac{\partial u_j}{\partial x_i})] + S_u \tag{5}$$

where S_u may now contain terms such as those expressing the effects of dilatation or turbulent stresses, as well as body forces.

The equations for the transport of a scalar quantity ϕ (which may stand for temperature, species concentration, turbulence kinetic energy etc.) may also be put in a single form as:

$$\frac{\partial}{\partial t}(\rho\phi) + \frac{\partial}{\partial x_j}(\rho u_j\phi) = \frac{\partial}{\partial x_j}(\frac{\mu}{Pr_\phi}\frac{\partial\phi}{\partial x_j}) + S_\phi \tag{6}$$

where Pr_ϕ is an effective Prandtl number for the diffusion of property ϕ and S_ϕ is the source or sink of ϕ which may also include turbulent fluxes when a turbulence model solving for these is employed.

If the Mach number of the flow is significant, the relationship between pressure and density is important and must therefore be included in the analysis. Generally, an equation of state may be written as:

$$\rho = f(p, T) \tag{7}$$

Typically, for a perfect gas, this relationship becomes:

$$p = R\rho T \tag{8}$$

For very low Mach numbers which is the case in many if not most practical situations, the pressure and density are weakly related and in the idealised limit of truly incompressible fluid, the density becomes completely decoupled from pressure, hence relation (7) is replaced by its equivalent:

$$\rho = f(T) \tag{9}$$

For uniform temperature flow which is often the case, ρ becomes constant. This has a notable effect on the continuity equation (1) in that the time derivative term vanishes thus reducing the equation to the form:

$$\frac{\partial u_i}{\partial x_i} = 0 \tag{10}$$

This feature of the incompressible continuity equation introduces difficulties into the solution procedure in that a main dependent variable can no longer be identified in the equation. Rather, the continuity relation (10) now assumes the role of a compatibility condition on the velocity field.

2.2 Problems Associated with Using the Density as Main Dependent Variable

Many of the existing methods in fluid dynamics invoke the fully compressible fluid equations to compute flows which are essentially incompressible. Such an approach is attactive because unlike the incompressible form (see eqn. (10)), the compressible continuity equation (eqn. (1)) contains the temporal derivative of density. This enables the treatment of that equation as one in terms of a single dependent variable, i.e. density, via which the equation can be coupled to those of momentum.

At low Mach numbers, the pressure and density become weakly related and the variations of the latter become negligible. Despite this fact, many examples can be found in the literature of applications of methods using density as a dependent variable to practically incompressible situations. The question of validity of such approaches is dealt with here.

In reality, fluids are compressible and the pressure is always related to the density no matter how weak their connection is. The limit of truly incompressible flow in which variations of density exactly vanish is only a theoretical state. Hence it ought to be legitimate to use the compressible form of the equations even for incompressible flow regimes. However, because the equations are solved numerically, constraints are imposed by numerical factors such as truncation and round-off errors. In what follows, it is demonstrated that such limitations, at least, arise from machine round-off errors.

Consider for example the case of a fluid obeying an equation of state such as eqn. (8), (a consideration which need not affect the generality of the analysis). This equation can be written as:

$$p = \frac{c^2}{\gamma}\rho \tag{11}$$

where C is the speed of sound and γ is the specific heat ratio. In a method using ρ as a main dependent variable, equation (11) is invoked to calculate the pressure from the density and the pressure-gradient terms in the momentum equations (5) are replaced by their equivalents in terms of density. If the operator Δ represents a spatial-difference analogue of a spatial derivative ($\frac{\partial}{\partial x}$ for e.g.), the finite-difference representation of that particular pressure gradient then becomes:

$$\frac{\partial p}{\partial x} = \Delta p$$

When p is replaced by ρ, this term becomes:

$$\Delta p = \Delta(\frac{C^2}{\gamma} \rho)$$

If the flow is nearly incompressible, the spatial variations of C become small and the above term can be represented by:

$$\Delta p = \frac{C^2}{\gamma} \Delta\rho$$

Let M_0, u_0 and ρ_0 be typical reference values, over the flow field, of the Mach number, velocity and density. Introduction of these quantities into the preceding expression leads to:

$$\frac{1}{\rho_o u_o^2} \Delta p = \frac{1}{\gamma M_o^2} \Delta(\rho/\rho_o) \tag{12}$$

Clearly, for low Mach number flows, although density gradients are very small (identically zero in the limit of truly incompressible flow), pressure gradients are finite and can be large. This is reflected by equation (12) where if $\Delta(\rho/\rho_o)$ is small, division by M_0^2 which is also small leads to a value of $(\frac{1}{\rho_o u_o^2} \Delta p)$ which need not be small. It is evident, that the accuracy of evaluating the pressure gradients is very sensitive to that of the density gradients.

On a computer, quantities including ρ/ρ_0 can be evaluated only to a given accuracy (say 10^{-m}) due to machine round-off. Hence, density gradients less than 10^{-m} cannot be discerned by the machine, and meaningful pressure gradients can be set up only if:

$$\Delta(\rho/\rho_o) > 10^{-m} \tag{13}$$

Furthermore, the evaluation of the pressure gradient itself is required to at least the same degree of accuracy as that desired for the momentum equation as a whole - say to 10^{-n}. Hence the minimum value of pressure gradient considered to be of significance is given by:

$$\frac{1}{\rho_o u_o^2} \Delta p = 10^{-n} \tag{14}$$

If conditions (13) and (14) are matched subject to relation (12), the following is obtained:

$$M_o^2 > 10^{-m+n}$$

or

$$M_o > 10^{(m+n)/2} \tag{15}$$

in order that the pressure gradients can be resolved to the required accuracy.

Condition (15) shows that a lower limit on the Mach number exists, if the influence of round-off errors on the solution is to be curtailed. The values of m and n are dependent on the machine used as well as on the overall solution method. However, realistic values of 8 and 4 respectively may be assumed here for the purpose of illustration. Relation (15) then gives a lower limit on the Mach number of .01, a value which, it is worth noting, is higher than is encountered in a great many cases.

The conclusion is reached that if the compressible form of the flow equations are to be used either for wholly incompressible flows or for mixed compressible and incompressible regimes*, the density must not be taken to stand as a main dependent variable in the equations. The alternative is to use the pressure, which does not give rise to any of the above pecularities throughout the Mach number range+.

Other considerations which affect the application of the compressible fluid equations to incompressible flow calculations are related to the mathematical character of these equations. For, whereas the compressible equations are of the hyperbolic type relating to initial-value problems, their incompressible counterpart are of elliptic/parabolic nature. Incompatability between equation-type and physical problem arising from application of the former equations to the latter situations can result in computational difficulties; however, this issue cannot be pursued here further.

2.3 Spatial Discretization

The choice of practices for expressing the spatial fluxes and derivatives in the differential equations in terms of difference formulae is very wide, and an adequate examination of these options is outside the objectives of the lectures. It suffices to draw the readers attention to this wide choice, and point out that the treatment presented in the following sections is independent of the choice of spatial differencing, although obviously, this choice affects the final form of the difference equations to be solved and hence has a bearing on the choice of method of solution. The only topic which relates to spatial discretization that is discussed here, is the choice between conservative and other difference schemes, and the way in which the former is implemented.

In finite-difference formulation, two main options arise. In the first, the spatial derivative terms in the transport equations are expressed in difference form and the equation is enforced at each point on the mesh thus satisfying the laws of transport at these discrete points. For the purpose of illustration, consider the model one-dimensional scalar conservation equation:

$$\frac{d\phi}{dx} - \frac{d}{dx}\left(A\,\frac{d\phi}{dx}\right) = 0 \qquad (16)$$

in which A is constant.

---

*An example of this case is the flow in the cylinder of a reciprocating combustion engine at different parts of its cycle.

+Including supersonic and hypersonic regimes, see e.g. [27].

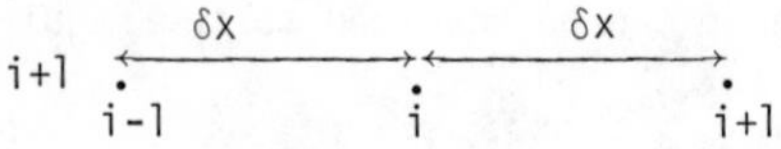

δx δx

i+1 i-1 i i+1

Consider also an equally spaced mesh with intervals δx. The finite-difference analogue of the derivatives in equation (16) evaluated at point i may be chosen for example as the second order formulae:

$$\left(\frac{d\phi}{dx}\right)_i = (\phi_{i+1} - \phi_{i-1})/2\delta x$$

and

$$\left(\frac{d^2\phi}{dx^2}\right)_i = (\phi_{i+1} + \phi_{i-1} - 2\phi_i)/\delta x^2$$

Insertion of these into equation (16) gives the difference equation for node i as:

$$\phi_i = \frac{1}{2}\left[\left(1 - \frac{\delta x}{2A}\right)\phi_{i+1} + \left(1 + \frac{\delta x}{2A}\right)\phi_{i-1}\right] \tag{17}$$

This formulation, however, does not guarantee that conservation is maintained over the whole of the domain except, of course, in the limit of vanishingly small grid spacing. This is because continuity of fluxes between grid points is not enforced and hence summation of the difference equations over all points such as i may not yield the required balance of fluxes of property ϕ entering and leaving the domain of solution (see [11] for a more thorough discussion of the topic). It is a coincidence that equation (17) above does in fact satisfy overall conservation.

The alternative formulation is to consider the domain as discretized into contiguous cells or control-volumes each surrounding a nodal point. The transport equations are first expressed in integral form expressing the balance of fluxes over each of the cells. These fluxes are then represented by finite-difference analogues, ensuring that continuity of these fluxes is maintained across each cell face. The example given above may be used again to illustrate the process. Cells may be considered surrounding each nodal point such as i. Equation (16) can be represented in integral form over the cell as:

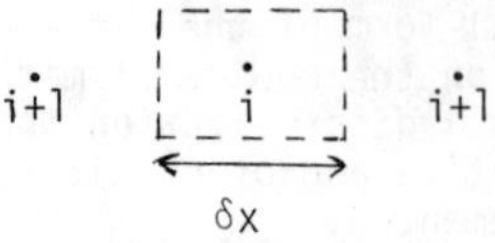

i+1 i i+1

δx

$$\int_{i-1/2}^{i+1/2} \left\{ \frac{d\phi}{dx} - \frac{d}{dx}\left(A\frac{d\phi}{dx}\right)\right\} dx = 0$$

which gives:

$$(\phi_{i+1/2} - \phi_{i-1/2}) - \left\{\left(A\frac{d\phi}{dx}\right)_{i+1/2} - \left(A\frac{d\phi}{dx}\right)_{i-1/2}\right\} = 0$$

Values of ϕ and its derivative at cell faces i+1/2 and i-1/2 are now required. If second order formulae similar to those used in arriving at equation (17) are introduced here, the following is obtained:

$$\phi_{i+1/2} = \frac{1}{2}(\phi_i + \phi_{i+1})$$

$$\phi_{i-1/2} = \frac{1}{2}(\phi_i + \phi_{i-1})$$

$$(\frac{d\phi}{dx})_{i+1/2} = (\phi_{i+1} - \phi_i)/\delta x$$

$$(\frac{d\phi}{dx})_{i-1/2} = (\phi_i - \phi_{i-1})/\delta x$$

Substitution of these relations into the preceeding equation yields:

$$\phi_i = \frac{1}{2}[(1 - \frac{\delta x}{2A})\phi_{i+1} + (1 + \frac{\delta x}{2A})\phi_{i-1}]$$

which happens for this particular case to be identical to equation (17) which was derived via the first approach. Overall conservation can now be ensured if the values of ϕ and its derivatives at cell faces such as that at i+1/2 are expressed by the same difference formulae when constructing the difference equations for both cells i and i+1.

Clearly, the conservative property of difference schemes, though not essential, is highly desirable, particularly so for engineering applications where overall balance of e.g. mass, energy and species are often of prime importance. In view of the practical orientation of the present work, conservative difference schemes are therefore recommended.

The example given above for the formulation of difference equations is for a simple model. In what follows, the general procedure for effecting conservative spatial discretization in multi-dimensions for a general transport equation is presented.

Consider the discrete control-volume surrounding node 0 and shown shaded in Fig. 1. The integrated form of any of the conservation equations (say eqn. (6) for ϕ which from here onwards is taken to stand for any of the field variables*) in for example, two dimensions+ for Cartesian co-ordinates, is:

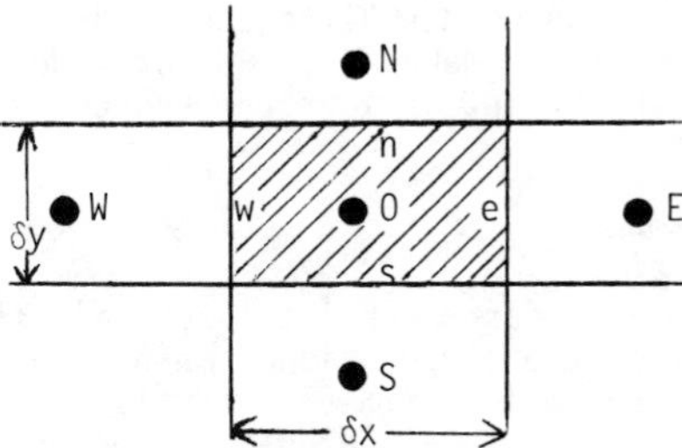

Fig. 1 Typical control-volume

*This includes the velocities in the momentum equations (5) as these are also expressible in the form of eqn. (6).

+Derivation in three dimensions following very similar lines.

$$\int \frac{\partial}{\partial t}(\rho\phi)dV = - \left[(\rho u\delta y\phi)_e - (\rho y\delta u\phi)_w + (\rho v\delta x\phi)_n - (\rho v\delta x\phi)_s\right]$$

$$+ \left(\frac{\mu\delta y}{Pr_\phi}\frac{\partial\phi}{\partial x}\right)_e - \left(\frac{\mu\delta y}{Pr_\phi}\frac{\partial\phi}{\partial x}\right)_w + \left(\frac{\mu\delta x}{Pr_\phi}\frac{\partial\phi}{\partial y}\right)_n - \left(\frac{\mu\delta x}{Pr_\phi}\frac{\partial\phi}{\partial y}\right)_s$$

$$+\int S_\phi dV \tag{18}$$

where subscripts n,s,e and w refer to space-averaged values prevailing over the faces of the control-volume. The values of ϕ and its derivatives at these faces are not immediately available; they must be evaluated by reference to nodal values at points such as O, E, W etc., where variable ϕ is stored. The variations of ϕ between these nodes is prescribed by the desired finite-difference scheme which makes such an evaluation possible. No particular differencing practice is selected presently and the difference equations presented below are of a general form in which any spatial difference formula might be incorporated.

Thus, if the appropriate terms in eqn. (18) are grouped as fluxes of ϕ at the appropriate cell faces, the flux across each cell face (say face e of the control-volume shown) may be represented by a general linear function of the form:

$$\text{flux at } e = a_O\phi_O + a_E\phi_E + a_N\phi_N + a_S\phi_S + \ldots \text{ etc.} \tag{19}$$

where the a-coefficients are in general functions of velocity, density, diffusivity etc. and where the number of terms in the expression (and consequently the number of nodes involved) is dictated by the type and order of the difference formula. When the expressions for the fluxes prevailing at all cell faces are introduced into equation (18), the following is obtained:

$$\int \frac{\partial}{\partial t}(\rho\phi)dV = A_i\phi_i + S_O \tag{20}$$

where the A's are finite-difference coefficients combining the effects of convection and diffusion, and the repeated suffix i indicates summation in which i stands for each and every nodal suffix such as O,E,W, etc. of the nodes involved in the difference representation of the spatial fluxes. For example, a five point second order difference scheme for the two-dimensional control volume above, makes i stand for nodes O,E,W, N and S. The term S_O in eqn. (20) is the integral of the source term over the control volume.

2.4 Choice of Grid Alignment

At first sight, it may appear natural to store all the field variables at each and the same node on the computational mesh; indeed, this has been the practice adopted for a long time. Recently developed methods, however, have for reasons to be explained below, elected to store the velocity components at locations staggered with respect to those at which the pressures and densities are stored. Figure 2 compares two such staggered arrangements with the standard one: in fig. 2(b), the grid employed by the MAC [12] and SIMPLE [13] methods is illustrated, while fig. 2(c) shows the arrangement favoured in the ALE [15] scheme.

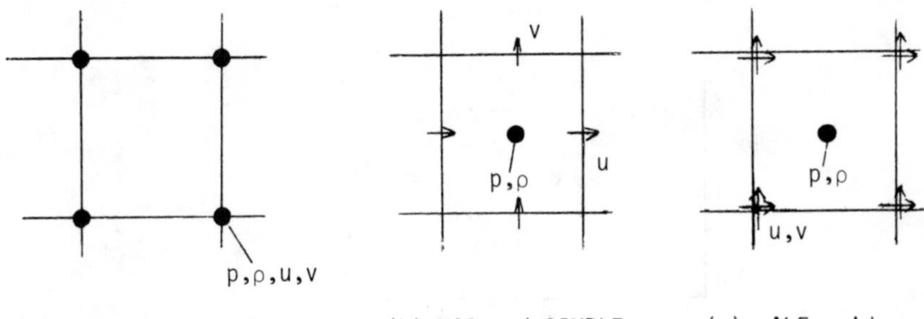

(a) Non-staggered (b) MAC and SIMPLE grids (c) ALE grid

Fig. 2 Alternative grid arrangements

One of the main advantages in using a staggered grid is that the pressure-gradient terms in the momentum equations can be centred about the velocities which they drive without resulting in the decoupling of adjacent pressure nodes (sometimes called checkerboard effect in the pressure field). This effect can be demonstrated by reference to a simple one-dimensional model. Consider the momentum equation for the case of uniform steady-state flow in the absence of body forces:

$$\dot{m}\,\frac{du}{dx} - \mu\frac{du^2}{dx^2} = -\,\frac{dp}{dx} \tag{21}$$

where $\dot{m}$ is the mass flow rate per unit area.

The finite-difference analogue of equation (21) cn a non-staggered grid for node i is for example:

δx δx
i+1 i i+1

$$\frac{\dot{m}}{2\delta x}(u_{i+1} - u_{i-1}) - \frac{\mu}{\delta x^2}(u_{i+1} + u_{i-1} - 2u_i) = \frac{1}{2\delta x}(p_{i-1} - p_{i+1})$$

which because of uniformity of the flow reduces to:

$$p_{i+1} = p_{i-1} \tag{22}$$

Similarly, the equation for node i+1 is:

$$p_{i+2} = p_i \tag{23}$$

It is evident from equations (22) and (23) that coupling in pressure betweeen alternate nodes is obtained. Hence, a pressure distribution such as that illustrated in Fig. 3 satisfies these equations and results in a uniform flow. Clearly this does not correspond with the physical situation.

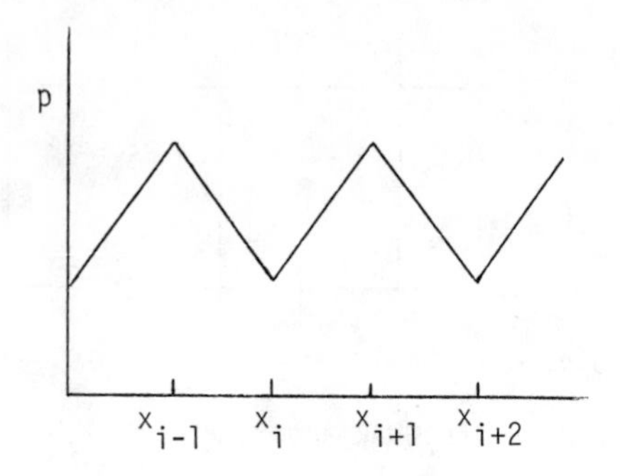

Fig. 3 Pressure field for uniform flow (non-staggered grid)

Alternatively, if equation (21) is discretized on a staggered grid such as that used in [13] for example in which the velocities are stored midway between nodes where pressures are located, then the following equation for u_i results:

$$\overset{\leftarrow\ \delta x\ \rightarrow}{\vec{u}_{i-1}\quad \dot{p}_{i-1}\quad \vec{u}_i\quad \dot{p}_i\quad \vec{u}_{i+1}}$$

$$\frac{\dot{m}}{2\delta x}(u_{i+1} - u_{i-1}) - \frac{\mu}{\delta x^2}(u_{i+1} + u_{i-1} - 2u_i) = \frac{1}{\delta x}(p_{i-1} - p_i)$$

which for uniform flow reduces to:

$$p_i = p_{i-1} \tag{24}$$

Thus, the pressure at each node is related by eqn. (24) to its immediate neighbours, and only a uniform pressure field which corresponds with the physical reality can satisfy this relationship.

Similar arguments to the above also apply when the continuity equation is considered. Centering of the difference expressions for the mass fluxes on a suitably staggered grid results in the desired coupling between adjacent velocities, whereas the same is not obtained with a non-staggered mesh.

A further benefit gained from staggered meshes, but which is not immediately obvious, is the elimination of the need for boundary conditions on pressure, since the domain boundaries can be chosen to fall on velocity nodes, thus making boundary conditions on velocity the only requirement; the same cannot be said of non-staggered meshes. It is not the task here to rigorously prove that pressure-boundary conditions are not required for closing the set of incompressible flow equations and that only the boundary conditions on velocity are sufficient for that purpose. However, an argument that is offered here in support of this contention is as follows. The momentum equations can be differentiated and combined to eliminate the pressure altogether from the resulting equation (which can be construed as one for vorticity). Since the equation thus obtained together with the continuity relation contain no reference to the pressure, no boundary conditions for the latter are necessary. Clearly this is possible solely because of the special role played by the pressure in the momentum equations where it appears in the form of gradients. It is interesting to note that when the equivalent manipulation is carried out using the difference form of the momentum equations, the resultant equation can be shown to be free from pressure if a staggered mesh is used, but the same is not obtained in the case of non-staggered grids. It may be argued, therefore, that grid staggering is the more natural arrangement as it leads to difference equations which not only correspond to the differential equations, but also conform to the same rules of manipulation and operation.

2.5 Temporal Discretization

To complete the discretization, equation (20) must now be discretized with respect to time such that the field variables can be advanced from their state at time t to that prevailing at time $t + \delta t$ (these time levels are denoted herein by superscripts k and k+1 respectively). Here again the choice of scheme for this purpose will not be dwelt upon for limitations on scope. It suffices to point out that there exist many temporal differencing schemes as inspection of the literature (e.g. refs. [11],[17] and [20]) will reveal. Some of these schemes invoke information available at just two time levels t and $t + \delta t$, while others rely on more than two time levels to achieve high order of temporal accuracy; some, which although ostensibly use two time levels only, involve more than one step in the integration from t to $t + \delta t$ (e.g. the two-step Lax-Wendroff method [17], and the time-splitting scheme of [21]). In any event, the incorporation of any of these difference schemes has a direct bearing on the form of the resulting difference equations and hence on the method to be chosen for their solution. For the sake of clarity of presentation, attention will be limited here to discretization schemes involving only two time levels which needs not affect the generality of the discussion.

The time derivative term in eqn. (20) may be expressed in the general form:

$$\int \frac{\partial}{\partial t}(\rho\phi)dV = (B\phi_0)^{k+1} + (B\phi_0)^k$$

where B is a quotient containing the density, volume of the cell and δt. The right-hand side of equation (20) must now be referenced to a particular time level, e.g. time t, time $t + \delta t$ etc. A convenient and general method of representing this group of terms is through the use of a linear combination of the spatial fluxes prevailing at the various time levels, suitably weighted by factors which take different values for the different temporal differencing schemes. Thus, for the two time levels considered here, the equation can be written in a general form as:

$$(B\phi_0)^{k+1} - (B\phi_0)^k = \{\alpha(A_i\phi_i)^{k+1} + (1 - \alpha)(A_i\phi_i)^k\} + S_0 \tag{25}$$

where α is the weighting factor defining the temporal difference scheme used e.g.: forward difference is given by $\alpha = 0$, backward difference by $\alpha = 1$ and centred difference by $\alpha = 1/2$.

There are as many equations such as (25) as there are conservation equations and the equation set is non-linear being coupled through the dependence of the A's, B's and S's on each of the dependent variables which ϕ represents.

3. SOLUTION METHODS

3.1 Preamble

In this section, methods for solving the discretized equations presented earlier are examined. The discussion commences with an evaluation of explicit versus implicit schemes, followed by a review of options for the handling of non-linearities. The main thrust of the section is to explore the means by which the continuity equation is coupled to the momentum ones. In this respect, attention is restricted to methods using the pressure as a main depedent variable rather than the density. The reason is that, as highlighted earlier, density based methods are liable to breakdown at very low Mach numbers, while such flows constitute the major field of interest in the present contribution.

3.2 Explicit Versus Implicit Schemes

Explicit schemes in which α in equation (25) takes the value of zero are the simplest to work with. Inspection of the equation reveals that the value of ϕ^{k+1} is determinable directly from known values at old time levels. However, it is well known that, for stability reasons, explicit methods (unlike implicit ones) suffer from severe limitations on the time step size δt (see |17|), especially for flows where diffusion is present. Explicit computations can therefore prove to be prohibitively expensive, particularly if only the steady-state solution is required. An illustration of this inefficiency is given in |21| where a comparison was made of the computing effort required by two schemes, one explicit and the other implicit, both using the same spatial discretization, and both applied to the same problem for which the steady-state solution was sought. For a moderately high Reynolds number flow, the explicit technique was shown to require up to 1000 times the effort expended by the implicit one, thus putting the former at a grave disadvantage.

For transient flows, temporal accuracy is called for, and this imposes its own restrictions on the time step size that can be taken, the limit being of the same order of magnitude as that demanded by stability considerations. It may, therefore, be argued that at least for transient flows, explicit techniques offer some advantage. However, the stability limit on δt is a local criterion and must be satisfied everywhere in the field, where large variations in velocities as well as mesh density often arise. The overall limit imposed on δt must, therefore, be the minimum occuring in the field which is likely to be significantly lower than what is dictated by temporal accuracy considerations. Hence, even in time dependent calculations, explicit methods can be inefficient.

A further drawback with explicit techniques is that each equation solved must contain a time derivative term. In the incompressible flow limit such a term is absent from the continuity equation (see qn. (10)), and a special treatment is therefore warranted such as the introduction of artificial compressibility (e.g. [31]) or implicit formulation of the equation (e.g. [28] to [30]), both practices of which detract from the advantages of the method.

Implicit methods on the other hand, do not suffer from time-step restrictions. They do, however, require the solution of sets of non-linear simultaneous equations like eqn. (25) for each of the dependent variables at each time step. It is the solution of such systems that the rest of the present paper is devoted to.

3.3 Treatment of Non-linearities

The discretized equations (25) are in general coupled and non-linear for which no direct solution is possible when considered in implicit form. Each of the A and B coefficients as well as the term S in the equation may be a function of the dependent variable ϕ for which the equation stands, as well as other dependent variables in the set. Two options exist for handling the non-linearity of the equations; these are: (i) linearisation and (ii) iteration, both of which are now assessed.

(i) Linearisation: It is possible to relate the coefficients at time level k+1 where their values are unknown to the values prevailing at time level k where they are readily determined. The relationship normally used for the purpose is the Taylor series expansion in time, truncated to first or second order terms. Thus for example:

$$A_i^{k+1} = A_i^k + \left(\frac{dA_i}{dt}\right)^k \delta t + O(\delta t^2) \tag{26}$$

First order linearisation is obtained by ignoring terms of order (δt) and higher in expression (26), hence:

$$A_i^{k+1} = A_i^k \tag{27}$$

Clearly, this type of linearisation is simple and easy to implement, the only drawback being its low order of accuracy. However, if the temporal discretization involved in the derivation of eqn. (25) is itself of first order accuracy (e.g. Euler implicit scheme, $\alpha = 1$), which is often the case, then a first order linearisation such as that in eqn. (27) suffices.

Alternatively, second order linearisation which has recently become popular (e.g. [18] to [20]) may be obtained by retaining the term $O(\delta t)$ in eqn. (26). However in this case, the whole flux term $(A\phi)$ must be considered for complete linearisation. Hence:

$$(A_i\phi_i)^{k+1} = (A_i\phi_i)^k + \left[\frac{d(A_i\phi_i)}{dt}\right]^k \delta t \quad \text{(no summation)} \tag{28}$$

Now, A_i is a function of many variables, say ϕ^m, where each value assigned for m signifies a certain variable. Hence, equation (28) can be replaced by:

$$(A_i\phi_i)^{k+1} = (A_i\phi_i)^k + \left[\frac{\partial(A_i\phi_i)}{\partial\phi^m}\right]^k \delta\phi^m \quad \text{(summation on m)}$$

or

$$(A_i\phi_i)^{k+1} = (A_i\phi_i)^k + \left[\frac{\partial(A_i\phi_i)}{\partial\phi^m}\right]^k \{(\phi^m)^{k+1} - (\phi^m)^k\} \tag{29}$$

(summation on m)

It is clear, that whereas the original flux term contains only one dependent variable (unsuperscripted ϕ), its replacement which is the right hand side of eqn. (29) contains as many dependent variables as the coefficient A_i depends on. To illustrate this, consider for example the convective flux term $(\rho vu)^{k+1}$ in the u-momentum equation, linearised in accordance with eqn. (29). It becomes:

$$(\rho vu)^{k+1} = (\rho vu)^k + (\rho v)^k(u^{k+1} - u^k) + (\rho u)^k(v^{k+1} - v^k)$$
$$+ (uv)^k(\rho^{k+1} - \rho^k) \tag{30}$$

Insertion of expression (30) into the finite-difference equation for u^{k+1}, introduces the additional variables v^{k+1} and ρ^{k+1}, thus making the original equation lose its role as one for a single variable i.e. for u^{k+1}. When all the difference equations are likewise linearised, the equations become simultaneous with respect to variables, and the set is called block simultaneous in which each element in the coefficient matrix is itself a matrix of m x m variables.

Evidently, the set of equations resulting from this type of linearisation requires special solution procedures - a subject which will be examined shortly. It should also be noted that, should the present conservation equations be modified for any reason, or if additional equations are introduced, substantial alterations to the finite-difference equations and to the solution procedure, which involve much effort, must be effected. Thus, second order accuracy is gained at the high cost of complexity and inflexibility. Moreover, unless second order (or higher) accurate temporal differencing scheme is employed in arriving at eqn. (25), implementation of such linearisation schemes becomes pointless.

ii) Iteration: The alternative to linearisation, is to effect an iterative process, in which the equations are solved sequentially and repeatedly while the coefficients are updated in between one iteration and the next, the process being carried out until convergence of the field variables is obtained. This approach is useful for methods which only solve for steady-state flows (e.g. [13] and [14]). In these methods, time is replaced by an equivalent iteration parameter, and the converged solution obtained corresponds to the final steady-state of an equivalent transient calculation. Intermediate stages in steady-state calculations are of no interest, and therefore the updating of coefficients can lag the determination of the field variables on which they depend. In the final converged solution, all the equations such as (25) are satisfied implicitly inclusive of all non-linear dependency.

For time-dependent methods, when iteration is employed to update the coefficients (e.g. |5|), the iterative process must be effected at each time step. Since a number of iterations have to be performed to arrive at the converged, solution these calculations can be very costly, and it is for this reason, that many existing time-dependent techniques avoid iteration and resort to linearisation instead. Here again, it is not certain whether precise evaluation of the coefficients at the k+1 time level is necessary, when the order of accuracy of the temporal difference scheme is the decisive factor in determining overall accuracy.

3.4 Solution of the Linearised Equations

When the implicit equations (such as (25), with $\alpha > o$) are either linearised, or when their coefficients and source terms are held constant in between iterations, linear simultaneous sets of equations are obtained, which may or may not be coupled through the variables appearing in them. The task now is to review the methods suitable for the solution of these systems, although the linear coupling between the equations of continuity and momentum through the pressure-velocity linkage will be overlooked for the moment, as it will be the subject of a separate treatment later.

The sets of equations may be cast in the convenient matrix form:

$$[B + A][\phi]^{k+1} = [S] \tag{31}$$

for systems with single variables ϕ, or:

$$[B + A][\phi^m]^{k+1} = [S] \tag{32}$$

for block simultaneous systems in which ϕ^m is the variable vector. The matrices [A], [B] and [S] are those whose elements are the corresponding terms appearing in equation (25). In eq. (32) each element of [A] is itself a matrix having the dimension m x m, where m is the number of variables. Such a system is obtained for example as a consequence of the second order linearization discussed in section 3.3 or when a compact differencing scheme is used for the representation of the spatial fluxes (see e.g. [33]). The main notable features of matrix [A] for both systems (31) and (32) are: the matrix is sparse and in general the elements are diagonally banded*.

The three main options available for the solution of systems (31) and (32) are: (i) direct methods, (ii) iterative schemes and (iii) split-operations algorithms.

(i) Direct methods: It is not unusual to find in the literature examples of the application of direct methods to the solution of single variable systems (31). However, most of these methods do not implement the elimination process directly to the set, as this involves much effort and storage (see, e.g. [34]). It is normally the practice to manipulate the matrices first such that the bands are rearranged into a form which minimizes the demands on intermediate storage in the implementation of the Gaussian elimination; they fail, however, to reduce the number of operations required (see [22]). Furthermore, these demands are still high when large matrices are involved as is the case with fine grids or three-dimensional configurations. No general direct method is viable for block systems such as (32) with the exception of block bi-diagonal and tridiagonal ones. These, however, are not obtained in general in multi-dimensional flows unless special practices are introduced such as the splitting of operations during the solution (e.g. [18]), or the time-splitting of spatial differences as in [21].

(ii) Iterative schemes: These are fairly widely used as they offer the decisive advantage over direct methods of not requiring the storage of any of the null elements in the matrices which are, as mentioned earlier, sparse. Their advantage is further enhanced by the requirement of a fewer number of operations(for an assessment of operation counts see e.g. [22]). Since such iterative schemes are well covered in the

---

*Provided that the domain is discretized into quadrilaterals with common vertices.

literature, the reader is referred to standard texts such as [22], [24] and [34] for a comprehensive treatment. It suffices here to mention the best known amongst them which are: the point Gauss-Seidel iteration, successive over-relaxation, line Gauss-Seidel iteration*, and incomplete factorization methods such as the "strongly implicit" technique of [23].

Experience shows that of the above schemes, the incomplete factorization method of [23] is fastest in terms of number of iterations, although this is offset by requiring a greater number of operations per iteration. The scheme is particulary superior to the others for a 'truly elliptic'+ equation such as that for pressure. Its rate of convergence is also insensitive to the type of boundary conditions and to non-homogeneity of the coefficients. Of the others, the line Gauss Seidel scheme, is quite powerful for equations having large coefficients in a preferred direction (i.e. parabolic-type equations).

Also shown by experience, is that for equations with coefficients which are direction-wise homogeneous, the rate of convergence slows down dramatically after a certain number of iterations. This is attributed to the ineffectiveness of iterative schemes to rapidly purge the low frequency errors residing in the domain. Multi-grid techniques, e.g. [35], have been developed to improve the performance of iterative schemes in eliminating these low frequency errors. Whether the additional complexity involved in introducing such convergence-accelerating pracrice is justified by the returns is open to question when considered in the present context, since the solution of the linear sets of equations occupies a smaller part of the overall computing effort required for the type of flows under consideration.

(iii) Split-operation schemes: It is sometimes possible to approximately factorise the operators in the original set of equations such that each of the new constituent operators is amenable to direct solution. The combination of the split operations then gives the final solution which is a close approximation to the exact one. The best known example of such schemes is the ADI method [25] and [26], in which the factorization is accomplished as follows. Equation (31) may be written as:

$$[B + A_x + A_y + A_z] \quad [\phi]^{k+1} = [S] \tag{33}$$

where A_x, A_y and A_z are coefficient matrices each relating to spatial fluxes in one of the spatial directions.

Equation (33) is factorised into:

$$[B + A_x] \ [\phi]^* = [S] - [A_y + A_z][\phi]^k \tag{34a}$$

$$[B + A_y] \ [\phi]^{**} = [S] - [A_x][\phi]^* - [A_z][\phi]^k \tag{34b}$$

and

$$[B + A_z] \ [\phi]^{***} = [S] - [A_x][\phi]^* - [A_y][\phi]^{**} \tag{34c}$$

*This is the equivalent of the point Gauss-Seidel iteration, but with the the solution accomplished over a whole grid line at a time rather than for a single node. The scheme takes advantage of the tridiagonal character of the coefficient matrix resulting from considering a single line, for a direct solution to such a system is possible.

+An equation in which off-diagonal coefficients are nearly homogeneous direction-wise.

where both ϕ^* and ϕ^{**} are intermediate fields and ϕ^{***} is the final field which is used to approximate the exact solution ϕ^{k+1}. It can be shown that by combining equations (34) together, the solution ϕ^{***} differs from the exact one ϕ^{k+1} (given by eqn. (33)) by terms of the order (δt^2) i.e. the factorization is second order accurate. The advantage of the above factorization is that each of equations (34) is a tridiagonal system which can be solved directly, thus obviating the need for iteration.

Extensions of the ADI technique have been made, in e.g. [18] to [20], to block simultaneous systems described by eqn. (32) in which, it may be recalled, ϕ^m is a vector. Factorization on the same lines as above results in equations like (34), except that each of the equations now becomes a block tridiagonal one which is also solvable by a direct method, such as that in [22].

Unfortunately, the ADI technique breaks down where a time derivative term is absent in the parent equation. The failure can be easily demonstrated by putting term B (which is the time derivative quotient) in eqns. (34) to zero with the result that both equations (34b) and (34c) become inoperative. Thus, only parabolic type equations, and not elliptic ones such as the incompressible flow continuity equation, can be handled by the method (which may still be adopted for the solution of the momentum equations) unless special practices, like the introduction of artificial compressibility, are resorted to (see, e.g. [31]).

3.5 Coupling of the Continuity and Momentum Equations

Earlier, it was outlined why the pressure should be retained as a main dependent variable in the equations in preference to the density, if the method is to be valide for the whole spectrum of Mach numbers including incompressible flows. Although pressure-based schemes were mostly developed for the incompressible regime, the extension to compressible flow is straightforward and is free of difficulties, as has been shown in [27].

The main obstacle arising from the choice of the pressure as a working variable is to devise a mechanism by which the continuity and momentum equations are linked together. The difficulty stems from the presence of the pressure in the momentum equations (which may be regarded as equations for the velocities) in the form of source terms, while its appearance in the continuity equation is conditional upon the fluid being compressible. Thus, the pressure does not, in general, possess its own equation amongst the raw set of equations. The path normally followed in surmounting this difficulty is by deriving a pressure (or pressure-like) equation from a joint manipulation of the continuity and momentum equations. The derivation may be achieved either by using the parent differential equations (details of which may be found in e.g. [11]) or by invoking the finite-difference analogues of these equations directly as is done in e.g. [12] to [14] and [30].

The differential approach can be demonstrated by considering the incompressible form of the momentum equations* (5) which when differentiated by $\frac{\partial}{\partial x_i}$ become:

---

*Only the incompressible equations will from here onwards be used. This is purely for reasons of convenience and clarity of presentation and does not affect the applicability of the analysis to the compressible flow equations.

$$\frac{\partial}{\partial t}\left(\frac{\partial u_i}{\partial x_i}\right) + \frac{\partial^2}{\partial x_i \partial x_j}(u_i u_j) = -\frac{1}{\rho}\frac{\partial^2 p}{\partial x_i \partial x_i} + \frac{\partial^2 \sigma_{ij}}{\partial x_i \partial x_j} + \frac{\partial S_u}{\partial x_i} \qquad (35)$$

In order for the velocity and pressure fields to satisfy the continuity relation (10), the latter is inserted into equation (35) so that all velocity divergence terms (i.e. $\frac{\partial u_i}{\partial x_i}$) vanish. The resulting equation may be rewritten as:

$$\frac{\partial^2 P}{\partial x_i \partial x_i} = f(u_i, u_j) \qquad (36)$$

Equation (36) is of the Poisson type and may be regarded as one for pressure, to be solved in conjunction with the momentum equations (5) to which it is linearly coupled. It may be discretized in the manner employed for discretizing the rest of the equations.

There are two main pitfalls in using the differential form of the equations. The first is that in the process of discretization of the pressure equation, there is no direct way of ensuring that the mass fluxes, which are implicitly included in the equation, are differenced in a conservative manner; overall mass balance may thus be infringed. Also, some fluxes appearing in both the momentum and pressure equations may not be differenced consistently, hence adversely affecting both accuracy and stability. The second pitfall is that unless the initial velocity field is divergence free* ($\frac{\partial u_i}{\partial x_i} = 0$), it is not correct to introduce equation (10) into every term in equation (35) to make the velocity divergence vanish. Thus, if $(\frac{\partial u_i}{\partial x_i})^k$ is not zero, then $\frac{\partial}{\partial t}(\frac{\partial u_i}{\partial x_i})$ is also not zero. Similarly if the spatial fluxes in the momentum equations are represented in the difference equations by weighted averages of old and new time level values (see e.g. eqn. (25)), then the value of $(\frac{\partial u_i}{\partial x_i})$ at time level k must be retained in the equation, a practice which is not immediately obvious when the differential form of the equations is used. Failure to account for a non-vanishing velocity divergence arising at any stage in the calculation leads to erroneous prediction of the pressure field, the effects of which persist for long in the rest of the calculation.

It is therefore evident that in order to circumvent the drawbacks outlined above, derivation of the pressure equation must proceed with the discretized form of the continuity and momentum equations. In what follows, the pressure equation is derived in discrete form on a two-dimensional staggered mesh a portion of which is shown in Fig. 4. Consider the control volume surrounding node 0, for whcih the continuity relation (10) may be expressed in discretized form as:

$$D_o = (\rho\delta yu)_e - (\rho\delta yu)_w + (\rho\delta xv)_n - (\rho\delta xv)_s = 0 \qquad (37)$$

where D_o is the divergence of the velocities, and all non-superscripted quantities are taken to prevail at the k+1 time level. The velocities u_e etc. are obtainable from the discretized momentum equations which are of the same form as (25). Let the equation for u_e, for example, be written as:

---

* In general the initial velocity field need not be divergence free, and to produce such a field would require a major effort in itself. More significantly, in an iterative procedure the velocity field obtained after each pass through the momentum equations, and which is subsequently used in the pressure equation, will also in general not be divergence free.

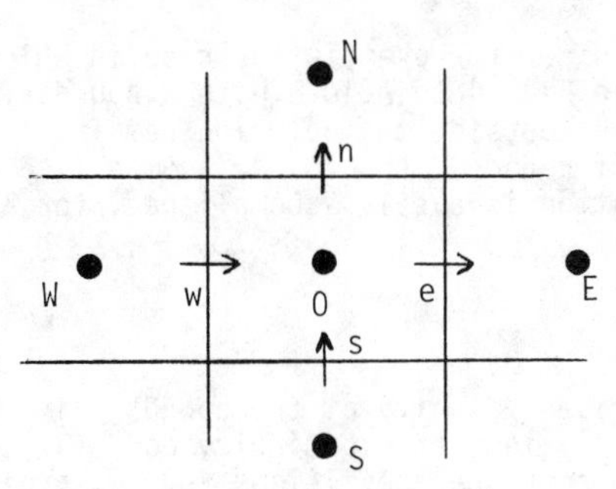

Fig. 4 Portion of mesh

$$u_e - u_e^k = \alpha(A_i u_i) + C_e(P_o - P_E) + S_e \tag{38}$$

where the summation index i, and the weighting factor α, retain the definitions given to them in eqn. (25). Here, again the non-superscripted velocities and pressures are assumed to prevail at the k+1 time level, while all k time level spatial derivatives are lumped into the S-term; also, the equation has been divided by the time step quotient B. The equation for v_n can similarly be written as:

$$v_n - v_n^k = \alpha(A_i v_i)_n + C_n(P_o - P_N) + S_n \tag{39}$$

Similar equations for u_w and v_s may also be obtained. When the velocities in equation (38), (39) etc. are substituted into the continuity equation (37), the following is obtained:

$$\begin{aligned}[(\rho\delta yC)_e + (\rho\delta yC)_w + (\rho\delta xC)_n + (\rho\delta xC)_s]\, p_o &= (\rho\delta yC)_e P_E + (\rho\delta yC)_w P_W \\ &+ (\rho\delta xC)_n P_N + (\rho\delta xC)_s P_S + D_o - D_o^k \\ &+ \alpha\{ -(\rho\delta y A_i u_i)_e + (\rho\delta y A_i u_i)_w - (\rho\delta x A_i v_i)_n - (\rho\delta x A_i v_i)_s\} \\ &+ S_o^P \end{aligned} \tag{40}$$

where the velocity divergence term D_o at level k+1 which is supposed to vanish is retained for reasons to be clarified later. The term S^P in eqn. (40) represents the lumped source terms appearing in the momentum equations (38) etc. The elliptic equation (40) is the sought Poisson equation for pressure in discrete form, and can be solved by any of the iterative schemes outlined in section 3.4 provided that the expression in { } brackets is given.

It was stated earlier, that when a staggered mesh is employed, only the boundary conditions on the velocity are required*; this may be

*This applies for cases with prescribed velocities at the boundaries. For cases where the pressure is specified, boundary conditions on the pressure are obviously necessary.

demonstrated now. Consider for example the case in which face e of the control volume shown in Fig. 4 coincides with a boundary; node E becomes redundant as it now lies outside the solution domain. The derivation of the pressure equation for node C should now be modified to take this into account. The modification is easily accomplished, for all is required is to replace eqn. (38) by:

$$u_e = U_B \tag{41}$$

where U_B is the prescribed velocity at the boundary in question. The result of substituting eqn. (41) into the continuity eqn. (37), is an equation like (40) but in which both the coefficient $(\rho\delta yC)_e$ and the term $(\rho\delta yA_i u_i)_e$ vanish. Thus, the equation will not contain any reference to pressures or velocities at nodes lying outside the domain, and no value for the pressure at the boundary is needed.

Evidently, equation (40) is linearly coupled to the momentum equations through the appearance of the {} bracket term in (40), which vanishes only when $\alpha = 0$, i.e. when an explicit temporal difference scheme is used for the discretization of the momentum equations. Attention will shortly be turned to methods of dealing with this coupling between the pressure and velocity equations. However before doing so, equation (40) is re-cast in a more compact form, for the sake of convenience, as:

$$A^P_o P_o = A^P_i P_i + (D_o - D^k_o) + \alpha H(\underline{u}) + S_P \tag{42}$$

where the A^P's are now the finite-difference coefficients of p, with i having the same significance as in eqn. (25), and $H(\underline{u})$ is a linear operator operating on the velocity vector field $\underline{u}$ and is defined by the expression in the {} bracket in (40). The diagonal coefficient A^P_o is given by:

$$A^P_o = \sum_i A^P_i \qquad (i = E,W,N,S) \tag{43}$$

Similarly, the momentum equations may be rewritten as:

$$\underline{u}_o - \underline{u}^k_o = \alpha A_i \underline{u}_i + \Delta p + S_u \tag{44}$$

where the Δ operator denotes the finite-difference equivalent of grad, i is the summation index over the cluster of nodes E, W ... etc. and S_u is the source term.

3.6 Algorithms for the Pressure-velocity Coupled System

Many schemes have been devised for the solution of the linearly coupled system in pressure and velocity; a description of the major alternatives is now presented.

The first notable feature of eqn. (42) is that if α is set to zero and with D_0 made to vanish, the equation becomes independent of the new time level velocity field and can hence be solved for the p^{k+1} field directly. Such methods are sometimes called semi- or pressure-implicit and a whole group of these were developed for incompressible (as in [12] and [30]) as well as for compressible flows (as in [28] and [29]). The implication of putting $\alpha = 0$ is that the momentum equations are discretized explicitly, although the continuity relation can still be satisfied in an implicit manner. Typically, methods based on this technique effect the solution as follows:

(a) Based on the given velocity field at time level k, the pressure equation is solved for p^{k+1}.

(b) The explicit momentum equations are solved for the new velocities using the new pressure.

(c) Other transport equations are solved.

(d) All fields are time advanced and the procedure can be repeated commencing with step (a) again.

The above class of methods suffers from the step size restrictions associated with explicit schemes as discussed earlier, although to their advantage, the need for iteration is obviated. The restriction may be quite severe in highly turbulent flows as the stability limit imposed by the diffusion terms becomes prominent, which makes the methods particularly unsuitable for computing steady state flows.

When fully implicit discretization is used for the equations including the momentum ones, the pressure equation (42) must be satisfied simultaneously with those of momentum. The most common approach to achieve that goal is by iteration. The simplest of iterative schemes would be to:

(a) Guess the pressure field at time k+1.

(b) Solve the momentum equations for the new velocities.

(c) Solve the pressure equation in which D_o is set to zero and the operator H is based on the latest velocities.

(d) Solve other equations.

(e) Repeat from step (b) if the solution is not converged.

(f) Advance by δt and repeat the procedure.

The procedure may be represented mathematically, with the aid of an iteration counter n by writing the momentum equation (44) as:

$$\underline{u}_o^{n+1} - \underline{u}_o^k = \alpha A_i \underline{u}_i^{n+1} + \Delta p^n + S_u \qquad (45)$$

and the pressure equation (42) as:

$$A_o^p p^{n+1} = A_i^p p_i^{n+1} - D_o^K + \alpha H(\underline{u}^{n+1}) + S_p \qquad (46)$$

where D_o^{n+1} has been set to zero. It is clear, that iteration is required since each approximation p^n yields a velocity $\underline{u}^{n+1}$ which does not satisfy the zero divergence condition, whereas the determination of p^n is actually based on the assumption that it yields a divergence free velocity field.

Many methods based in essence on the above principle exist. However, most solve for pressure-correction rather than for pressure itself and the solution is achieved by its decomposition into predictor and corrector steps. The central idea was first developed in [32], and was applied to the semi-implicit scheme of [30], although it is really useful for fully implicit ones such as in [13], [14] and [16]. The basis of this type of method will now be analysed. If the superscript * denotes an intermediate value, the momentum equation (44) can be written as:

$$\underline{u}_o^* - \underline{u}_o^k = \alpha A_i \underline{u}_i^* + \Delta p^n + S_u \tag{47}$$

which constitutes a predictor phase for $\underline{u}^*$, and where the pressure p^n is taken to be that at the previous iteration. Clearly, if equation (47) is used as the basis for forming the divergence of the intermediate velocity field i.e. D_o^* (as defined by eqn. (37)), the resulting pressure equation is obtained†:

$$A_o^p p_o^n = A_i^p p_i^n + (D_o^* - D_o^k) + \alpha H(\underline{u}^*) + S_p \tag{48}$$

which is automatically satisfied when eqn. (47) is solved.

The corrector phase is formulated by writing a corrected momentum equation as:

$$\underline{u}_o^{n+1} - \underline{u}_o^k = \alpha A_i \underline{u}_i^* + \Delta p^{n+1} + S_u \tag{49}$$

where it should be noted, and for reasons given shortly, only the velocity on the left-hand side of the equation is updated with a corresponding update of the pressure. Subtraction of equation (49) from (47) yields:

$$\delta \underline{u}_o = \Delta(\delta p) \tag{50}$$

where the correction quantities $\delta\underline{u}$ and δp defined by:

$$\underline{u}^{n+1} = \underline{u}^* + \delta\underline{u} \tag{51}$$

$$\underline{p}^{n+1} = p^n + \delta p \tag{52}$$

have been introduced. When equation (49) is employed in formulating a pressure equation†, the result is:

$$A_o^p p_o^{n+1} = A_i^p p_i^{n+1} - D_o^k + \alpha H(\underline{u}^*) + S_p \tag{53}$$

where D_o^{n+1} has been made to vanish. Equation (53) is the operative pressure equation and can be solved straigh away. However, a similar equation may be obtained, as is often done, by subtracting equation (48) from (53) to give the so called pressure-correction equation:

$$A_o^p \delta p_o = A_i^p \delta p_i - D_o^* \tag{54}$$

Equation (54) is solvable directly to yield the δp field, which can then be used to evaluate a $\delta\underline{u}$ field from eqn. (50). The velocity field is then

†The derivation follows the same steps taken in arriving at eqn. (42).

updated according to eqn. (51) and so is the pressure field† according to eqn. (52). Schemes which are based on the predictor-corrector method defined by eqns. (47), (50), (51), (52) and (54) are those of [13],[14].

It can be shown that the above predictor-corrector process is equivalent to solving equations (45) and (46) in succession, with the exception that with the pressure-correction approach, the velocity field $\underline{u}^{n+1}$ at the end of the iteration is divergence free and this is advantageous if the coefficients of the finite-difference equations are to be based on the new velocity field. Furthermore, as under-relaxation is most likely to be used, the relaxation will be achieved by reference to a divergence free velocity field which should enhance the rate of iteration convergence. The advantage of using a pressure-correction equation such as (54) rather than the full pressure equation (53) is its simplicity and the ease of application of the boundary conditions.

The reason behind formulating the corrector phase in the manner shown in eqn. (49) is that if the $\underline{u}$ field on the right-hand side of (49) is also taken to be at the n+1 iteration level, the resulting pressure equation will be the saem as in eqn.(46), which unlike its counterpart eqn. (53) contains unknown velocities. The corresponding pressure-correction equation will therefore be:

$$A^p_p \delta p_o = A^p_i \delta p_i - D^*_o + \alpha H(\delta \underline{u}) \tag{55}$$

which unlike eqn. (54) cannot be solved directly as it contains the unknown term $H(\delta \underline{u})$, which is a function of the δp field.

Experience has shown that the above predictor-corrector iteration scheme is seldom stable without heavy under-relaxation and is often slow to converge. This is atrributable to the approximation made in formulating eqn. (49) which has led to the vanishing of the term $H(\delta \underline{u})$ from eqn. (54). To overcome this shortcoming a two stage corrector phase is proposed here. In the first corrector stage, exactly the same steps as in the single-corrector approach, where the pressure-correction equation (54) is solved, are followed. The second stage comprises the solution of another pressure-correction equation i.e. (55), with the term $H(\delta \underline{u})$ constructed from the approximate $\delta \underline{u}$ field calculated from eqn. (51) in the first corrector stage. This amounts to the break up of the solution of the exact equation (55) into two stages which yields a pressure field that is much closer to the final solution than that obtained by the single corrector stage, hence resulting in a substantial improvement in the stability and rate of convergence of the iteration process.

The SIMPLER technique proposed in [16] shares some of the features of the two-stage corrector scheme outlined above. However, it involves the nearly twice the amount of work needed. Nevertheless, considerable saving in computing effort is reported to have been gained.

---

†In some methods such as [36], the pressure field is not updated by eqn. (52) but is evaluated from the integral of the momentum equations in which the $\underline{u}$ field is corrected by eqn. (51).

Finally, a new non-iterative scheme has recently been developed [37] to solve the pressure-velocity system. In this scheme, the predictor-corrector concept just outlined, is re-formulated into a series of split operations such that the final solution obtained is a close approximation to the exact solution of the original equations. Thus, the momentum equation (44) is operator-split into:

$$\underline{u}_o^* - \underline{u}_o^k = A_i \underline{u}_i^* + \Delta p^k + S_u \tag{56}$$

$$\underline{u}_o^{**} - \underline{u}_o^k = A_i \underline{u}_i^* + \Delta p^* + S_u \tag{57}$$

and

$$\underline{u}_o^{***} - \underline{u}_o^k = A_i \underline{u}_i^{**} + \Delta p^{**} + S_u \tag{58}$$

where *, ** and *** denote fields at various stages of the solution. Further splitting is of course possible but it is sufficient to take u^{***} and p^{**} to be the approximate solutions for $\underline{u}^{k+1}$ and p^{k+1}. The pressure fields p^* and p^{**} are obtained from their respective pressure-equations which are derived by requiring that both D_0^{**} and D_0^{***} vanish†. It can be shown that the approximate final velocity and pressure fields are accurate to $O(\delta t^3)$ and $O(\delta t^2)$ respectively, which is sufficient considering the approximate nature of the finite-difference equations themselves. Thus, iteration is obviated leading to substantial reductions of computing efforts (by factors of 3 or 4 if not greater).

4. CONCLUSIONS

In what preceded, the factors which must enter into consideration when finite-difference methods are to be applied to practical engineering problems, are stated. The limitation of using the density as a working variable is investigated and methods for handling turbulence effects are selected. Spatial and temporal discretizations are briefly examined including the choice of staggered mesh arrangement but no particular difference scheme was chosen, so as not to restrict the generality of the presentation. The choice between explicit and implicit methods is then assessed. The treatment of non-linearity is discussed and the various options are presented. Also examined is the choice of methods for solving the linearised sets of equations. Finally, the techniques for formulating and solving the equations for the coupled pressure-velocity system are reviewed.

REFERENCES

1. Gosman, A.D., Pun, W.M., Runchal, A.K., Spalding, D.B. and Wolfstein, M. 1969. Heat and mass transfer in recirculating flows. Academic Press.

2. Launder, B.E. and Spalding, D.B. 1972. Mathematical models of turbulence. Academic Press.

3. Monin, A.S. and Yaglom, A.M. 1977. Statistical fluid mechanics. MIT Press.

4. Hinze, J.O. 1959. Turbulence. McGraw-Hill.

---

†Similar to the derivation of eqn. (42).

5. Gosman, A.D., Johns, R.J.R. and Watkins, A.P. 1980. Development of prediction methods for in-cylinder processes in reciprocating engines. Combustion Modelling in Reciprocating Engines, Plenum Press.

6. Ahmadi-Befrui, B., Gosman, A.D., Lockwood, F.C. and Watkins, A.P. 1981. Multidimensional calculation of combustion in an idealized homogeneous charge engine: a progress report. SAE Paper 810151, Detroit.

7. Demirdzic, I., Gosman, A.D. and Issa, R.I. 1980. A finite-volume method for the prediction of turbulent flow in arbitrary geometries. Proc. of the Seventh Int. Conf. on Numerical Methods in Fluid Dynamics, Stanford.

8. Antonopoulos, K., Gosman, A.D. and Issa, R.I. 1978. Flow and heat transfer in tube assemblies. Proc. of the First Int. Conf. on Numerical Methods in Laminar and Turbulent Flow, Swansea.

9. Antonopolous, K., Gosman, A.D. and Issa, R.I. 1978. A prediction method for laminar and turbulent flow in tube assemblies. Proc. of the Sixth Int. Conf. on Numerical Methods in Fluid Dynamics, U.S.S.R.

10. Issa, R.I. and Gosman, A.D. 1981. The computation of three-dimensional turbulent two-phase flows in mixer vessels. Proc. of the Second Int. Conf. on Numerical Methods in Laminar and Turbulent Flow, Venice.

11. Roach, P. 1976. Computational fluid dynamics. Hermosa Publishers.

12. Harlow, F.H. and Welch, J.E. 1965. Numerical calculation of time-dependent viscous incompressible flow of fluid with free surface. The Physics of Fluids, Vol. 8, p. 2182.

13. Patankar, S.V. and Spalding, D.B. 1972. A calculation procedure for heat, mass and momentum transfer in three-dimensional parabolic flows. Int. J. Heat Mass Transfer, Vol. 15, p. 1787.

14. Caretto, L.S., Gosman, A.D., Patankar, S.V. and Spalding, D.B. 1972. Two calculation procedures for steady three-dimensional flows with recirculation. Proc. of Third Int. Conf. on Numerical Methods in Fluid Dynamics, Paris, p. 60.

15. Hirt, C.W., Amsden, A.A. and Cook, J.L. 1974. An arbitrary Lagrangian-Eulerian computing method for all flow speeds. J. Comp. Phys., Vol. 14, p. 227.

16. Patankar, S.V. 1980. Numerical heat transfer and fluid flow. McGraw-Hill.

17. Richtmyer, R.D. and Morton, K.W. 1967. Difference methods for initial-value problems. Interscience Publishers.

18. Briley, W.R. and McDonald, H. 1977. Solution of the multi-dimensional compressible Navier-Stokes equations by a generalised implicit method. J. Comp. Phys., Vol. 24, p. 372.

19. Beam, R.M. and Warming, R.F. 1978. An implicit factored scheme for the compressible Navier-Stokes equations. AIAA Journal, Vol. 16, p. 393.

20. Warming, R.F. and Beam, R.M. 1978 On the construction and application of implicit factored schemes for conservative laws. SIAM-AMS Proc., Vol. 11, p. 85.

21. MacCormack, R.W. 1981. A numerical method for solving the equations of compressible viscous flow. AIAA-81-0110, 19th Aerospace Sci. Meeting, St. Louis.

22. Isaacson, E. and Keller, H.B. 1966. Analysis of numerical methods. John Wiley & Sons Inc.

23. Stone, H.L. 1968. Iterative solution of implicit approximations of multidimensional partial differential equations. SIAM J. Num. Anal., Vol. 5, p. 530.

24. Ames, W.F. 1977. Numerical methods for partial differential equations. Academic Press.

25. Peaseman, D.W. and Rachford, H.H. 1955. The numerical solution of parabolic and elliptic differential equations. J. SIAM, Vol. 3, p. 28.

26. Douglas, J. and Gunn, J.E. 1964. A general formulation of alternating direction implicit methods, Part I, Parabolic and hyperbolic problems. Numerische Mathematik, Vol. 6, p. 428.

27. Issa, R.I. and Lockwood, F.C. 1977. On the prediction of two-dimensional supersonic viscous interactions near walls. AIAA J., Vol. 15, p. 182.

28. Harlow, F.H. and Amsden, A.A. 1968. Numerical calculation of almost incompressible flow. J. Comp. Phys., Vol. 3, p. 80.

29. Harlow, F.H. and Amsden, A.A. 1971. A numerical fluid dynamics calculation method for all flow speeds. J. Comp. Phys., Vol. 8, p. 197.

30. Amsden, A.A. and Harlow, F.H. The SMAC method: A numerical technique for calculating incompressible fluid flows. Los Alamos Sci. Lab. Report, LA-4370.

31. Chorin, A.J. 1967. A numerical method for solving incompressible viscous flow problems. J. Comp. Phys., Vol. 2, p. 12.

32. Chorin, A.J. 1968. Numerical solution of the Navier-Stokes equations. Maths. of Comp., Vol. 22, p. 745.

33. Ciment, M., Leventhal, S.H. and Weinberg, B.C. 1978. The operator compact implicit method for parabolic equations. J. Comp. Phys., Vol. 28, p. 135.

34. Smith, G.D. 1979. Numerical solution of partial differential equations: Finite difference methods. Clarendon Press.

35. Brandt, A. and Dinar, N. 1979. Multigrid solutions to elliptic flow problems. Numerical Methods for Partial Differential Equations. Academic Press.

36. Mazhar, Z. and Raithby, G.D. 1981. A refined PUMPIN (Pressure Update by Multiple Path Integration) method for updating pressures in the numerical solution of the incompressible fluid flow equations. Proc. of Second Int. Con. on Numerical Methods in Laminar and Turbulent Flow, p. 255.

37. Issa, R.I. Split-operator solution of the pressure-velocity coupled equations. To be published.

The Computation of Transonic Potential Flow

TIMOTHY J. BAKER

1. INTRODUCTION

In these lecture notes we shall review the numerical computation of transonic flow. The progress in this area has been startling and the discipline now encompasses a range of separate and differing activities. It follows that the task of reviewing current progress is a formidable one and any comprehensive account can only provide a superficial discussion of the theoretical details. To avoid this fate and make a reasonable attempt at describing the present state of the art, it is necessary to restrict oneself to a narrowly defined part of the total activity. In the following we shall therefore discuss the currently available methods for computing steady, inviscid, potential flow by finite difference techniques. By potential flow we mean the solution of the full potential equation and we shall make only occasional reference to the Transonic Small Perturbation or TSP equation. This is justified by the general recognition that TSP methods are less accurate and rely too heavily on empirical fiddle factors to be satisfactory for prediction purposes. The reason that TSP methods have enjoyed popularity in the past has been the relative simplicity of the grids they employ. This has enabled the TSP approach to be applied to configurations of greater geometric complexity than the corresponding full potential technique. Indeed, the only finite difference methods that are at present available for calculating transonic flow over the combination of wing + fuselage + stores appear to be those of Boppe [1] and Albone [2]; both these codes are based on the TSP equation. However, the current progress in grid generation techniques (see chapter 4) should soon erode this advantage and in due course open the way to the application of the full potential equation to compute transonic flow over complete aircraft.

On the other hand, a complete description of the flowfield can only be obtained from the Navier Stokes equations and many physical phenomena (eg buffet and other separated flow behaviour) cannot be properly modelled until such methods are available. For the present and foreseeable future it appears that computer codes based on the Navier Stokes equations do not offer the prospect of being low cost and routine methods for aerodynamic design purposes. It is therefore to potential flow methods that we look for the routine computation of transonic flow. It should nevertheless be borne in mind that the present emphasis on potential flow is only a transient affair and the ultimate goal remains the solution of the Navier Stokes equations about complete aircraft.

Keeping within our narrow framework it is apparent that the numerical computation of steady transonic flow is an activity that has been widely researched in recent years. Potential methods have been developed for a wide variety of geometric shapes and are now used extensively throughout the aerospace industry. The next generation of aircraft are currently in the design stage and the success of their aerodynamic design will depend much on the accuracy or otherwise of the computer codes that are now available.

In parallel with the increasing variety of geometric applications, there has been tremendous progress in the numerical algorithms used to solve the flow equation. The most obvious example of this phenomenon is the improvement in the calculation of steady transonic flow over an aerofoil. The first viable computer code appears to be that of Yoshihara [3] who solved the unsteady equations for isentropic flow on a rather complex set of overlapping grids (in today's jargon we would probably call it grid embedding). The quoted computer times were several hours on a CDC 6400. Roughly a decade later Jameson with his MAD method [4] was able to obtain converged solutions after only ten or twenty iterations. It is difficult to obtain a precise comparison of the computing effort required to produce a converged result in each case. However, it seems reasonable to estimate that this represents an improvement of between two and three orders of magnitude, and that, by any standards, is quite spectacular. Between these two codes lie a few key breakthroughs and the frustrations and triumphs of those who have struggled with the problem of calculating transonic flow.

The progress we have mentioned is in the numerical techniques for solving the flow problem. If one takes into account the equally dramatic improvement in computer technology that has also taken place over the last decade, then it is not difficult to see why computational fluid dynamics is considered to be a field of rapidly growing importance.

And yet the present state of the art leaves much to be desired. The use of the potential equation rests on the assumption that the flow is isentropic and whenever shock waves exist this ceases to be true. It is argued that the entropy rise through a shock wave is proportional to the third power of the shock strength which is measured by M^2-1 where M is the Mach number just ahead of the shock. If this Mach number is less than about 1.4 then the corresponding entropy increase should be small. Since we are mainly interested in calculating flows at or near design conditions, where the flow is expected to be attached and any shock waves fairly weak, it seems at first sight plausible that the isentropic flow assumption is not too far from the truth. But leaving aside the objection that we also want to know what happens at off-design conditions (ie with strong shock waves and regions of separated flow) one still has to admit that the potential equation can only provide an approximate description of transonic flow. What is particularly disturbing is the recognition that we have no quantitative feel for the effect of an apparently small entropy increase on the complete flowfield. Experience indicates that surprisingly good predictions of transonic flows are obtained from the potential equation but, with the present state of knowledge, thus must be regarded as being somewhat fortuitous. When we carry out a transonic flow calculation by a potential method we can never be sure just how close this result will be to the correct inviscid solution. In other words, the potential equation is an inadequate model for transonic flow and worse still we do not know how inadequate it is.

This uncertainty is compounded by the inevitable appearance of other errors that arise in a numerical calculation. A particularly difficult question concerns the effect of truncation errors introduced by the discretisation of the differential equation. On the stretched and contorted grids that are required for the more complex geometric configurations, truncation errors can be quite large. We rarely have much idea of their size and hardly any knowledge of the way these truncation errors can interact and hence modify the equation and its solution. We have to conclude that we are using a physical model that is inadequate and justifying our numerical computations with a mathematical theory that at best is of doubtful rigour. Seen from this point of view, it is perhaps ironic that potential flow methods should have made such a profound impact. We make this point not out of despair but rather to strike a mood of sober realism amidst the euphoria that surrounds much of the computational fluid dynamics scene.

Perhaps the best attitude to adopt is one of subdued optimism. Despite their deficiencies the potential flow methods are at present the best developed and most widely used techniques for predicting transonic flow. We shall discuss much of the theory on which these methods are based and try to convey an impression of the scope and versatility of the many finite difference codes that are available. The subject falls into three loosely related areas, grid generation, discretisation of the potential equation and the iterative solution of the finite difference equations. Discretisation will be considered in the next chapter. This involves the concepts of artificial viscosity, artificial compressibility and a discussion of conservative and nonconservative differencing. Conceptually, the problem of grid generation precedes that of solving the finite difference equations. However, recent advances in iterative algorithms for computing transonic flow have brought this area to a state of relative maturity. It therefore seems sensible to describe the current status of the solution algorithms before discussing grid generation techniques.

Although it has always been regarded as an important element in the total package, it is only fairly recently that grid generation has been generally recognised as a discipline in its own right [5]. Throughout most of the development of flowfield methods, grids have been produced in an almost ad hoc manner to suit the particular geometry under consideration. Although this approach was adequate for simple shapes of aerodynamic interest, the application of potential flow methods to more complex configurations requires considerable ingenuity. In fact, it seems that the problem of grid generation has reached a crisis in which ingenuity alone is unable to define a global coordinate transformation for complicated geometries. The appearance of numerical and hence semi-automated grid generation [6] is one response to this difficulty. So too is the realisation that different parts of the flowfield must be treated separately and then patched together [7,8,9] if one is to compute the flow around complete aircraft. Indeed, it is probable that grid generation will now be the main area of interest and progress as far as the potential equation is concerned. The discussion of this area that is presented here is in some ways rather vague and inconclusive but this is perhaps inevitable bearing in mind that grid generation is a new and burgeoning field.

Finally, we note that the emergence of grid generation as a subject in its own right, represents a rather belated recognition of its importance by those who are involved in finite difference techniques. In the finite element method the importance of geometry was recognised from the start. Differential equations which are elliptic in character can be recast as a variational problem in which the required solution follows from the minimisation of a certain functional. The theoretical basis of this method is now well developed and the success of its application to several areas of engineering is well known. Unfortunately, the relevance of these ideas to equations of a mixed elliptic and hyperbolic character is less well understood. Nevertheless, developments along these lines [10,11] are to be welcomed even if the computed shock waves are not always convincing. Particularly significant is the work of Periaux et al [11,12] who can now compute transonic flow over a configuration consisting of wings, fuselage and rear mounted nacelles. This is a remarkable achievement and it is quite possible that further developments in finite element methods may supersede much of the present work using finite difference techniques.

2. DISCRETISATION OF THE POTENTIAL EQUATION

2.1. The Potential Equation

The equation of mass conservation in a steady flow can be written in the following divergence form

$$\nabla.(\rho \underline{u}) = 0 \tag{2.1}$$

We assume the existence of a velocity potential Φ so that the velocity can be represented as

$$\underline{u} = \nabla\Phi$$

Since the flow has been assumed irrotational and isentropic (these conditions are necessary for the existence of a potential function) we can integrate the momentum equation to obtain Bernoulli's equation

$$\frac{q^2}{2} + \frac{a^2}{\gamma-1} = \text{const} \tag{2.2}$$

which relates the sound speed a to the flow speed $q = |\underline{u}|$ at any point in the fluid. The sound speed is defined by the equation

$$a^2 = \frac{dp}{d\rho}$$

where p and ρ are the local static pressure and density respectively. The relation

$$\frac{p}{\rho^\gamma} = \text{const}$$

holds for the isentropic flow of a perfect gas and completes the set of governing equations. After normalising the velocities and sound speed by the freestream velocity and the density by its freestream value we can write the potential equation in either divergence form

$$\nabla.(\rho\nabla\Phi) = 0 \tag{2.3}$$

or quasilinear form

$$\nabla^2\Phi - \frac{\nabla\Phi}{2a^2}.\nabla(|\nabla\Phi|^2) \tag{2.4}$$

where

$$a^2 = \frac{1}{M_\infty^2} + \frac{(\gamma-1)}{2}(1-q^2) \tag{2.5}$$

and

$$\rho = (M_\infty^2 a^2)^{1/\gamma-1} = \{1 + \frac{(\gamma-1)}{2} M_\infty^2 (1-q^2)\}^{1/\gamma-1} \tag{2.6}$$

In a Cartesian coordinate system it follows that for two dimensional flow eqn (2.3) becomes

$$(\rho\Phi_x)_x + (\rho\Phi_y)_y = 0 \tag{2.7}$$

and the corresponding quasilinear form (2.4) becomes

$$(1 - \frac{u^2}{a^2})\, \Phi_{xx} - \frac{2uv}{a^2}\, \Phi_{xy} + (1 - \frac{v^2}{a^2})\, \Phi_{yy} = 0 \tag{2.8}$$

with $u = \Phi_x$ and $v = \Phi_y$

2.2. Finite Difference Formulae

In the following sections we shall make frequent use of difference formulae and in order to keep the notation reasonably concise it is convenient to introduce certain difference operators. For example the first order central difference operator in the x direction is defined by

$$\delta_x f_{j,k} = f_{j+\frac{1}{2},k} - f_{j-\frac{1}{2},k}$$

This extends in an obvious way to a second order operator δ_{xx} which has the effect

$$\delta_{xx} f_{j,k} = \delta_x(\delta_x f_{j,k}) = f_{j+1,k} - 2f_{j,k} + f_{j-1,k}$$

The following backward and forward difference formulae are frequently used

$$\overleftarrow{\delta}_x f_{j,k} = f_{j,k} - f_{j-1,k}$$

$$\overrightarrow{\delta}_x f_{j,k} = f_{j+1,k} - f_{j,k}$$

The operators $\overleftarrow{\delta}_x$ and $\overrightarrow{\delta}_x$ are usually written as ∇ and Δ in the numerical analysis literature. In adopting the arrowed notation we follow the accepted usage in recent papers on transonic flow computation and reserve the symbol Δ for other purposes. Finally, we introduce the averaging operator μ_x defined by the formula

$$\mu_x f_{j,k} = \tfrac{1}{2}(f_{j+\frac{1}{2},k} + f_{j-\frac{1}{2},k})$$

and remark that identical operators can be defined with respect to the y coordinate.

Consider a uniform grid with increment Δx between grid lines in the x direction and Δy in the y direction. Next write $\Phi_{j,k}$ for the discrete approximation to $\Phi(j\Delta x, k\Delta y)$ and assume that we can replace the differential equation for $\Phi(x,y)$ by a set of difference equations for $\Phi_{j,k}$ in such a way that the discrete solution $\Phi_{j,k}$ will approximate the continuous potential $\Phi(x,y)$ at the point $(j\Delta x, k\Delta y)$ arbitrarily closely as the grid increments Δx and Δy tend to zero. The component of velocity u at the point j,k can be approximated in the following way:

$$u_{jk} = \frac{\mu_x \delta_x}{\Delta x}\, \Phi_{j,k} = \frac{\Phi_{j+1,k} - \Phi_{j-1,k}}{2\Delta x} \tag{2.9}$$

If $\Phi_{j,k}$ is replaced by the corresponding value of the continuous potential $\Phi(j\Delta x, k\Delta y)$ and we assume that it can be expanded as a Taylor series we find that

$$\frac{\Phi_{j+1,k} - \Phi_{j-1,k}}{2\Delta x} = \Phi_x + \frac{\Delta x^2}{6}\, \Phi_{xxx} + O(\Delta x^4) \tag{2.10}$$

The truncation error is $\Delta x^2/6\ \Phi_{xxx}$ which is $O(\Delta x^2)$ in smooth regions of flow and tends to zero with Δx as the grid increment is reduced. Similarly we can approximate the second x derivative of Φ at the point j,k by the formula

$$\frac{\delta_{xx}}{\Delta x^2}\cdot\Phi_{j,k} = \frac{\Phi_{j+1,k} - 2\Phi_{j,k} + \Phi_{j-1,k}}{\Delta x^2}$$

$$= \Phi_{xx} + \frac{\Delta x^2}{12}\Phi_{xxxx} + O(\Delta x^4) \qquad (2.11)$$

which is also second order accurate. The formulae (2.10) and (2.11) together with similar formulae for the derivatives with respect to y and the following finite difference approximation for the cross derivative,

$$\frac{\mu_x\mu_y\delta_x\delta_y}{\Delta x\Delta y}\Phi_{j,k} = \frac{1}{4\Delta x\Delta y}\left[\Phi_{j+1,k+1} - \Phi_{j-1,k+1} - \Phi_{j+1,k-1} + \Phi_{j-1,k-1}\right]$$

$$= \Phi_{xy} + \frac{\Delta x^2}{6}\Phi_{xxxy} + \frac{\Delta y^2}{6}\Phi_{yyyx} + O(\Delta^4)$$

are then used to obtain a discrete approximation to the differential euqation (2.8).

2.3. Perturbation Potential

For an external flow problem we expect the flow to be at essentially freestream conditions far upstream or downstream of any disturbance. Thus if the freestream flow is aligned with the x axis, we expect that

$$\Phi \sim x \quad \text{for large } x.$$

The uniform flow $\Phi = x$ is clearly a solution of eqn (2.8) and since

$$\delta_{xx}\Phi = x + \Delta x - 2x + x - \Delta x = 0$$

we see that it is also a solution of the difference equations when a uniform Cartesian grid is used. In practice we would normally introduce a stretching transformation $x = x(\xi)$ to take the x axis into a finite computational interval (-1,1) say. Equal increments $\Delta\xi$ are then taken in the computation space to define a grid that is nonuniform in physical space. Under this transformation we find that

$$\Phi_{xx} = (\xi_x)^2\ \Phi_{\xi\xi} + \xi_{xx}\Phi_\xi$$

If central difference formulae are used to approximate Φ_ξ and $\Phi_{\xi\xi}$ the term Φ_{xx} will now be approximated by the formula

$$D_{xx}\Phi = (\xi_x)^2\ \frac{\delta_{\xi\xi}}{\Delta\xi^2}\Phi + \xi_{xx}\ \frac{\mu_\xi\delta_\xi}{\Delta\xi}\Phi$$

$$= (\xi_x)^2\left[\Phi_{\xi\xi} + \frac{\Delta\xi^2}{12}\Phi_{\xi\xi\xi\xi}\right]$$

$$= \xi_{xx}\left[\Phi_\xi + \frac{\Delta\xi^2}{6}\Phi_{\xi\xi\xi}\right] + O(\Delta\xi^4)$$

$$\text{ie}\quad D_{xx}\Phi = \Phi_{xx} + \frac{\Delta\xi^2}{6}\left[\xi_{xx}\ \Phi_{\xi\xi\xi} + \frac{1}{2}(\xi_x)^2\ \Phi_{\xi\xi\xi\xi}\right] + O(\Delta\xi^4) \qquad (2.12)$$

The difference approximation is formally second order accurate. However, if we consider a stretching transformation like

$$x = -\log(1-\xi) \quad \text{or } \xi = 1 - e^{-x}$$

which takes the interval $[0,\infty)$ in physical space into $[0,1)$ in ξ space, the following transformation derivatives are obtained

$$\xi_x = 1 - \xi, \qquad \xi_{xx} = \xi - 1$$

If we now put $\Phi = x$ it follows that

$$\Phi_{\xi\xi\xi} = \frac{2}{(1-\xi)^3} \text{ and } \Phi_{\xi\xi\xi\xi} = \frac{6}{(1-\xi)^4}$$

so that the truncation error in eqn (2.12) becomes

$$\frac{1}{6}\frac{\Delta\xi^2}{(1-\xi)^2} + O(\Delta\xi^4)$$

Provided ξ is not too close to 1 we are justified in claiming that the truncation error is second order. On the other hand, if

$$1 - \xi = O(\Delta\xi)$$

then the truncation error is O(1). It follows that, at large values of x, the truncation error arising from our approximation to Φ_{xx} will swamp the other terms in the difference equation and we therefore cannot expect to obtain an accurate solution of the form $\Phi = x$ on a stretched grid.

The simplest way out of this difficulty is to introduce a perturbation potential defined by

$$\Phi = x + \phi \tag{2.13}$$

The potential equation (2.8) then becomes

$$(1 - \frac{u^2}{a^2})\,\phi_{xx} - \frac{2uv}{a^2}\,\phi_{xy} + (1 - \frac{v^2}{a^2})\,\phi_{yy} = 0 \tag{2.14}$$

with $u = \Phi_x = 1 + \phi_x$ and $v = \Phi_y = \phi_y$

The perturbation ϕ is small at large x values and the truncation error arising from the difference formula for ϕ_{xx} should now be sufficiently small to avoid any adverse effects on the numerical computation.

2.4. Stability Analysis

The potential equation and the set of difference equations that approximate it are non-linear. It is therefore necessary to solve the difference equations by an iterative procedure in which the approximate solution after n iteration cycles is given by ϕ^n_{jk} and the error is defined by

$$e^n_{jk} = \phi^n_{jk} - \phi_{jk} \tag{2.15}$$

where ϕ_{jk} is the exact solution of the difference equations. The question arises as to whether the iterative scheme converges, that is whether

$$||e^n|| \to 0 \qquad \text{as } n \to \infty$$

for some suitably defined norm. If the difference equation had constant coefficients then the error e^n_{jk} could be written as a superposition of Fourier components

$$e^n_{jk} = \sum_{p=1}^{N} \sum_{q=1}^{M} \rho^n(p,q)\, e^{ipj\Delta x}\, e^{iqk\Delta y} \tag{2.16}$$

where $N = 2\pi/\Delta x$ and $M = 2\pi/\Delta y$ are the number of grid points in the x and y directions respectively. According to the von Neumann stability criterion, a necessary condition for the iteration scheme to converge is that all Fourier coefficients $\rho^n(p,q)$ should remain bounded. In other words we require

$$|G(p,q)| \leq 1, \text{ for all } p,q$$

where the amplification factor is

$$G(p,q) = \frac{\rho^{n+1}}{\rho^n}$$

Although this harmonic or modal analysis is not strictly valid for the potential equation, it is often applied on the assumption that the equation coefficients are approximately constant over distances of the order of a grid increment. The local stability bounds that are obtained in this way often provide a good indication of the stability of an iterative scheme. As an example, we consider the potential equation (2.14) with the flow direction aligned with the x axis. In this case the equation can be written

$$(1 - M^2)\,\phi_{xx} + \phi_{yy} = 0 \tag{2.17}$$

where M is the local Mach number. The derivatives are replaced by the centred difference formulae (2.11) and if we consider a Line Gauss Seidel iteration (ie SLOR with the relaxation factor $\omega = 1$) we obtain the equation

$$(1 - M^2)(e^n_{j+1,k} - 2e^{n+1}_{jk} + e^{n+1}_{j-1,k}) + \left(\frac{\Delta x}{\Delta y}\right)^2 (e^{n+1}_{j,k+1} - 2e^{n+1}_{jk} + e^{n+1}_{j,k-1}) = 0 \tag{2.18}$$

It follows that the amplification factor is

$$G = \frac{(1 - M^2)e^{i\xi}}{(1 - M^2)(2 - e^{-i\xi}) + 2(\Delta x/\Delta y)^2\,(1 - \cos\eta)} \tag{2.19}$$

where $\xi = p\Delta x$ and $\eta = q\Delta y$

When $M \leq 1$

$$|G| \leq \frac{1 - M^2}{1 - M^2 + 2(\Delta x/\Delta y)^2(1 - \cos\eta)} \leq 1$$

so that the iterative scheme should be stable for subsonic flow. On the other hand, when $M > 1$, consider a frequency component such that $\xi = O(\Delta x)$. In this case

$$G = \frac{M^2 - 1}{M^2 - 1 - 2(\Delta x/\Delta y)^2(1 - \cos\eta)} + O(\Delta x)$$

It follows that there exist frequency components for which

$$|G| > 1$$

so that the iterative scheme becomes unstable when the local Mach number exceeds one.

2.5. Type Dependent Differencing

The use of centred difference formulae for ϕ_{xx} and ϕ_{yy} in eqn (2.17) mimics the behaviour of an elliptic system for which disturbances at any position influence, and are influenced by, all neighbouring points. In a hyperbolic system the field value at any point depends on data within a domain bounded by the two characteristics which pass through that point. To simulate this in the finite difference scheme we replace the centred difference approximation for ϕ_{xx} by the following upwind formula

$$\frac{\overleftarrow{\delta}_x \overleftarrow{\delta}_x}{\Delta x^2} \phi_{jk} = \frac{\phi_{jk} - 2\phi_{j-1,k} + \phi_{j-2,k}}{\Delta x^2}$$

$$= \phi_{xx} - \Delta x \, \phi_{xxx} + O(\Delta x^2) \qquad (2.20)$$

The truncation error is now $O(\Delta x)$ and the finite difference equation actually approximates the differential equation

$$(M^2 - 1)\, \phi_{xx} - \phi_{yy} = (M^2 - 1)\, \Delta x \, \phi_{xxx}$$

The term on the right hand side can be interpreted as an artificial viscosity which vanishes in the limit $\Delta x \to 0$. This viscosity has the effect of enforcing a type of numerical entropy condition which ensures that the only shocks to appear will be compression shocks, behaviour which is of course consistent with that of a real fluid.

If we replace the centred approximation for ϕ_{xx} by the upwind formula (2.20) when $M > 1$, we obtain a marching scheme for which the amplification factor is zero, indicating stability.

We thus arrive at the concept of a type dependent difference scheme first introduced by Murman and Cole [13]. In regions of subsonic flow centred differencing is used. This is switched to a combination of centred differencing in the y direction and upwind differencing in the x direction whenever the flow becomes locally supersonic. The upwind formula used for the cross derivative term ϕ_{xy} in eqn (2.14) is

$$\frac{\mu_y \delta_y \overleftarrow{\delta}_x}{\Delta x \Delta y} \phi_{jk} = \phi_{xy} - \frac{\Delta x}{2} \phi_{xxy} + O(\Delta^2) \qquad (2.21)$$

so that the artificial viscosity introduced by the simple upwind differenced approximation to eqn (2.14) is

$$\Delta x \left[\left(\frac{u^2}{a^2} - 1\right) u_{xx} + \frac{uv}{a^2} v_{xx} \right] \qquad (2.22)$$

2.6. Rotated Difference Scheme

The artificial viscosity introduced by the upwind formula for ϕ_{xx} is $\Delta x(u^2/a^2 - 1)\, u_{xx}$ and when the flow is not aligned with the x axis it is possible to have regions of flow where $q^2 > a^2 > u^2$. In such cases the artificial viscosity introduced by ϕ_{xx} is negative and can have a destabilising effect on the numerical scheme. To overcome this limitation and ensure that the difference approximation has the correct domain of dependence it is necessary to adopt a rotated difference scheme [14]. Consider an orthogonal coordinate system with one coordinate s aligned with the flow direction and the other coordinate n normal to the flow direction. In this coordinate system eqn (2.14) becomes

$$(1 - q^2/a^2)\ \phi_{ss} + \phi_{nn} = 0 \qquad (2.23)$$

When the flow is locally supersonic, upwind differencing must be used for ϕ_{ss} to maintain computational stability. We therefore identify the form of ϕ_{ss} in terms of the x and y derivatives of ϕ in order to determine the correct combination of upwind and central differencing to be used in eqn (2.14). The transformation from the x,y system to the s,n system can be thought of as a local rotation of the coordinate axes through an angle θ where

$$\cos\theta = u/q \text{ and } \sin\theta = v/q$$

It follows that the transformation is defined by the equations

$$x = \frac{u}{q}\, s - \frac{v}{q}\, n \quad \text{and} \quad y = \frac{v}{q}\, s + \frac{u}{q}\, n$$

and hence, on retaining only the highest derivatives of ϕ, that

$$\phi_{ss} = \frac{1}{q^2}\ (u^2 \phi_{xx} + 2uv\phi_{xy} + v^2 \phi_{yy}) \qquad (2.24)$$

and

$$\phi_{nn} = \frac{1}{q^2}\ (v^2 \phi_{xx} - 2uv\phi_{xy} + u^2 \phi_{yy})$$

A simpler way of obtaining the same result is to compare the canonical form (2.23) with (2.14) and extract the ϕ_{ss} contribution from the terms multiplying $1/a^2$ in (2.14); ϕ_{nn} then follows from the remaining terms. The advantage of the second approach is its simplicity, particularly when one considers a general nonorthogonal coordinate transformation (see section 2.12). The upwind difference formula for ϕ_{xy} when $u > 0$ and $v > 0$ is now

$$\frac{\overleftarrow{\delta}_x \overleftarrow{\delta}_y}{\Delta x \Delta y}\ \phi_{jk} = \frac{1}{\Delta x \Delta y}\ \{\phi_{jk} - \phi_{j-1,k} - \phi_{j,k-1} + \phi_{j-1,k-1}\}$$

$$= \phi_{xy} - \frac{\Delta x}{2}\ \phi_{xxy} - \frac{\Delta y}{2}\ \phi_{xyy} + O(\Delta^2)$$

It follows that under a rotated difference scheme the artificial viscosity introduced at supersonic points is

$$\frac{1}{q^2}\ (\frac{q^2}{a^2} - 1)\ \left[\Delta x|u|(u\ u_{xx} + v\ v_{xx}) + \Delta y|v|(u\ u_{yy} + v\ v_{yy})\right] \qquad (2.25)$$

which is symmetric in x and y and positive for any flow alignment.

2.7 Conservation Form

The equations governing fluid flow are mathematical statements derived from the physical laws of conservation of mass, momentum and energy. Although the laws are expressed in integral form, the divergence theorem allows one to replace the integral conditions by equivalent differential forms provided the flow is sufficiently smooth. For example, eqn (2.1) is the differential form expressing conservation of mass for a steady flow. The required solution is therefore one which satisfies eqn (2.1) where the flow is smooth and the following jump condition across a discontinuity

$$[\rho v]\ \cos\theta - [\rho u]\ \sin\theta = 0$$

where $\tan\theta$ is the slope of the discontinuity. It is clearly desirable that the numerical scheme should mirror the physics and hence obey the same conservation law. In other words the numerical solution should approximate a proper weak solution [15].

If we expect the numerical approximation to eqn (2.1) to conserve mass then the mass flux into each cell of the finite difference grid should be balanced by the mass outflow. The nonconservative differencing described above makes no attempt to ensure mass conservation. This can be illustrated by considering the one dimensional form of eqn (2.1)

$$(\rho u)_x = 0 \tag{2.26}$$

which in quasilinear form becomes

$$A\, u_x = 0$$

where $A = \rho(1 - u^2/a^2)$

The velocity at point j is given by

$$u_j = \frac{\phi_{j+1} - \phi_{j-1}}{2\Delta x}$$

and a test is made to determine whether the value of A_j is positive or negative. In regions of subsonic flow where $A > 0$ the centred difference approximation is

$$A_j \frac{(u_{j+\frac{1}{2}} - u_{j-\frac{1}{2}})}{\Delta x} = 0, \text{ where } u_{j+\frac{1}{2}} = \frac{\vec{\delta}_x \phi_j}{\Delta x} = \frac{\phi_{j+1} - \phi_j}{\Delta x}$$

When $A < 0$ which is the case at a supersonic point the following upwind differencing is used

$$A_j \frac{(u_{j-1/2} - u_{j-3/2})}{\Delta x} = 0 \qquad \text{if } u_j > 0$$

$$A_j \frac{(u_{j+3/2} - u_{j+1/2})}{\Delta x} = 0 \qquad \text{if } u_j < 0$$

In smooth regions of flow these formulae approximate eqn (2.26) to within the truncation error and so the consequent error in mass conservation should be of the same order. The important exception occurs in the vicinity of a shock where the flow undergoes an abrupt deceleration. If j is the first subsonic point following a supersonic region in which $u > 0$ then we have a contribution of

$$\frac{A_{j-1}\, u_{j-3/2} - A_j\, u_{j-1/2}}{\Delta x}$$

across the shock wave. Since the flow conditions ahead of the shock at point j-1 and behind the shock at point j are quite different, this quantity need not be small. Its precise value depends on the values of the flow variables at the grid points j-1 and j and hence on the local grid spacing. The net effect is that extra mass is introduced at a shock. The amount of mass and hence the shock that is predicted by the numerical scheme can therefore be expected to vary according to the local grid spacing. It is this apparent dependence of the shock strength and position on nonphysical properties such as grid spacing that is the main weakness of a nonconservative solution.

To overcome this limitation and ensure the conservation of mass, it is necessary to return to the equation in divergence form [16], ie eqn (2.26). Centred differencing is used to give the following discretised form at the point j,

$$\frac{(\rho u)_{j+\frac{1}{2}} - (\rho u)_{j-\frac{1}{2}}}{\Delta x} = 0$$

The outflow from the cell at j-1 now cancels the inflow to cell j and mass is therefore conserved.

2.8. Explicit Artificial Viscosity

To extend this idea to regions of supersonic flow we must add the artificial viscosity in divergence form as well [16]. Specifically we write the equation as

$$(\rho u + P)_x = 0$$

where P is zero in subsonic regions. When the flow is locally supersonic we choose $P = O(\Delta x)$ such that P_x approximates the artificial viscosity which arises from the quasilinear form. Thus we select P so that a centred difference approximation to P_x gives roughly the value

$$\mp \Delta x \, A u_{xx} \quad \text{for } u \gtrless 0$$

To retain the type dependent switching which produces a sharp shock we first construct a switching function

$$\mu = \min\,[0,A]$$

and then define

$$T_j = \frac{P_{j+\frac{1}{2}} - P_{j-\frac{1}{2}}}{\Delta x} \tag{2.27}$$

where

$$P_{j+1/2} = \begin{cases} -\mu_j(u_{j+1/2} - u_{j-1/2}), & u_{j+1/2} > 0 \\ \mu_{j+1}(u_{j+3/2} - u_{j+1/2}), & u_{j+1/2} < 0 \end{cases} \tag{2.28}$$

It can be seen that the artificial viscosity T_j is a centred difference approximation to

$$\mp \Delta x (\mu u_x)_x \quad \text{for } u \gtrless 0$$

in which the term with the highest derivative of u is

$$\mp \Delta x A u_{xx} \quad \text{for } u \gtrless 0$$

as required. The introduction of an explicit artificial viscosity was first suggested by Jameson [16,17] and this scheme ensures mass conservation throughout the flow.

2.9. An Alternative Form for Artificial Viscosity

Since

$$\rho_x = -\frac{\rho u u_x}{a^2}$$

we can write

$$Au_x = -(1 - a^2/u^2)\, \rho \frac{u^2}{a^2} u_x = (1 - a^2/u^2)\, u\rho_x$$

It follows that an equivalent form for the artificial viscosity is

$$-\Delta x(\nu|u|\rho_x)_x$$

where the switching function is now defined as

$$\nu = \max\,[0,(1 - a^2/u^2)]$$

The artificial viscosity that is added still has the form (2.27) but now

$$P_{j+\frac{1}{2}} = \begin{cases} -u_j\nu_j(\rho_{j+\frac{1}{2}} - \rho_{j-\frac{1}{2}}), & u_{j+\frac{1}{2}} > 0 \\ u_{j+1}\nu_{j+1}(\rho_{j+3/2} - \rho_{j+1/2}), & u_{j+\frac{1}{2}} < 0 \end{cases} \tag{2.29}$$

One advantage of this form for the artificial viscosity is that the extension to two and three dimensions is straightforward. Thus, in order to construct a conservative difference scheme for eqn (2.7) we form the centred difference formula

$$S_{jk} + T_{jk} = 0 \tag{2.30}$$

where

$$S_{jk} = \frac{[(\rho u)_{j+\frac{1}{2},k} - (\rho u)_{j-\frac{1}{2},k}]}{\Delta x} + \frac{[(\rho v)_{j,k+\frac{1}{2}} - (\rho v)_{j,k-\frac{1}{2}}]}{\Delta y} \tag{2.31}$$

and T_{jk} represents the artificial viscosity. The velocity components are required at points half way between grid points. They are usually evaluated as

$$u_{j+\frac{1}{2},k} = \frac{\Phi_{j+1,k} - \Phi_{j,k}}{\Delta x}$$

$$v_{j+\frac{1}{2},k} = \frac{[\Phi_{j+1,k+1} + \Phi_{j,k+1} - \Phi_{j+1,k-1} - \Phi_{j,k-1}]}{4\Delta y}$$

with similar formulae for $v_{j,k+\frac{1}{2}}$ and $u_{j,k+\frac{1}{2}}$. The switching function is now

$$\nu = \max\,[0,(1 - a^2/q^2)] \tag{2.32}$$

and we add an artificial viscosity in the form

$$-\Delta x(\nu|u|\rho_x)_x - \Delta y(\nu|v|\rho_y)_y \tag{2.33}$$

The terms containing the highest derivatives of u and v are, apart from a factor ρ, the same as expression (2.25) for the artificial viscosity introduced by the nonconservative form. In the numerical scheme the artificial viscosity is represented by the centred difference approximation

$$T_{jk} = \frac{(P_{j+\frac{1}{2},k} - P_{j-\frac{1}{2},k})}{\Delta x} + \frac{(Q_{j,k+\frac{1}{2}} - Q_{j,k-\frac{1}{2}})}{\Delta y} \tag{2.34}$$

The extension of eqn (2.29) to two dimensions is

$$P_{j+\frac{1}{2},k} = \begin{cases} -u_{jk}\nu_{jk}\,(\rho_{j+\frac{1}{2},k} - \rho_{j-\frac{1}{2},k}),\; u_{j+\frac{1}{2},k} > 0 \\ -u_{j+1,k}\,\nu_{j+1,k}\,(\rho_{j+\frac{1}{2},k} - \rho_{j+3/2,k}),\; u_{j+\frac{1}{2},k} < 0 \end{cases} \tag{2.35}$$

with a similar formula for $Q_{j,k+\frac{1}{2}}$. Other forms of artificial viscosity have been reported. These generally differ only slightly in detail and give almost identical results but the existence of these variants does lead to some confusion. For example, the form of artificial viscosity reported by Jameson [17] uses

$$P_{j+\frac{1}{2},k} = \begin{cases} -u_{j+\frac{1}{2},k}\,\nu_{j,k}\,(\rho_{j+\frac{1}{2},k} - \rho_{j-\frac{1}{2},k}),\; u_{j+\frac{1}{2},k} > 0 \\ -u_{j+\frac{1}{2},k}\,\nu_{j+1,k}\,(\rho_{j+\frac{1}{2},k} - \rho_{j+3/2,k}),\; u_{j+\frac{1}{2},k} < 0 \end{cases}$$

However, Holst and Ballhaus [19] remark that Jameson actually uses eqn (2.35) to determine $P_{j+\frac{1}{2},k}$. On the other hand, Holst [18] adopts the above form for $P_{j+\frac{1}{2},k}$ but redefines the switching function (see eqn 2.39) and writes

$$P_{j+\frac{1}{2},k} = -u_{j+\frac{1}{2},k}\,\nu_{j+\frac{1}{2},k}\,(\rho_{j+\frac{1}{2},k} - \rho_{j+r+\frac{1}{2},k}) \tag{2.36}$$

$$\text{where } r = \begin{cases} -1,\; u_{j+\frac{1}{2},k} > 0 \\ 1,\; u_{j+\frac{1}{2},k} < 0 \end{cases}$$

The extension of these ideas to three dimensions is quite straightforward.

2.10. Holst Ballhaus Arrangement

A neat rearrangement of the difference equations in conservation form has been introduced by Holst and Ballhaus [19]. The difference approximation is given by eqn (2.30) where S_{jk} is given by (2.31) which can be written as

$$S_{jk} = \frac{\overleftarrow{\delta}_x(\rho u)}{\Delta x}\,{}_{j+\frac{1}{2},k} + \frac{\overleftarrow{\delta}_y(\rho v)}{\Delta y}\,{}_{j,k+\frac{1}{2}}$$

The artificial viscosity is defined by eqn (2.34) and (2.36) which can be written as

$$T_{jk} = \frac{-\overleftarrow{\delta}_x[u_{j+\frac{1}{2},k}\nu_{j+\frac{1}{2},k}(\rho_{j+\frac{1}{2},k} - \rho_{j+r+\frac{1}{2},k})]}{\Delta x}$$

$$\frac{-\overleftarrow{\delta}_y[v_{j,k+\frac{1}{2}}\nu_{j,k+\frac{1}{2}}(\rho_{j,k+\frac{1}{2}} - \rho_{j,k+s+\frac{1}{2}})]}{\Delta y}$$

where r is defined above and

$$s = \begin{cases} -1,\; v_{j,k+\frac{1}{2}} > 0 \\ 1,\; v_{j,k+\frac{1}{2}} < 0 \end{cases}$$

Combining these expressions for S_{jk} and T_{jk} gives the following compact difference form for eqn (2.30)

$$\frac{\overleftarrow{\delta}x(\tilde{\rho}u)}{\Delta x}_{j+\frac{1}{2},k} + \frac{\overleftarrow{\delta}y(\bar{\rho}v)}{\Delta y}_{j,k+\frac{1}{2}} = 0 \tag{2.37}$$

where

$$\tilde{\rho}_{j+\frac{1}{2},k} = [(1-\nu)\rho]_{j+\frac{1}{2},k} + \nu_{j+\frac{1}{2},k}\,\rho_{j+r+\frac{1}{2},k} \tag{2.38}$$

and

$$\bar{\rho}_{j,k+\frac{1}{2}} = [(1-\nu)\rho]_{j,k+\frac{1}{2}} + \nu_{j,k+\frac{1}{2}}\,\rho_{j,k+s+\frac{1}{2}}$$

The addition of artificial viscosity is thus seen to be equivalent to retarding the density by an amount which is controlled by the switching function (2.32) which can also be written

$$\nu = \max\,[0,\ 1 - 1/M^2]$$

where M is the local Mach number. For flows containing strong shocks, oscillations in the solution can appear ahead of the shock. These arise because of insufficient artificial viscosity and can quickly lead to numerical instability. To damp down the preshock oscillations, the amount of artificial viscosity can be increased by using a different definition of the switching function. Various alternatives have been suggested [18,20]. For example,

$$\nu = \max\,[0,\ C\,M^n\,(1 - 1/M^2)]$$

where C is a user defined constant.

Deconinck and Hirsch [20] determine a value for n by numerical experiment while Holst [18] appears to favour the value $n = 2$. It should be noted that, in order to achieve some economy in notation, Holst [18] defines his switching function as

$$\nu_{j+\frac{1}{2},k} = \begin{cases} \max\,[0,\ C(M^2_{j,k}-1)], & u_{j+\frac{1}{2},k} > 0 \\ \max\,[0,\ C(M^2_{j+1,k}-1)], & u_{j+\frac{1}{2},k} < 0 \end{cases} \tag{2.39}$$

2.11. Artificial Compressibility

Equation (2.37) can be considered as a finite difference approximation to the equation

$$(\tilde{\rho}u)_x + (\bar{\rho}v)_y = 0$$

where

$$\tilde{\rho} = \rho - \Delta x\ \nu\ \rho_x$$

and

$$\bar{\rho} = \rho - \Delta y\ \nu\ \rho_y$$

It is tempting to replace $\tilde{\rho}$ and $\bar{\rho}$ by a single modified density, say

$$\hat{\rho} = \rho - \underline{\nu\Delta s}.\underline{\rho_s}$$

where

$$\underline{\Delta s} = (\Delta x,\ \Delta y)$$

and

$$\underline{\rho_s} = \left(\frac{u}{q}\rho_x,\ \frac{v}{q}\rho_y\right)$$

so that

$$\hat{\rho} = \rho - \nu\left(\frac{u}{q}\Delta x\rho_x + \frac{v}{q}\Delta y\rho_y\right) \qquad (2.40)$$

The equation then becomes

$$(\hat{\rho}u)_x + (\hat{\rho}v)_y = 0 \qquad (2.41)$$

This is formally equivalent to eqn (2.7) with the true density modified according to eqn (2.40). When the flow direction is aligned with either of the coordinate directions this is the same as the Holst Ballhaus scheme (2.37) and (2.38) which is also very similar to Jameson's use of artificial viscosity. The form of eqn (2.41) suggests a link with finite element methods in which solution techniques applicable to elliptic equations could be used to solve the flow equations even in regions of mixed flow. The theoretical justification for this approach is not entirely clear but the idea has recently become quite popular [10,20,21]. The term artificial compressibility has been coined for such methods. It appears that these pseudo finite element methods require fairly hefty doses of artificial viscosity in order to compute flows containing strong shocks and the predicted pressure distributions do not always look convincing. Nevertheless, the potential benefit of an accurate and reliable transonic code based on the finite element approach is so great that this is certainly an area worth pursuing.

2.12. General Coordinate Systems

The extension of the preceding differential and difference equations to an arbitrary non-orthogonal coordinate system is straightforward although the number of terms increases and the coding of such problems can become quite messy. The difficulties can be eased considerably if one resorts to tensor manipulations. We assume a general transformation

$$x^m(\xi^1,\xi^2,\xi^3),\ m = 1,2,3$$

from a Cartesian system (x^1,x^2,x^3) to a general coordinate system (ξ^1,ξ^2,ξ^3). The vector dx^m is related to $d\xi^m$ by the formula

$$dx^m = \sum_{\ell=1}^{3} \frac{\partial x^m}{\partial \xi^\ell}\, d\xi^\ell$$

and we therefore call the matrix H whose entries are

$$h_{m\ell} = \frac{\partial x^m}{\partial \xi^\ell}$$

the transformation matrix. The metric tensor is defined by the matrix

$$G = H^T H$$

which has the entries

$$g_{m\ell} = \sum_{k=1}^{3} \frac{\partial x^k}{\partial \xi^m} \frac{\partial x^k}{\partial \xi^\ell}$$

The determinant of G is equal to J^2 where J, which is the absolute value of the determinant of H, is known as the Jacobian of the transformation.

The contravariant components of the metric tensor are written as $g^{m\ell}$ and defined as the entries of the inverse matrix G^{-1}. Any vector $\underline{a}$ can be expressed in either covariant form a_m or contravariant form a^m. The metric tensor provides the means for changing from covariant to contravariant form and vice versa. The rules are

$$a^m = g^{m\ell} a_\ell \;, \; a_m = g_{m\ell} a^\ell$$

In the above equations we have made use of the summation convention which states that the appearance of the same symbol for a suffix and superscript is to be interpreted as implying summation with respect to that symbol. In other words

$$a^m = g^{m\ell} a_\ell = g^{m1} a_1 + g^{m2} a_2 + a^{m3} a_3$$

We note that for a transformation to orthogonal coordinates, the only non zero entries in the metric tensor are the diagonal terms. In this case the distinction between covariant and contravariant components of a vector disappears since the vectors have the same direction and differ only in magnitude.

The utility of tensor notation becomes apparent when one wishes to write an equation such as (2.1) in a general coordinate system. The scalar product of two vectors $\underline{a}$ and $\underline{b}$ can be written in the following equivalent ways

$$\underline{a}.\underline{b} = a^m b_m = g^{m\ell} a_m b_\ell = a_m b^m$$

The divergence of a vector $\underline{u}$ generalises as

$$\nabla.\underline{u} = \frac{1}{J} \frac{\partial}{\partial \xi^m} (J u^m)$$

or equivalently

$$\nabla.\underline{u} = \frac{1}{J} \frac{\partial}{\partial \xi^m} (J g^{m\ell} u_\ell)$$

where $u_m = \Phi_{\xi^m}$ are the covariant velocity components and $u^m = g^{m\ell} u_\ell$ are the contravariant components. The nonconservative or quasilinear form of the potential equation is given by (2.4) which in tensor notation becomes

$$\frac{1}{J} \frac{\partial}{\partial \xi^m} \left(J g^{m\ell} \frac{\partial \Phi}{\partial \xi^\ell}\right) - \frac{u^m}{2a^2} \frac{\partial}{\partial \xi^m} \left(g^{k\ell} \frac{\partial \Phi}{\partial \xi^k} \frac{\partial \Phi}{\partial \xi^\ell}\right) = 0$$

This replaces eqn (2.8). The sound speed and density are given by eqns (2.5) and (2.6) with

$$q^2 = u^m u_m$$

If we define a perturbation potential by eqn (2.13) we can make use of the identity

$$\nabla^2 x = 0$$

to obtain

$$\nabla^2\Phi = \nabla^2\phi = g^{m\ell}\,\phi_{\xi^m\xi^\ell} + \frac{1}{J}\,\phi_{\xi^m}\,\frac{\partial}{\partial\xi^\ell}\,(Jg^{m\ell})$$

After some further manipulation we find that the perturbation potential satisfies the following equation

$$\left(g^{m\ell} - \frac{u^m u^\ell}{a^2}\right)\phi_{\xi^m\xi^\ell} + \frac{1}{J}\,\phi_{\xi^m}\,\frac{\partial}{\partial\xi^\ell}\,(Jg^{m\ell}) - \frac{u^m}{2a^2}\,u_k u_\ell\,\frac{\partial g^{k\ell}}{\partial\xi^m} - \frac{u^m u^\ell}{a^2}\,\frac{\partial^2 x}{\partial\xi^m\partial\xi^\ell} = 0 \tag{2.42}$$

where $u_m = \phi_{\xi^m} + \partial x/\partial\xi^m$

This replaces equation (2.14) in a general coordinate system. If a rotated difference scheme is used it is necessary to identify the form of ϕ_{ss} and ϕ_{nn} that appear in (2.23). A comparison of eqn (2.23) with eqn (2.42) enables one to extract ϕ_{ss} from those second order derivative terms which are multiplied by $1/a^2$. We obtain

$$\phi_{ss} = \frac{u^m u^\ell}{q^2}\,\phi_{\xi^m\xi^\ell} \tag{2.43}$$

and also

$$\phi_{nn} = \left(g^{m\ell} - \frac{u^m u^\ell}{q^2}\right)\phi_{\xi^m\xi^\ell}$$

These formulae replace (2.24).

To obtain the general form for the conservation equation we return to eqn (2.1) which in a general coordinate system becomes

$$\frac{1}{J}\,\frac{\partial}{\partial\xi^m}\,(J\rho u^m) = 0 \tag{2.44}$$

where $u^m = g^{m\ell}\,\dfrac{\partial\Phi}{\partial\xi^\ell}$

A suitable generalisation of eqn (2.33) for the artificial viscosity is now

$$\frac{-\Delta\xi^1}{J}\,\frac{\partial}{\partial\xi^1}\left(J\nu|u^1|\,\frac{\partial\rho}{\partial\xi^1}\right) - \frac{\Delta\xi^2}{J}\,\frac{\partial}{\partial\xi^2}\left(J\nu|u^2|\,\frac{\partial\rho}{\partial\xi^2}\right) - \frac{\Delta\xi^3}{J}\,\frac{\partial}{\partial\xi^3}\left(J\nu|u^3|\,\frac{\partial\rho}{\partial\xi^3}\right) \tag{2.45}$$

This is the form used by Holst [18] although we note that he defines the transformation matrix as H^{-1} rather than H. It follows that the Jacobian he uses is equal to J^{-1} in our notation.

Finally, we note that use of the conservation form (2.1) or (2.44) precludes the extraction of the term $\nabla^2 x = 0$ which is the main benefit of defining a perturbation potential (see section 2.3). It follows that truncation error can be expected to cause problems on stretched grids. This is noted by Jameson [16] who overcomes the difficulty by using the quasilinear form (2.14) or (2.42) in the farfield. Provided no shock waves extend that far, this strategy should be effective in avoiding the problem of truncation error while maintaining conservation from in regions where shocks are present. Interestingly the problem does not appear to be reported elsewhere in the literature on conservative potential flow methods. Most of these codes use a grid that covers only a finite region. In such cases, the grid will not be as stretched as it would be if it covered the entire flowfield. It is therefore possible that those who use grids of finite extent have not experienced such difficulties. It should be noted, though, that Steger [83] has met an essentially similar problem when solving the Euler equations.

2.13. Conservative versus Nonconservative Controversy

In the preceding sections we have described both nonconservative and fully conservative differencing without making any attempt to discuss their relative merits. The question is now examined and it is important to note that the author's opinions expressed in this section represent a minority viewpoint. The reader who is new to the subject is therefore advised to keep this in mind and to read what others have to say before forming his own opinion.

The Euler equations are accepted as providing an adequate model of inviscid flow in which mass, momentum and energy are conserved. When solving the Euler equations there is no obvious reason for adopting a nonconservative approach and use of conservation form would appear to be mandatory. The situation with the potential equation is rather different. In this case the flow is required to be isentropic; an assumption that is not valid when shocks are present. It is then not possible to conserve both mass and momentum. The choice of a conservative scheme which ensures conservation of mass [16,17] and relates the deficiency in momentum to wave drag [22] has the appeal of being consistent in the sense that at least one physical property is always conserved. Nevertheless, this is still an incomplete and hence inadequate model of inviscid flow. It is therefore arguable whether a conservative potential method that conserves mass but not momentum is necessarily a better approximation to physical reality than a nonconservative scheme which makes no attempt to conserve either property. The main arguments in favour of the conservative form are first the claim that nonconservative differencing can be shown to generate nonunique solutions [23] and second that the strength and position of the shock can depend on the local grid spacing.

The discussion of conservative and nonconservative differencing that is presented in Reference [23] is based on solutions of an ordinary differential equation with overspecified boundary conditions. The conclusions are suggestive but they do not constitute a proof that the same behaviour applies to a partial differential equation such as (2.7) or (2.14). It is still an open question whether or not a unique solution is guaranteed by either form of differencing.

The second allegation that the solution may depend on the grid spacing is perhaps the most serious. In defence of nonconservative differencing we can only argue that this does not seem to be a serious problem in practice. For example, in Figure 1 we compare results [24] using a grid obtained from the circle plane mapping with those calculated by Carlson [25] who uses a nonaligned Cartesian grid. The predicted results are in surprisingly good agreement considering the wide disparity between the grids that have been used. Similarly, the use of a finer grid will certainly sharpen up the shock jump but this does not significantly affect either its strength or position.

We conclude that despite the obvious appeal of conserving at least something, there is no compelling theoretical justification for preferring a conservative solution of the potential equation. Both nonconservatrive and conservative solutions of the potential equation are incorrect models of an invicsid flow containing shock waves. The only reasonable basis from which to make a preference is a comparison of predicted results with those obtained by solving the Euler equations.

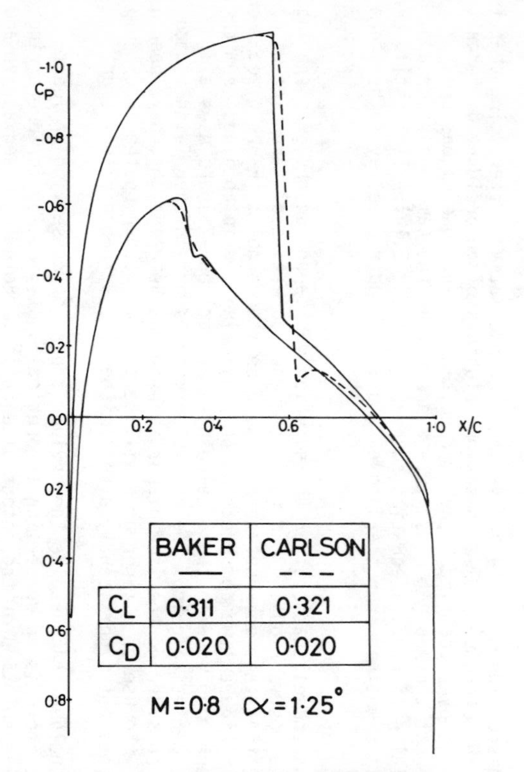

FIG 1 PRESSURE DISTRIBUTION COMPARISON FOR AEROFOIL NACA 0012

	FC (----)	NC (——)	EULER (-·-)
C_L	0·337	0·335	0·327
C_D	0·0003	0·0000	0·0014

M=0·63 α=2°

FIG. 2(a) PRESSURE DISTRIBUTION COMPARISON FOR AEROFOIL NACA 0012

Unfortunately, this is not as easy as it sounds since there appears to be no clear consensus about what is an accurate solution of Euler equations. However, an important step in this direction was taken at the GAMM Workshop [26] where results for several test cases were presented for both Euler and potential methods. The Euler results that are shown in Figures 2a,b,c,d,e are those of Rizzi [26,27] since his method appears to be the best developed and probably most accurate of the Euler codes presently in existence. We note that the predictions of Sells' Euler method [26,28] generally agree closely with Rizzi. The nonconservative (NC) and fully conservative (FC) potential results are those produced by Baker and Holst respectively. These potential flow predictions were obtained using a very similar grid and distribution of grid points and in each case are probably the most accurate of their genre.

The first comparison shown in Figure 2a compares the three methods at a subcritical condition. As one would expect the pressure distributions as well as the lift and drag coefficients all agree closely. The comparison in Figure 2b shows excellent agreement between the conservative potential and Euler solutions with the nonconservative shock too weak and too far forward. At a higher freestream Mach number (Figure 2c) the nonconservative scheme still underestimates the shock but the conservative method predicts a shock that is too strong and too far back. If we consider a lifting condition (eg Figure 2d) then the conservative prediction shows an upper surface shock at the trailing edge and the result bears little resemblance to the Euler solution. At the higher Mach number (Figure 2e) the disagreement between conservative potential and Euler is less pronounced and, like the non-lifting case at this Mach number (Figure 2c), the Euler solution lies roughly half way between the conservative and nonconservative solutions.

On this empirical evidence it is still arguable whether either type of difference scheme produces a more satisfactory result. However, the level of agreement between conservative potential and Euler varies considerably while the nonconservative solution appears to underestimate the Euler result in a fairly consistent manner. To this extent the nonconservative scheme can be regarded as being the more reliable potential method for predicting flow development at transonic conditions.

3. ITERATIVE ALGORITHMS

3.1 Historical Perspective

In this chapter we discuss the numerical solution of the difference equations. The iterative algorithms that are described can be applied with only minor differences in detail, to either the nonconservative or fully conservative forms of the potential equation. The feature common to all the techniques which have appeared is that they first evolved as methods for solving elliptic equations. Not all have made a successful transition to equations of mixed type. In fact, successive line overrelaxation (SLOR) which was first applied to a mixed flow problem by Murman and Cole [13] has endured a surprisingly long time. Only recently has this method been surpassed by faster and more reliable algorithms. SLOR, semi direct solvers and eigenvalue extrapolation have been adequately described elsewhere [17] so we shall concentrate on the approximate factorisation and multigrid methods.

Alternating Direction Implicit (ADI) or approximate factorisation techniques were originally developed for equations of parabolic and elliptic type [29,30]. The first application to transonic flow problems is reported by Ballhaus and Steger [31] who used these techniques to remove the time step

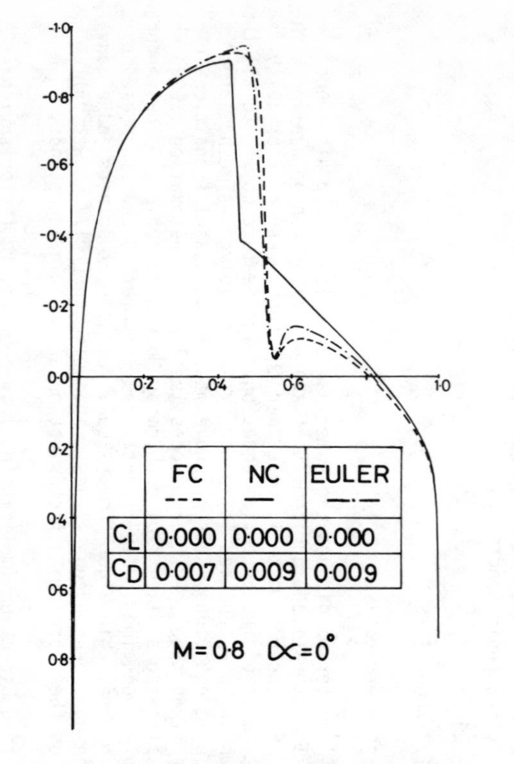

FIG.2(b) PRESSURE DISTRIBUTION COMPARISON FOR AEROFOIL NACA 0012

	FC	NC	EULER
C_L	0·000	0·000	0·000
C_D	0·059	0·036	0·047

M = 0·85 α = 0°

FIG 2(c) PRESSURE DISTRIBUTION COMPARISON FOR AEROFOIL NACA 0012

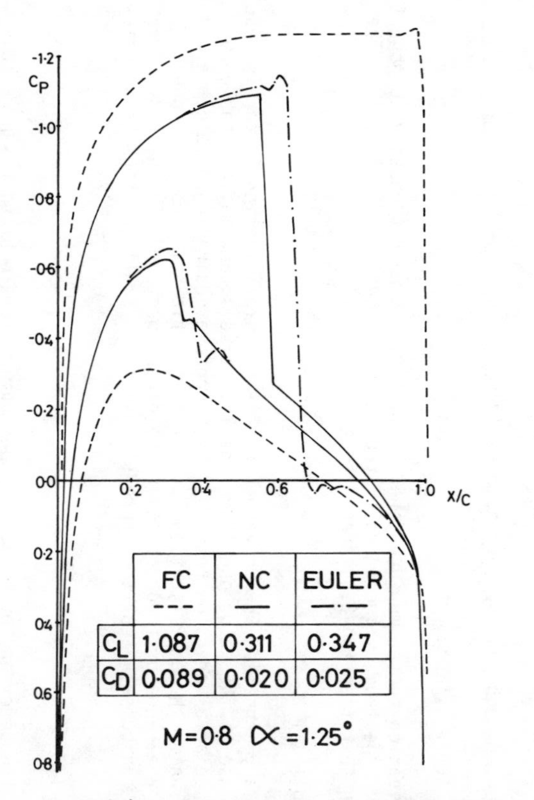

	FC (- - -)	NC (——)	EULER (—·—)
C_L	1·087	0·311	0·347
C_D	0·089	0·020	0·025

FIG 2(d) PRESSURE DISTRIBUTION COMPARISON FOR AEROFOIL NACA 0012

	FC (- - -)	NC (——)	EULER (—·—)
C_L	0·569	0·237	0·360
C_D	0·084	0·042	0·063

M=0·85 α=1°

FIG 2(e) PRESSURE DISTRIBUTION COMPARISON FOR AEROFOIL NACA 0012

limitation inherent in a relaxation solution of the unsteady TSP equation. The extension to steady TSP came quickly [32] and the AF2 scheme proposed by Ballhaus, Jameson and Albert formed the basis for further developments aimed at solving the exact potential equation. This followed two independent but similar lines; one for the fully conservative equation [18], the other for the nonconservative equation [33]. The factorisations of Holst and Baker have been extremely successful and as a result a number of existing SLOR codes have already been refurbished with an approximte factorisation routine (see for example section 3.7).

In parallel with these developments the multigrid method was being investigated. The introduction of the multigrid concept is credited to Fedorenko [34] but it is probably Brandt [35,36] who has done most to develop and popularise the technique for a variety of flow problems. Transonic flow with shock waves presents particular difficulties and work in this area by Brandt, South, Jameson and Arlinger [37,38] produced encouraging but not entirely satisfactory results. The recent intense interest in the method stems from Jameson's synthesis [4] of multigrid with an ADI smoothing routine. This has produced a remarkable improvement in convergence rate which must surely be at or at least close to the ultimate limit on the computational effort required to carry out a potential flow calculation.

3.2 The ADI Method

In order to motivate the development of ADI and approximate factorisation techniques we first consider the typical convergence properties of point and line iterative methods. These can be illustrated by examining a line Gauss Seidel scheme applied to Laplace's equation. The amplification factor in this case is given by equation (2.19) with $M = 0$. If we make the further simplifying assumption that $\Delta x = \Delta y = h$ then

$$G\,(p,q) = \frac{e^{i\xi}}{2 - e^{-i\xi} + 2(1 - \cos\eta)} \tag{3.1}$$

where $\xi = ph$, $\eta = qh$ and the wave numbers are

$$p,q = 1,\ \dots\ \pi/h$$

The amplification factor measures the extent to which each frequency component of the error spectrum is increased or reduced after one iteration cycle. The von Neumann test (section 2.4) states that

$$|G(p,q)| \leq 1$$

is a necessary condition for convergence. If the amplification factor is less than but close to the value one, then the iterative scheme may be stable but only converge at a very slow rate. It is therefore desirable to select a method for which the amplification factor is significantly less than one throughout the range of frequencies in the error spectrum. In the above example it can be seen that the expression (3.1) is small for high frequencies. For example, if $\xi \sim \pi$ and $\eta \sim \pi$ then

$$|G| \sim 1/7$$

At the low frequency end of the error spectrum where $\xi \sim h$ and $\eta \sim h$ we have

$$G \sim 1 - 2h^2$$

which implies a very slow convergence rate. For optimum SLOR this can be improved so that

$$G = 1 - O(h)$$

for low frequency components. Nevertheless, the convergence rate of a relaxation scheme can be expected to become progressively worse as the number of grid points is increased since this causes a corresponding reduction in the size of the grid increment h. In order to alleviate this difficulty it is usual to obtain a relaxation solution on a coarser grid and then interpolate this result onto the fine grid for use as a starting solution for the fine grid calculation. The coarse grid calculation is faster, first because the grid increment h is larger and second since the computation time per iteration cycle is less. This is a useful strategy for decreasing the overall computation time but the convergence rate of a relaxation method still leaves much to be desired. To illustrate this point we first show convergence histories for the calculation of subcritical lifting flow over an aerofoil by relaxation.

Note that all the computed results presented in this chapter have been obtained from methods which use a grid generated by the circle plane mapping (ie by conformally mapping the region exterior to the aerofoil onto the interior of a circle). By taking equal increments Δr in the radial direction and $\Delta\theta$ in the circumferential direction one produces a grid that is nonuniform in physical space. The relaxation code mentioned here is very similar to the familiar Garabedian and Korn method [39] which also uses the circle plane mapping. The approximate factorisation codes which produced the results shown later in this chapter use the same grid. The example is that shown in Figure 2a and the convergence histories for relaxation with and without grid refinement are presented in Figure 3. In each case it can be seen that the residual drops rapidly at first, corresponding to the removal of high frequency components. Eventually, the residual is dominated by low frequency errors and the asymptotic rate of decay is seen to be very slow.

Loosely speaking, high frequency errors are associated with localised deviations from the converged solution. On the other hand, low frequency errors reflect global discrepancies, for example, an incorrect circulation. It is therefore not surprising that a good starting solution, such as that provided by a coarse grid calculation, is effective in reducing the initial low frequency content. This behaviour is reflected in the convergence histories which show that grid refinement leads to a lower overall residual even though the asymptotic rate of decay is the same. In practice, one continues the iterative solution until the residual or some other measure of convergence reaches a sufficiently small level. The convergence parameter that is plotted in Figure 3 is the maximum residual multiplied by h^2. A value of 10^{-5} is usually required to achieve adequate convergence and it can be seen that the slow convergence rate of relaxation prevents this level from being reached until a great many iterations have been carried out.

When computing a transonic flow in which strong shocks are present the convergence rate of the relaxation scheme can be much slower and the manner in which the solution converges can be quite misleading. This is well illustrated by the example shown in Figure 4. The convergence history with grid refinement is presented in Figure 5 and the circulation growth is shown in Figure 6. Between about 200 and 400 iterations the circulation remains almost constant and only reaches the final converged level much later. The final change in circulation and the associated blip in the redisual at around 420 iterations correspond with a shift in the shock position by one grid point (see Figure 4).

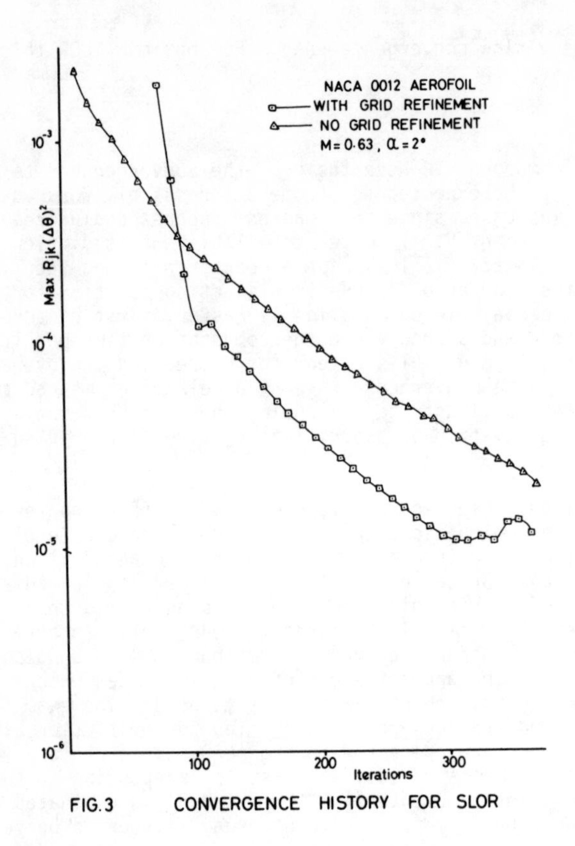

FIG. 3 CONVERGENCE HISTORY FOR SLOR

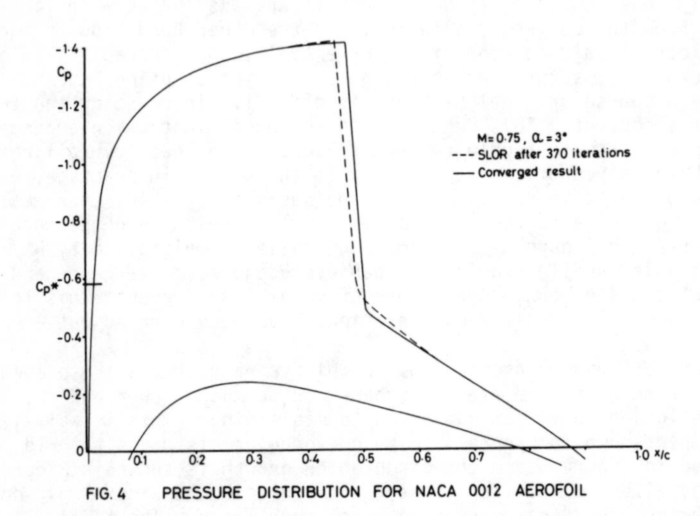

FIG. 4 PRESSURE DISTRIBUTION FOR NACA 0012 AEROFOIL

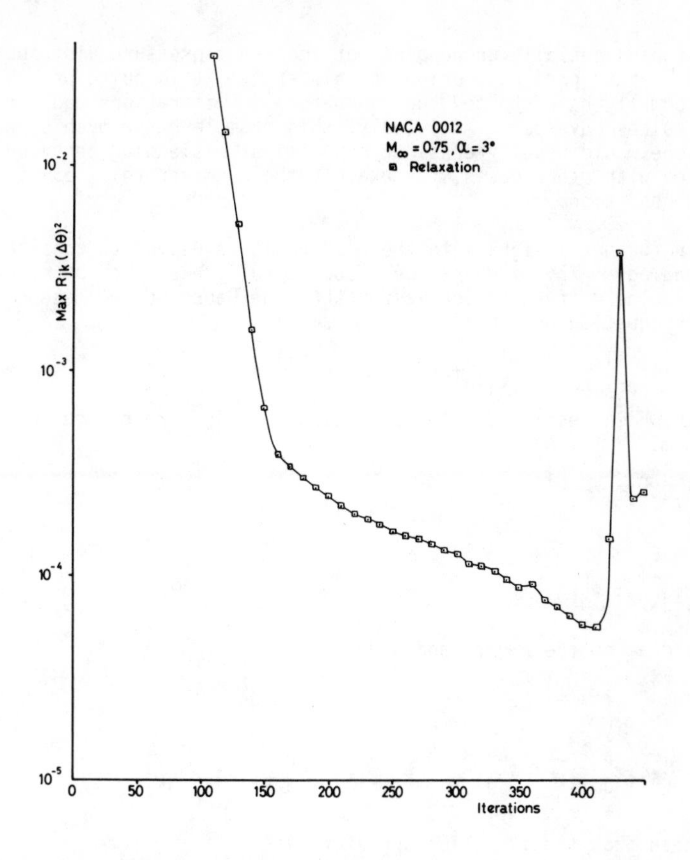

FIG. 5 CONVERGENCE HISTORY OF SLOR

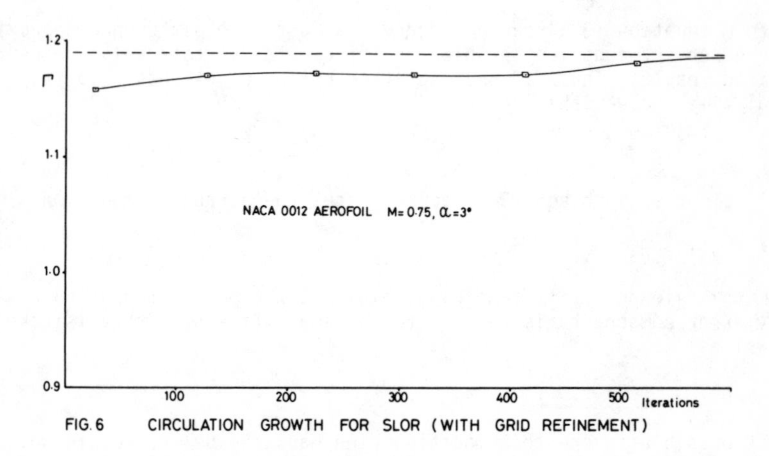

FIG. 6 CIRCULATION GROWTH FOR SLOR (WITH GRID REFINEMENT)

In view of the essentially unchanging but incorrect pressure distribution and circulation that is predicted prior to this shift, it is quite probable that one would normally have stopped the run before 400 iterations and considered the result to be converged. Admittedly, this example has a much stronger shock than one would usually consider but similar misleading behaviour has been observed with other cases, for example the KORN aerofoil near its shock free design condition.

To gain further insight into the problem it is useful to consider a time dependent analogy. This idea has been successfully exploited by several people [40,14,17]. The difference scheme for the line Gauss Seidel method applied to Laplace's equation on a uniform grid is

$$\phi^{n}_{j+1,k} + \phi^{n+1}_{j-1,k} + \phi^{n+1}_{j,k+1} + \phi^{n+1}_{j,k-1} - 4\phi^{n+1}_{jk} = 0$$

which was shown in section 2.4 to be stable. We first rearrange this equation into the form

$$N\Delta^{n}_{jk} = L\phi^{n}_{jk} \tag{3.2}$$

where

$$\Delta^{n}_{jk} = \phi^{n+1}_{jk} - \phi^{n}_{jk}$$

and the residual on the right hand side is

$$L\phi^{n}_{jk} = \left(\frac{\delta_{xx}}{h^2} + \frac{\delta_{yy}}{h^2}\right)\phi^{n}_{jk} \tag{3.3}$$

$$= \frac{1}{h^2}\left(\phi^{n}_{j+1,k} + \phi^{n}_{j-1,k} + \phi^{n}_{j,k+1} + \phi^{n}_{j,k-1} - 4\phi^{n}_{jk}\right)$$

It can be seen that the iteration operator is

$$N = \frac{1}{h^2}\left(1 + \overleftarrow{\delta}_x - \delta_{yy}\right)$$

In this form equation (3.2) can be regarded as a finite difference approximation to a time dependent equation in which the final steady state solution represents the required result. Thus, if one iteration cycle corresponds to a step Δt in artificial time and we take

$$\Delta t = h^2$$

then to first order in h eqn (3.2) approximates the parabolic equation

$$\phi_t = \phi_{xx} + \phi_{yy} \tag{3.4}$$

Seen from this viewpoint it seems plausible that the poor attenuation of low frequency error components is related to the small time step that is taken. If one now writes

$$\Delta t = rh^2$$

then it is possible to create a modified Line Gauss Seidel scheme in which the parameter r could be varied to change the effective time step. Unfortunately, a modal analysis shows that this explicit scheme is only stable for

$$r \leqslant 3/2$$

so that little is gained by this particular approach.

The remedy lies in the adoption of an implicit scheme which will remove the time step limitation. We again take the parabolic equation (3.4) as our prototype and following Mitchell [41], develop a finite difference approximation by using the formal relationship

$$\phi(x,y,t+\Delta t) = \left(1 + \Delta t \frac{\partial}{\partial t} + \frac{\Delta t^2}{2} \frac{\partial^2}{\partial t^2} + \ldots .\right) \phi(x,y,t)$$

$$= \exp\left(\Delta t \frac{\partial}{\partial t}\right) \phi(x,y,t)$$

Next replace the continuous function $\phi(x,y,t)$ by the finite difference form ϕ_{jk} and the differential operator $\partial/\partial t$ by the residual operator (3.3) to get

$$\phi_{jk}^{n+1} = \exp\,[r(\delta_{xx} + \delta_{yy})]\, \phi_{jk}^{n} \tag{3.5}$$

where as before

$$r = \frac{\Delta t}{h^2}$$

If one expands the exponential to first order in r, an explicit formula is obtained with the stability bound

$$r \leqslant \frac{1}{4}$$

If, however, eqn (3.5) is rearranged into the form

$$\exp\left(-\frac{r}{2}(\delta_{xx} + \delta_{yy})\right) \phi_{jk}^{n+1} = \exp\left(\frac{r}{2}(\delta_{xx} + \delta_{yy})\right) \phi_{jk}^{n} \tag{3.6}$$

then expanding the exponential to first order in r leads to the following implicit scheme

$$\left(1 - \frac{r}{2}(\delta_{xx} + \delta_{yy})\right) \phi_{jk}^{n+1} = \left(1 + \frac{r}{2}(\delta_{xx} + \delta_{yy})\right) \phi_{jk}^{n}$$

This difference scheme is stable for all positive r but, owing to the complexity of the left hand side, a large matrix inversion is required to obtain values at the new time level. An alternative procedure that leads to a set of straightforward tridiagonal inversions follows from the formula (cf eqn (3.6))

$$\exp\left(-\frac{r}{2}\delta_{xx}\right) \exp\left(-\frac{r}{2}\delta_{yy}\right) \phi_{jk}^{n+1} = \exp\left(\frac{r}{2}\delta_{xx}\right) \exp\left(\frac{r}{2}\delta_{yy}\right) \phi_{jk}^{n}$$

This can be expanded to give the ADI scheme

$$\left(1 - \frac{r}{2}\delta_{xx}\right)\left(1 - \frac{r}{2}\delta_{yy}\right) \phi_{jk}^{n+1} = \left(1 + \frac{r}{2}\delta_{xx}\right)\left(1 + \frac{r}{2}\delta_{yy}\right) \phi_{jk}^{n} \tag{3.7}$$

A modal analysis results in the following amplification factor

$$G(p,q) = \left(\frac{1 - r(1 - \cos\xi)}{1 + r(1 - \cos\xi)}\right) \left(\frac{1 - r(1 - \cos\eta)}{1 + r(1 - \cos\eta)}\right) \tag{3.8}$$

where $\xi = ph$ and $\eta = qh$ as before. It follows that

$$|G| \leqslant 1 \quad \text{for all } p \text{ and } q,$$

and hence the ADI method is stable for all positive r.

Rearranging eqn (3.7) into the form (3.2) and, for convenience, replacing the parameter r by α where

$$\alpha = \frac{2}{rh^2}$$

leads to the difference scheme

$$\left(\alpha - \frac{\delta_{xx}}{h^2}\right)\left(\alpha - \frac{\delta_{yy}}{h^2}\right)\Delta^n_{jk} = 2\alpha L\phi^n_{jk} \tag{3.9}$$

This is the AF1 scheme that was proposed by Ballhaus et al [32]. On examining the amplification factor (3.8) it can be seen that the choice

$$\alpha = \frac{2}{h^2}(1 - \cos ph)$$

will reduce the pth frequency component to zero. Thus the finite sequence

$$\alpha_p = \frac{2}{h^2}(1 - \cos ph), \; p = 1 \ldots \frac{\pi}{h} \tag{3.10}$$

should remove all error components and hence produce a converged result. The αs are called acceleration parameters and a sequence such as (3.10) is equivalent to the application of a series of different time steps. Large values of α correspond to small time steps which are effective against high frequency error components. On the other hand, small values of α which represent large time steps reduce the low frequency end of the error spectrum.

The above analysis is only strictly valid for the solution of an equation with constant coefficients on a uniform grid with periodic boundary conditions. In practice, we require the solution of a nonlinear equation on a nonuniform grid and so the above conclusion must be regarded as offering only a guideline for solving the full potential equation. if the equation to be solved is

$$C_1 \phi_{xx} + C_2 \phi_{xy} + C_3 \phi_{yy} + D = 0 \tag{3.11}$$

then the iterative scheme (3.9) can be generalised to

$$(\alpha - C_1 \frac{\delta_{xx}}{\Delta x^2})(\alpha - C_3 \frac{\delta_{yy}}{\Delta y^2})\, \Delta^n_{jk} = \sigma\alpha L\phi^n_{jk} \tag{3.12}$$

where L is now the residual operator corresponding to eqn (3.11). The difference equations (3.12) are solved in two steps. First we obtain the intermediate values F_{jk} by inverting

$$(\alpha - C_1 \frac{\delta_{xx}}{\Delta x^2})\, F_{jk} = \sigma\alpha L\phi^n_{jk}$$

on each constant y line. The second step requires a similar series of tridiagonal inversions on constant x lines

$$(\alpha - C_3 \frac{\delta_{yy}}{\Delta y^2})\, \Delta^n_{jk} = F_{jk}$$

whence

$$\phi^{n+1}_{jk} = \phi^n_{jk} + \Delta^n_{jk}$$

In practice it is best to use a small sequence of acceleration parameters running between the minimum and maximum values given by (3.10). The minimum value $\alpha_L = 1$ occurs for p = 1, while the maximum

$$\alpha_H = \frac{4}{h^2}$$

is obtained by taking $ph \sim \pi$. The following geometric sequence [32], which is repeated in a cyclic fashion, has proved very effective

$$\alpha_k = \alpha_H \left(\frac{\alpha_L}{\alpha_H}\right)^{k-1/M-1} \quad , \quad k = 1 \ .... \ M \tag{3.13}$$

where M typically has a value of around six to eight. The parameter σ in eqn (3.12) is usually taken to be two (cf (3.9)).

The improvement in convergence rate that can be obtained with the ADI or AF1 scheme is readily apparent if one considers its application to the example presented in Figure 2a. The convergence history for this case is shown in Figure 7 and it can be seen the residual drops much more quickly than for SLOR (cf Figure 3). Each iteration of the AF1 scheme takes about 1½ times as long as a relaxation iteration and so the convergence history presented in Figure 7 has been plotted against equivalent relaxation iterations. The AF1 scheme clearly offers a significant improvement over SLOR. The AF1 or ADI scheme described above is a fast procedure for solving elliptic problems. An important example is the set of elliptic equations that arise in numerical grid generation by Thompson's method [6]. Holst [18,42] for example, uses an AF1 algorithm to generate two dimensional grids by the Thompson technique.

The AF1 scheme extends readily to three dimensions and the following factorisation has been used to solve Laplace's equation

$$\left(\alpha - \frac{\delta_{xx}}{\Delta x^2}\right)\left(\alpha - \frac{\delta_{yy}}{\Delta y^2}\right)\left(\alpha - \frac{\delta_{zz}}{\Delta z^2}\right) \Delta^n_{jk\ell} = \sigma\alpha^2 L\phi^n_{jk\ell}$$

where

$$L\phi^n_{jk\ell} = \left(\frac{\delta_{xx}}{\Delta x^2} + \frac{\delta_{yy}}{\Delta y^2} + \frac{\delta_{zz}}{\Delta z^2}\right) \phi^n_{jk\ell}$$

In this case a modal analysis does not reveal a simple sequence for the acceleration parameters and a correct choice becomes rather obscure. However, Forsey [43] has used the above factorisation to generate three dimensional grids by the Thompson method. His numerical results indicate that the parameter sequence (3.13), with the above values for α_L and α_H, is still very effective. A value for σ of around 2.5 rather than 2 appeared to give the best convergence in his 3D computations.

3.3 The AF2 Factorisation

Although very effective for solving elliptic equations, the ADI or AF1 factorisation is not well suited to equations of mixed type. This problem has been carefully examined by Jameson [14,17] who shows that the reason lies in the form of the time like terms in the equivalent time dependent equation. In terms of the time dependent analogy the solution of an equation such as (2.17) by an AF1 method is equivalent to solving the equation

$$\phi_t = A\phi_{xx} + \phi_{yy}$$

where $A = 1 - M^2$. When the flow is locally subsonic A is positive and the above equation is parabolic. In this case the ϕ_t term introduces an effective damping that ensures a stable iterative scheme. When the flow is locally supersonic, A becomes negative and the ϕ_t term no longer provides the required damping.

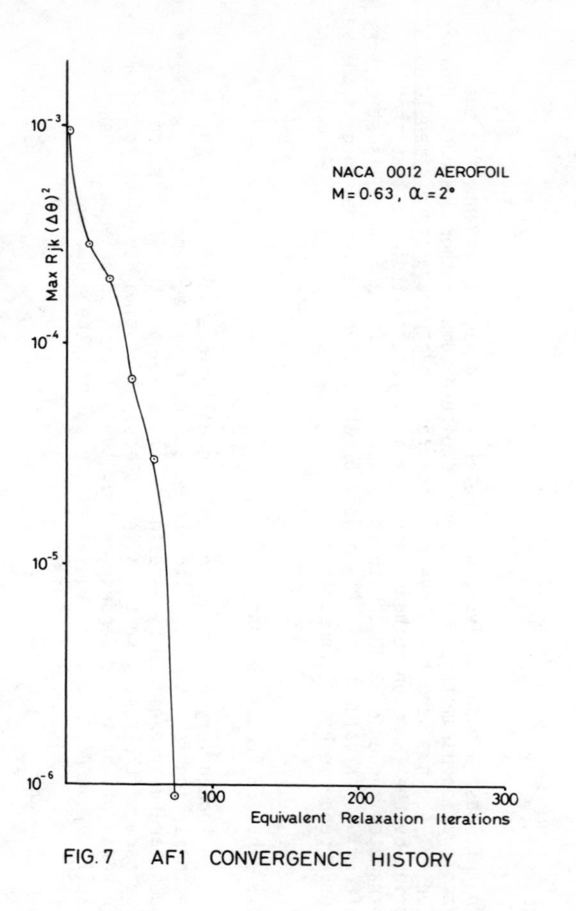

FIG. 7 AF1 CONVERGENCE HISTORY

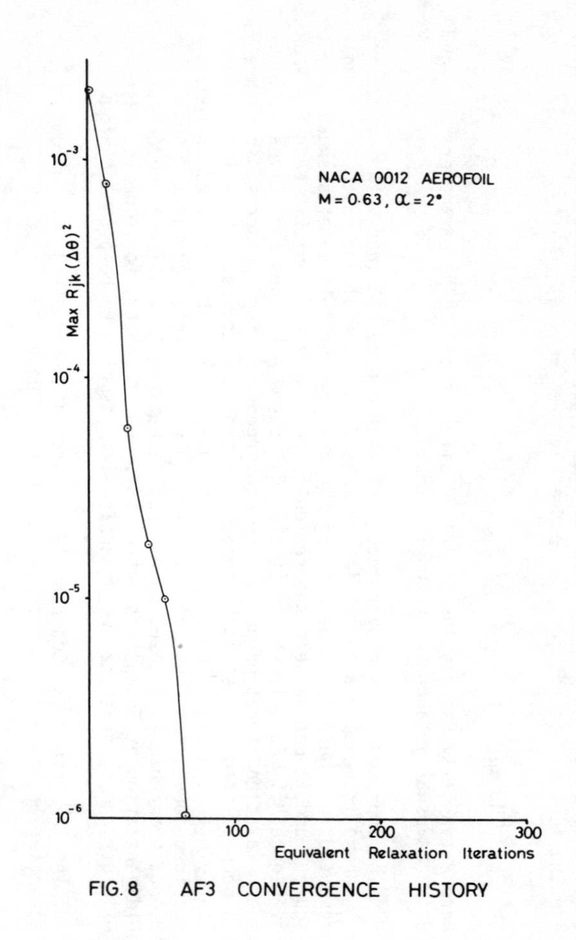

FIG. 8 AF3 CONVERGENCE HISTORY

Its presence can have a destabilising effect since the equation then admits solutions that grow exponentially. It is possible to stabilise the calculation by introducing terms like ϕ_{xt} and ϕ_{yt} [32] but a more satisfactory approach is the construction of an iterative scheme for which the equivalent time dependent equation contains no ϕ_t term. We therefore take the equation

$$\phi_{yt} = A\phi_{xx} + \phi_{yy}$$

as a prototype and seek a factored difference approximation of the form (3.2). One possibility is to use the identity (see section 2.2)

$$\delta_{yy} = \overset{\rightarrow}{\delta}_y \overset{\leftarrow}{\delta}_y$$

and alter the AF1 factorisation (3.9) so that the difference operator δ_{yy} is split between the two factors. For example

$$\left(-\alpha\frac{\overset{\rightarrow}{\delta}_y}{h} - A\frac{\delta_{xx}}{h^2}\right)\left(\alpha + \frac{\overset{\leftarrow}{\delta}_y}{h}\right)\Delta^n_{jk} = \sigma\alpha L\phi^n_{jk} \tag{3.15}$$

A modal analysis shows that this scheme is stable and the parameter sequence

$$\alpha_p = \frac{2\sin ph/2}{h}, \quad p = 1 \ldots. \frac{\pi}{h} \tag{3.16}$$

approximately minimises the amplification factor. Again it is usual to employ a smaller sequence of parameters such as (3.13) and repeat this in a cyclic manner. For the AF2 factorisation, however, we find that

$$\alpha_L = 1, \quad \alpha_H = \frac{2}{h} \tag{3.17}$$

and the parameter σ is given a value nearer one. The presence of the $\overset{\rightarrow}{\delta}_y$ operator in the first factor imposes a sweep restriction. To show this we write out the finite difference equation for the first step of the inversion

$$\left(-\alpha\frac{\overset{\rightarrow}{\delta}_y}{h} - A\frac{\delta_{xx}}{h^2}\right)F_{jk} = \sigma\alpha L\phi^n_{jk} \tag{3.18}$$

which expands to give

$$-\frac{AF_{j+1,k}}{h^2} + \left(\frac{\alpha}{h} + \frac{2A}{h^2}\right)F_{jk} - \frac{AF_{j-1,k}}{h^2} = \sigma\alpha L\phi^n_{jk} + \alpha\frac{F_{j,k+1}}{h} \tag{3.19}$$

When solving the tridiagonal system on the line $y = kh$ we require values of $F_{j,k+1}$. Hence we must sweep in the direction of decreasing k. Note that the minus sign in front of the $\overset{\rightarrow}{\delta}_y$ operator in (3.15) is imposed by the requirement of diagonal dominance. The factorisation in which the inversion of the first factor is swept in the direction of increasing k is

$$\left(\alpha\frac{\overset{\leftarrow}{\delta}_y}{h} - A\frac{\delta_{xx}}{h^2}\right)\left(\alpha - \frac{\overset{\rightarrow}{\delta}_y}{h}\right)\Delta^n_{jk} = \sigma\alpha L\phi^n_{jk} \tag{3.20}$$

When A is negative corresponding to locally supersonic flow the term ϕ_{xx} in the residual is backward differenced. The factorisation (3.15) is then replaced by

$$\left(-\alpha\frac{\overset{\rightarrow}{\delta}_y}{h} - A\frac{\overset{\leftarrow}{\delta}_x\overset{\leftarrow}{\delta}_x}{h^2}\right)\left(\alpha + \frac{\overset{\leftarrow}{\delta}_y}{h}\right)\Delta^n_{jk} = \sigma\alpha L\phi^n_{jk} \tag{3.21}$$

Apart from minor differences, the factorisations (3.15) and (3.21) are those used in the AF2 scheme proposed by Ballhaus et al [32]. In their paper they show several comparisons of AF2 and relaxation convergence rates for the TSP equation. The improvement they obtained is impressive and encouraged the expectation that a similar improvement would be forthcoming for the full potential equation.

3.4 Holst's Scheme for the Full Potential Equation

The extension of these ideas to the full potential equation is not obvious and there appear to be no sound theoretical criteria for deciding what should be a good factorisation. It is probably fair to say that the two schemes that have proved particularly successful, and which we now describe, owe their development more to inspired guesswork than theoretical insight. In order to extract some kind of guideline we note that there is a feature common to both the AF1 factorisation (3.9) and the AF2 factorisation (3.15). The terms in the iteration operator N which are linear in α correspond to the residual operator L. For want of anything better, we adopt the same general guideline when developing a factorisation for the full potential equation. In other words, on expanding the iteration operator as a polynomial in α, we require those terms which are linear in α to resemble closely the residual operator L.

On a Cartesian grid the residual for the full potential equation in conservation form is given by eqn (2.37). In view of the remarks about the AF1 scheme in section 3.3, it appears best to restrict our attention to factorisations of the AF2 type (cf eqn (3.15)). These considerations lead to Holst's AF2 scheme [18,42].

$$\left(-\alpha\frac{\vec{\delta}_y\,\rho^n_{j,k-\frac{1}{2}}}{\Delta y}-\frac{\vec{\delta}_x\,\tilde{\rho}^n_{j-\frac{1}{2},k}\,\overleftarrow{\delta}_x}{\Delta x^2}\right)\left(\alpha+\frac{\overleftarrow{\delta}_y}{\Delta y}\right)\Delta^n_{jk}=\sigma\alpha L\Phi^n_{jk} \tag{3.22}$$

where

$$L\Phi^n_{jk}=\vec{\delta}x\,\frac{(\tilde{\rho}u)^n}{\Delta x}_{j-\frac{1}{2},k}+\vec{\delta}y\,\frac{(\bar{\rho}v)^n}{\Delta y}_{j,k-\frac{1}{2}} \tag{3.23}$$

and the modified densities $\tilde{\rho}^n$ and $\bar{\rho}^n$ are given by eqn (2.38). This is a very neat factorisation and has some particularly attractive features. First we note that the form taken by the modified densities (see section 2.10) always maintains the correct upwind influence in the difference scheme when the flow is locally supersonic. Thus the effect of a rotated difference scheme is included in the factorisation as well as the residual. In addition the inversion of the first factor is accomplished by a straightforward tridiagonal solution while the second factor requires only a very simple bidiagonal inversion. It is, however, necessary to evaluate the modified densities $\tilde{\rho}$ and $\bar{\rho}$ at the previous iteration level n which suggests a possible loss of implicitness. As a result one might anticipate some stability problems. This does not appear to cause any difficulty in practice although, when strong shocks are present, Holst augments the factorisation by adding the term

$$-\alpha\frac{\beta\vec{\delta}_x}{\Delta x}\ \text{or}\ +\alpha\frac{\beta\overleftarrow{\delta}_x}{\Delta x} \tag{3.24}$$

to the first factor. This effectively introduces a ϕ_{xt} term into the equivalent time dependent equation and therefore has a stabilising influence on the iteration scheme [14]. The parameter β is a user defined constant which is set equal to zero in regions of subcritical flow. The choice of a forward or backward difference operator in (3.24) is of course dictated by the local flow direction.

The extension to a general two dimensional coordinate system follows from eqn (2.44). After multiplying this equation by the Jacobian J we obtain

$$\frac{\partial}{\partial \xi^1}(J\rho u^1) + \frac{\partial}{\partial \xi^2}(J\rho u^2) = 0$$

This is approximated by the residual

$$L\phi^n_{jk} = \overset{\rightarrow}{\delta}_{\xi^1}\frac{(J\tilde{\rho}u^1)^n_{j-\frac{1}{2},k}}{\Delta\xi^1} + \overset{\rightarrow}{\delta}_{\xi^2}\frac{(J\bar{\rho}u^2)^n_{j,k-\frac{1}{2}}}{\Delta\xi^2} \qquad (3.25)$$

where u^1 and u^2 are the contravariant velocity components and $\tilde{\rho}$ and $\bar{\rho}$ are defined as before by eqn (2.38). Equation (3.22) now generalises to the following

$$\left(-\alpha\overset{\rightarrow}{\delta}_{\xi^2}\frac{(J\bar{\rho}g^{22})^n_{j,k-\frac{1}{2}}}{\Delta\xi^2} - \overset{\rightarrow}{\delta}_{\xi^1}\frac{(J\tilde{\rho}g^{11})^n_{j-\frac{1}{2},k}}{(\Delta\xi^1)^2}\overset{\leftarrow}{\delta}_{\xi^1}\right)\left(\alpha + \frac{\overset{\leftarrow}{\delta}_{\xi^2}}{\Delta\xi^2}\right)\Delta^n_{jk} = \sigma\alpha L\phi^n_{jk} \qquad (3.26)$$

where the metric coefficients g^{11} and g^{22} have been defined in section 2.12. Full details of this scheme can be found in Reference [18] in which Holst shows several impressive examples. These demonstrate that his method is both reliable and very fast for all kinds of flow conditions including cases with extensive regions of supersonic flow.

It appears that Holst uses the parameter sequence (3.13) but he recommends the end points $\alpha_L = 0.07$ and $\alpha_H = 1.5$ which differ from those given in (3.17) by an order of magnitude. The apparent conflict arises because the modal analysis used to determine (3.17) assumes that the independent variables range between 0 and 2π so that $h \simeq 2\pi/N$ where N is the number of grid intervals. Holst normalises his independent variables so that the grid increment in each coordinate direction is unity. It follows that the end points (3.17) must be rescaled in a similar way.

Finally, we note that Holst has extended his method to three dimensions and has produced a code to compute transonic flow over a swept wing fixed between two walls. In place of eqn (3.25) the residual is now

$$L\phi^n_{jk\ell} = \overset{\rightarrow}{\delta}_{\xi^1}\frac{(J\tilde{\rho}u^1)^n_{j-\frac{1}{2},k,\ell}}{\Delta\xi^1} + \overset{\rightarrow}{\delta}_{\xi^2}\frac{(J\bar{\rho}u^2)^n_{j,k-\frac{1}{2},\ell}}{\Delta\xi^2} + \overset{\rightarrow}{\delta}_{\xi^3}\frac{(J\hat{\rho}u^3)^n_{j,k,\ell-\frac{1}{2}}}{\Delta\xi^3} \qquad (3.27)$$

and the factorisation he proposes is

$$\left[\left(\alpha - \frac{\Delta\xi^2}{A_k}\frac{\overset{\rightarrow}{\delta}_{\xi^3}A_\ell\overset{\leftarrow}{\delta}_{\xi^3}}{(\Delta\xi^3)^2}\right)\left(\frac{A_k}{\Delta\xi^2} - \frac{1}{\alpha}\frac{\overset{\rightarrow}{\delta}_{\xi^1}A_j\overset{\leftarrow}{\delta}_{\xi^1}}{(\Delta\xi^1)^2}\right) - \alpha\frac{E_{\xi^2}}{\Delta\xi^2}A_k\right]$$

$$\left(\alpha + \frac{\overset{\leftarrow}{\delta}_{\xi^2}}{\Delta\xi^2}\right)\Delta^n_{jk\ell} = \alpha\sigma L\phi^n_{jk\ell} \qquad (3.28)$$

where

$$A_j = (J\tilde{\rho}g^{11})^n_{j-\frac{1}{2},k,\ell},\quad A_k = (J\bar{\rho}g^{22})^n_{j,k-\frac{1}{2},\ell} \text{ and } A_l = (J\hat{\rho}g^{33})^n_{j,k,\ell-\frac{1}{2}}$$

When the flow is two dimensional with no dependence on the variable ξ^3, we may set the ξ^3 difference equal to zero in which case eqn (3.28) reduces to the two dimensional factorisation (3.26).

3.5 The AF3 Factorisation for the Nonconservative Equation

In principle it should be possible to use Holst's version of the AF2 scheme to solve the potential equation in nonconservative form. However, an alternative factorisation has been proposed by Baker [33,44] and for the nonconservative equation this is a more suitable extension of the original AF2 method. The name AF3 has been used to identify this particular scheme although it has much in common with Holst's AF2 algorithm.

In Cartesian coordinates the quasilinear or nonconservative form of the potential equation is given by eqn (2.14) which for convenience can be written as

$$A\,\phi_{xx} + B\,\phi_{xy} + C\,\phi_{yy} = 0 \tag{3.29}$$

where

$$A = 1 - \frac{u^2}{a^2}, \quad B = \frac{-2uv}{a^2} \quad \text{and} \quad C = 1 - \frac{v^2}{a^2}$$

When the flow is subcritical all ϕ derivatives are replaced by central difference formulae (see section 2.2) and eqn (3.29) is approximated by the residual

$$L\phi^n_{jk} = \left(A\,\frac{\delta_{xx}}{\Delta x^2} + B\,\frac{\mu_x\mu_y\delta_x\delta_y}{\Delta x\Delta y} + C\,\frac{\delta_{yy}}{\Delta y^2}\right)\phi^n_{jk} \tag{3.30}$$

An obvious extension of the AF2 scheme (3.15) is the following

$$\left(-\alpha C\,\frac{\vec{\delta}_y}{\Delta y} - A\,\frac{\delta_{xx}}{\Delta x^2}\right)\left(\alpha + \frac{\overleftarrow{\delta}_y}{\Delta y}\right)\phi^n_{jk} = \sigma\alpha L\phi^n_{jk} \tag{3.31}$$

We observe that no term corresponding to the cross derivative $B\phi_{xy}$ has been included in the factorisation. This omission does not appear to slow down the convergence rate of the scheme. To illustrate this and demonstrate the algorithm's effectiveness we show the AF3 convergence history in Figure 8 for the case presented previously in Figure 2a. A cycle of the iteration scheme (3.31) requires one set of tridiagonal and one set of bidiagonal inversions in place of the two sets of tridiagnonal inversions required by an AF1 iteration. One therefore finds that each AF2 or AF3 iteration takes slightly less computational effort than an AF1 iteration. In fact, one iteration of the AF3 scheme is equivalent in computational time to about 1.3 relaxation iterations. We have again plotted the convergence history in terms of equivalent relaxation iterations in order to get a fair comparison between convergence rates of the different solution algorithms. Figures 3 and 7 when compared with Figure 8 show that the AF3 scheme, like the AF1 scheme, is a considerable improvement over SLOR.

When regions of supersonic flow are present, we construct a residual that contains a combination of upwind and centrally differenced terms (see sections 2.5 and 2.6). For convenience we now write

$$A = A_u + A_c, \quad B = B_u + B_c \quad \text{and} \quad C = C_u + C_c$$

where the subscript u refers to the contribution of a coefficient to the upwind differenced term and the subscript c denotes the centrally differenced contribution. A comparison with eqns (2.23) and (2.24) indicates that

$$A_c = \frac{v^2}{q^2}, \quad B_c = \frac{-2uv}{q^2}, \quad C_c = \frac{u^2}{q^2} \tag{3.32}$$

$$A_u = \left(1 - \frac{q^2}{a^2}\right) \frac{u^2}{q^2}, \; B_u = 2\left(1 - \frac{q^2}{a^2}\right) \frac{uv}{q^2}, \; C_u = \left(1 - \frac{q^2}{a^2}\right) \frac{v^2}{q^2}$$

and, when the two velocity components u and v are both positive, the residual can be written as

$$L\phi^n_{jk} = \left\{A_c \frac{\delta_{xx}}{\Delta x^2} + A_u \frac{\overleftarrow{\delta}_x \overleftarrow{\delta}_x}{\Delta x^2} + B_c \frac{\mu_x \mu_y}{\Delta x \Delta y} \delta_x \delta_y \right.$$

$$\left. + B_u \frac{\overleftarrow{\delta}_x \overleftarrow{\delta}_y}{\Delta x \Delta y} + C_c \frac{\delta_{yy}}{\Delta y^2} + C_u \frac{\overleftarrow{\delta}_y \overleftarrow{\delta}_y}{\Delta y^2}\right\} \phi^n_{jk} \tag{3.33}$$

The factorisation (3.31) can then be modified as follows

$$\left(- \alpha C_c \frac{\overrightarrow{\delta}_y}{\Delta y} - A_c \frac{\delta_{xx}}{\Delta x^2} - A_u \frac{\overleftarrow{\delta}_x \overleftarrow{\delta}_x}{\Delta x^2}\right)\left(\alpha + \frac{\overleftarrow{\delta}_y}{\Delta y}\right) \Delta^n_{jk} = \sigma\alpha L\phi^n_{jk} \tag{3.34}$$

When u is negative the backward difference operator $\overleftarrow{\delta}_x \overleftarrow{\delta}_x$ in eqns (3.33) and (3.34) is replaced by the corresponding forward difference operator $\overrightarrow{\delta}_x \overrightarrow{\delta}_x$. In addition the cross derivative term

$$B_u \frac{\overleftarrow{\delta}_x \overleftarrow{\delta}_y}{\Delta x \Delta y}$$

in eqn (3.33) is replaced by

$$B_u \frac{\overrightarrow{\delta}_x \overleftarrow{\delta}_y}{\Delta x \Delta y}$$

The upwind differencing complicates the algorithm, and allowing for the possibility of both backward and forward differencing in the x direction, this requires the inversion of a pentadiagonal matrix. If we drop one of the backward difference operators $\overleftarrow{\delta}_x$, we obtain the alternative factorisation

$$\left(- \alpha C_c \frac{\overrightarrow{\delta}_y}{\Delta y} - A_c \frac{\delta_{xx}}{\Delta x^2} - A_u \frac{\overleftarrow{\delta}_x}{\Delta x^2}\right)\left(\alpha + \frac{\overleftarrow{\delta}_y}{\Delta y}\right) \Delta^n_{jk} = \sigma\alpha L\phi^n_{jk} \tag{3.35}$$

which only requires a tridiagonal inversion for the first factor. Removing one of the $\overleftarrow{\delta}_x$ operators has the effect of replacing the upwind formula

$$\Delta x^2 \, D_{xx} \, \phi = \phi^{n+1}_{jk} - 2\phi^{n+1}_{j-1,k} + \phi^{n+1}_{j-2,k}$$

for ϕ_{xx} by the following

$$\Delta x^2 \, D_{xx} \, \phi = \phi^{n+1}_{jk} - \phi^{n+1}_{j-1,k} - \phi^{n}_{j-1,k} + \phi^{n}_{j-2,k}$$

This alternative form is stable and a comparison of the two factorisations (3.34) and (3.35) for a number of numerical examples confirms that the convergence rates are similar. The upwind differenced term in (3.35) now corresponds to a ϕ_{xt} term in the associated time dependent equation and therefore has a stabilising influence on the iterative scheme. When u is negative the factorisation (3.35) is replaced by the following

$$\left(-\alpha C_c \frac{\overrightarrow{\delta}_y}{\Delta y} - A_c \frac{\delta_{xx}}{\Delta x^2} + A_u \frac{\overrightarrow{\delta}_x}{\Delta x^2}\right)\left(\alpha + \frac{\overleftarrow{\delta}_y}{\Delta y}\right) \Delta^n_{jk} = \sigma\alpha L\phi^n_{jk}$$

It will be noticed that the factorisation contains no upwind differenced term in the y direction. Thus the upwind differencing for ϕ_{yy} is evaluated at the previous level n, viz

$$\Delta y^2 D_{yy}\, \phi = \phi^n_{jk} - 2\phi^n_{j,k-1} + \phi^n_{j,k-2}$$

This does not appear to slow the convergence of the scheme, presumably because the ϕ_{xt} term is sufficient to maintain stability. If desired, however, the factorisation (3.35) can be modified to accommodate a forward difference approximation for ϕ_{yy} when v, the velocity component in the y direction is negative, viz

$$\left(-\alpha C_c \frac{\vec{\delta}_y}{\Delta y} - \alpha C_u \frac{E_y \overleftarrow{\delta}_y}{\Delta y} - A_c \frac{\delta_{xx}}{\Delta x^2} - A_u \frac{\overleftarrow{\delta}_x}{\Delta x^2}\right)\left(\alpha + \frac{\overleftarrow{\delta}_y}{\Delta y}\right) \Delta^n_{jk} = \sigma\alpha L \phi^n_{jk}$$

where E_y is the shift operator defined by

$$E_y\, \phi_{jk} = \phi_{j,k+1}$$

The AF3 scheme described above has been used to solve the potential equation over a wide range of flow conditions including flows containing strong shocks and with extensive regions of supersonic flow. The aerofoil calculations presented here have all been carried out on a grid of 160 x 32 points. The number of grid intervals in the radial or y direction is therefore N = 31. If we take

$$h \simeq \frac{2\pi}{N} \simeq \frac{1}{5}$$

the end point $\alpha_H \simeq 10$. In practice, a larger value, $\alpha_H = 20$ seems better, particularly for flows with strong shocks. We therefore use a geometric sequence such as (3.13) with the end points

$$\alpha_L = 1, \quad \alpha_H = 20 \tag{3.36}$$

and also take a value 1.3 for σ. This set of parameters has proved to be very successful.

To illustrate this we first consider the computed pressure distributions for a transonic lifting flow at various stages of convergence. Our example is the NACA 0012 aerofoil at a Mach number of 0.75 and an incidence of 2 degrees. Figure 9 compares the converged pressure distribution with results computed by SLOR after 100 coarse grid iterations (equivalent in computing time to 25 fine grid iterations) followed by a further 100 and 400 fine grid iterations. It is evident that the result after an effective computing time of 125 fine grid iterations is some way off convergence and even after 425 iterations the shock strength has still not reached its final converged value. The results can be contrasted with those shown in Figure 10 for AF3 after 5 and 10 AF iterations (equivalent in computing time to 7 and 13 relaxation iterations respectively). After 10 AF iterations the lower surface distribution is correct and the final shock position has been reached. After a further 10 AF iterations there is no plottable difference between the computed pressure distribution and the converged result. If the incidence is increased to 3 degrees we obtain the pressure distribution shown in Figure 4; the convergence history and circulation growth for SLOR have already been presented in Figures 5 and 6 respectively. These are now reproduced in Figures 11 and 12 together with the corresponding convergence information for AF3. In sharp contrast to the behaviour of the relaxation scheme, it an be seen that the AF3 algorithm quickly locks on to the converged solution. This rapid and assured convergence

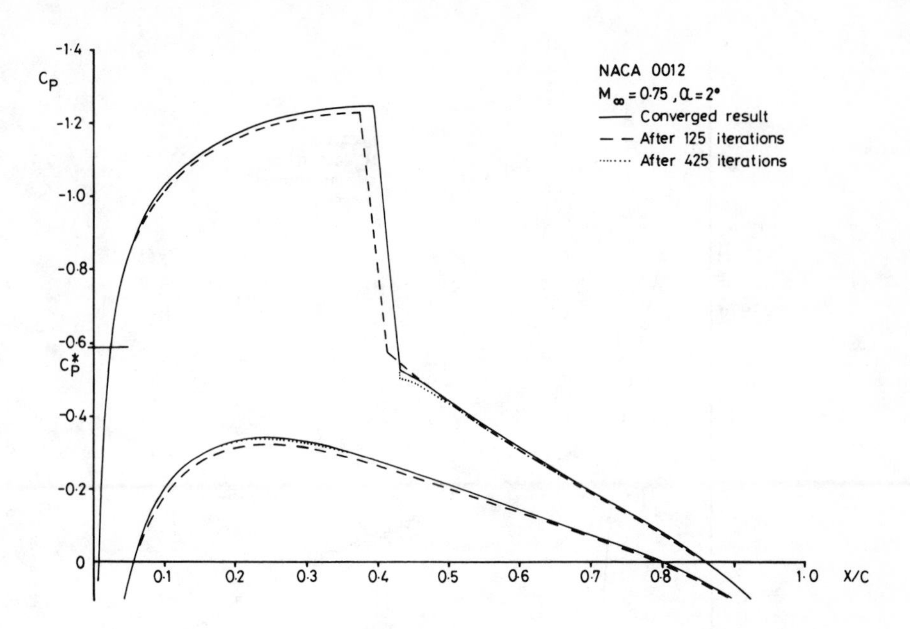

FIG. 9 RELAXION COMPUTATION OF PRESSURE DISTRIBUTION

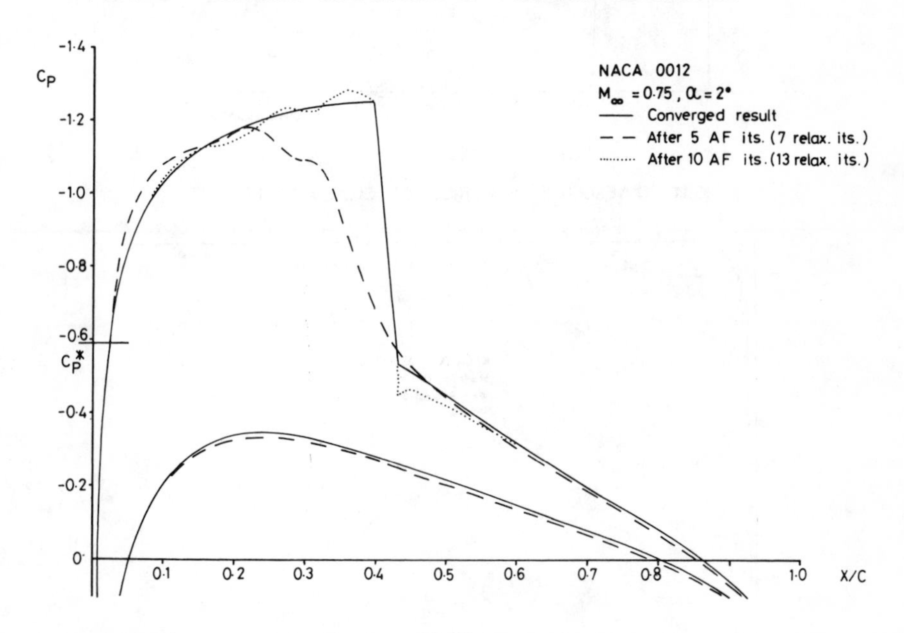

FIG.10 AF3 COMPUTATION OF PRESSURE DISTRIBUTION

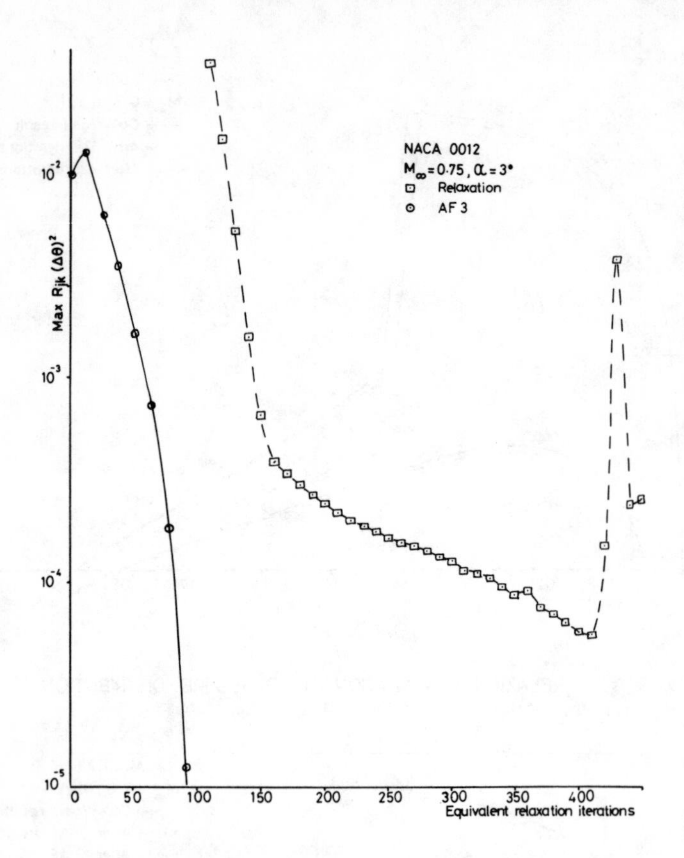

FIG.11 CONVERGENCE HISTORIES OF SLOR AND AF3

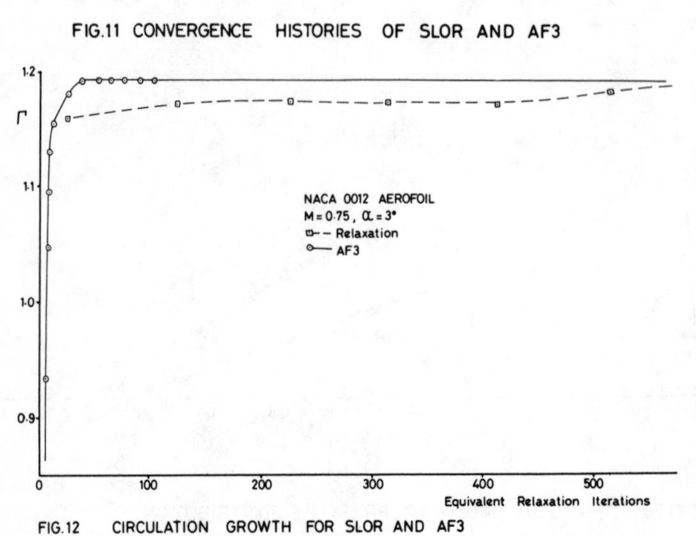

FIG.12 CIRCULATION GROWTH FOR SLOR AND AF3

is a particularly gratifying feature of approximate factorisation methods.

Unlike Holst's scheme it is not generally necessary to augment the factorisations by additional terms such as (3.24). However, this is required if one wishes to reduce the amount of dissipation introduced by the artificial viscosity in supersonic zones. For example, eqn (2.20) shows that the usual upwind formula for ϕ_{xx} is

$$\frac{\overleftarrow{\delta}_x \overleftarrow{\delta}_x}{\Delta x^2} \phi_{jk} = \phi_{xx} - \Delta x \, \phi_{xxx} + O(\Delta x^2)$$

If we now take the following term

$$\varepsilon \frac{\overleftarrow{\delta}_x \overleftarrow{\delta}_x \overleftarrow{\delta}_x}{\Delta x^2} \equiv \frac{\varepsilon \, (\phi_{jk} - 3\phi_{j-1,k} + 3\phi_{j-2,k} - \phi_{j-3,k})}{\Delta x^2}$$

$$= \varepsilon \Delta x \, \phi_{xxx} + O(\Delta x^2)$$

and add this to the above upwind formula for ϕ_{xx} we obtain

$$(1 + \varepsilon \overleftarrow{\delta}_x) \frac{\overleftarrow{\delta}_x \overleftarrow{\delta}_x}{\Delta x^2} \phi_{jk} = \phi_{xx} - (1 - \varepsilon) \, \Delta x \, \phi_{xx} + O(\Delta x^2) \qquad (3.37)$$

The artificial viscosity introduced by the modified formula (3.37) is

$$-(1 - \varepsilon) \, \Delta x \, \phi_{xxx}$$

which will be second order if

$$1 - \varepsilon = O(\Delta x)$$

In practice the flow computation, whether by SLOR or approximate factorisation, usually becomes unstable if ε is taken too close to one. It is therefore not possible, in general, to compute transonic potential flows with finite difference approximations that are second order accurate everywhere. It is possible to obtain converged results for modest values of ε up to about 0.8. The reduced amount of dissipation in supersonic zones tends to sharpen the shock profile and it is sometimes argued that use of this facility leads to more accurate solutions of the potential equation. If the modified upwind formula (3.37) is used in the residual on the right hand side of (3.35) then the first factor on the left hand side has to be augmented with terms like (3.24).

3.6 Approximate Factorisation on Stretched Grids

The AF3 factorisation (3.31) has been proposed for solving eqn (3.29) and in section 3.3, it was stated that the parameter sequence (3.16) approximately minimises the amplification factor generated by schemes of this type. The modal analsys, on which this result is based, holds for a uniform Cartesian grid but under a transformation to a stretched coordinate system the conclusion is no longer valid. It is then necesary to modify either the parameter sequence or preferably the form of the factorisation to achieve fast convergence on a stretched grid. This observation was brought to the author's attention by Catherall of RAE who has successfully applied approximate factorisation to calculate transonic flow over an aerofoil on a non-aligned grid [45].

In order to examine this question further we consider the solution of Laplace's equation

$$\phi_{xx} + \phi_{yy} = 0$$

where x and y are Cartesian coordinates. Next introduce a coordinate transformation to a new set of curvilinear coordinates X(x,y) and Y(x,y). To simplify the analysis we assume that the mapping is orthogonal. In the transformed plane the second derivative terms that appear in Laplace's equation can then be written as

$$C_1 \phi_{XX} + C_2 \phi_{YY} \tag{3.38}$$

where C_1 and C_2 are the metric components g^{11} and g^{22} (the off diagonal component g^{12} is of course zero for an orthogonal transformation). We now consider the following general factorisation of the AF2 or AF3 type

$$\left(-\alpha A_2 \frac{\vec{\delta}_Y}{\Delta Y} - A_1 \frac{\delta_{XX}}{\Delta X^2}\right)\left(B_1 \alpha + B_2 \frac{\overleftarrow{\delta}_Y}{\Delta Y}\right) \Delta^n_{jk} = \alpha L \phi^n_{jk} \tag{3.39}$$

where A_1, A_2, B_1 and B_2 are functions of X and Y which satisfy the conditions

$$A_1 B_1 = C_1 \text{ and } A_2 B_2 = C_2$$

but are otherwise arbitrary. Note that we have taken $\sigma = 1$ to simplify the analysis. We also assume that $\Delta X = \Delta Y = h$ and write

$$S_p = \frac{\sin ph/2}{h/2} \text{ and } e_p = e^{iph/2}$$

A von Neumann analysis leads to the following expression for the amplification factor [44],

$$|G(p,q)|^2 = \frac{S_q^{\,2}[\alpha^4 A^2 - 2\alpha^2 ABS_p^{\,2} \cos qh + B^2 S_p^{\,4}]}{[S_p^{\,4} + h\alpha AS_q^{\,2}S_p^{\,2} + \alpha^2 A^2 S_q^{\,2}][\alpha^2 + h\alpha BS_q^{\,2} + B^2 S_q^{\,2}]} \tag{3.40}$$

where $A = \frac{A_2}{A_1}$ and $B = \frac{B_2}{B_1}$

The combination

$$AB = \frac{A_2 B_2}{A_1 B_1} = \frac{C_2}{C_1}$$

is a constant for the purposes of our local stability analysis.

For the case

$$A_1 = B_1 = A_2 = B_2 = 1$$

the choice (3.16) or $\alpha = |S_p|$ approximately minimises $|G(p,q)|^2$. If we substitute this value of α into (3.40) we obtain the following expression for the amplification factor,

$$|G|^2 = \frac{S_q^{\,2} S_p^{\,2}[A^2 - 2AB \cos qh + B^2]}{[S_p^{\,2} + Ah|S_p|S_q^{\,2} + A^2 S_q^{\,2}][S_p^{\,2} + Bh|S_p|S_q^{\,2} + B^2 S_q^{\,2}]} \tag{3.41}$$

It can readily be seen that an inappropriate choice of the coefficients A_1, A_2, B_1 and B_2 can lead to slow convergence. For example, if we take the extreme case

$$\frac{A_2}{A_1} >> 1 >> \frac{B_2}{B_1}$$

in expressions (3.41) we find that

$$|G| \sim 1$$

It follows that a transformation from Cartesian coordinates to a stretched coordinate system can have an adverse effect on the convergence rate of an approximate factorisation scheme.

The best convergence rate will be obtained when the functions A and B are chosen to minimise $|G|^2$ subject to the constraint that AB is constant. Expression (3.41) is symmetric in A and B and, as one might expect, the amplification factor is minimised by the choice [44]

$$A = B,$$

in other words by selecting the coefficients so that

$$\frac{A_2}{A_1} = \frac{B_2}{B_1} = \sqrt{\frac{C_2}{C_1}} \tag{3.42}$$

On making this substitution in eqn (3.41) the following expression for the amplification factor is found

$$|G| = \frac{h|S_p|AS_q^2}{S_p^2 + h|S_p|A\,S_q^2 + A^2\,S_q^2}$$

On using the inequality

$$S_p^2 + A^2 S_q^2 \geq 2|S_p||S_q|A$$

we obtain

$$|G| \leq \frac{h|S_q|}{2 + h|S_q|}$$

But

$$h|S_q| = 2|\sin qh/2| \leq 2$$

whence we get the bound

$$|G| \leq \tfrac{1}{2} \tag{3.43}$$

Thus when the coefficients are chosen according to (3.42) and $\alpha = |S_p|$, the pth error mode is reduced in magnitude by at least one half. In order to reduce all error modes we use a sequence such as (3.13) with the end points (3.17). If we take a sequence of say, six acceleration parameters, then all modes and hence the complete error vector e^n should be reduced by at least one half every six iterations. Thus if we define an average convergence rate as

$$\lim_{n\to\infty}\left(\frac{e^n}{e^o}\right)^{1/n}$$

we can expect this to be bounded from above by

$$\left(\frac{1}{2}\right)^{1/6} \simeq 0.89$$

This ties in well with an observed residual reduction rate of about 0.87 which corresponds to a reduction of the residual by three orders of magnitude every 50 iterations (cf Figure 8).

Although the analysis was restricted to orthogonal transformations we shall assume that condition (3.42) still applies when the grid is mildly non-orthogonal. We observe that on expanding the factorisation (3.39), the operators appearing in the first factor will act on B_1 to give terms like

$$-\alpha^2 A_2 \frac{\partial B_1}{\partial Y} - \alpha A_1 \frac{\partial^2 B_1}{\partial X^2}$$

When B_1 is a function of X and Y, these extra terms will be non-zero and will introduce a ϕ_t term into the associated time dependent equation. We therefore make the restriction that the coefficient of α in the second factor is a constant which, without loss of generality, we may take equal to one. It follows that $B_1 = 1$ and hence $A_1 = C_1$. The condition (3.42) for optimum convergence then implies

$$\frac{A_2}{A_1} = \sqrt{\frac{C_2}{C_1}}$$

whence

$$A_2 = \sqrt{C_1 C_2}$$

and also

$$\frac{B_2}{B_1} = \sqrt{\frac{C_2}{C_1}}$$

whence

$$B_2 = \sqrt{\frac{C_2}{C_1}}$$

Under a general coordinate transformation we therefore require the following factorisation

$$\left(-\alpha\sqrt{C_1C_2}\,\frac{\overset{\rightarrow}{\delta}_Y}{\Delta Y} - C_1\frac{\delta_{XX}}{\Delta X^2}\right)\left(\alpha + \sqrt{\frac{C_2}{C_1}}\,\frac{\overset{\leftarrow}{\delta}_Y}{\Delta Y}\right)\Delta^n_{jk} = \sigma\alpha L\phi^n_{jk} \qquad (3.44)$$

If the transformation is conformal it can be written as

$$Z = f(z)$$

where $Z = X + iY$ is a point in computational space and $z = x + iy$ represents the corresponding point in physical space. Defining the mapping modulus as

$$H = \left|\frac{dZ}{dz}\right|$$

leads to the following coefficients in the differential equation

$$C_1 = C_2 = H^2$$

In this case the factorisation (3.44), apart from a multiplicative factor H^2, reduces to the basic AF2 or AF3 format.

3.7 Viscous Transonic Flow over Aerofoils

So far we have made no mention of viscous effects and in all the aerofoil calculations presented so far we have assumed the flow to be inviscid. In reality the boundary layer generated by lifting flow over an aerofoil can be quite thick and the actual pressure distribution can differ considerably from the corresponding inviscid solution. Provided the flow remains attached it is possible to allow for viscous effects by combining the inviscid algorithm with a boundary layer calculation. The boundary layer can be thought of as a displacement to the original aerofoil shape. It follows that the inviscid solution corresponding to the modified profile will depend on the properties of the boundary layer. Likewise, the calculated boundary layer will depend on the pressure gradients along the aerofoil surface. It is therefore necessary to link the inviscid calculation and boundary layer calculation in an iterative manner in order to arrive at a converged solution of the viscous flow. A number of such codes have been developed [23,46,47] and the reader is referred to the literature for further details. In all the codes the inviscid part of the calculation is based on SLOR. The purpose of this section is to mention briefly a recent application of the AF3 algorithm to the method of Reference [47].

In section 2.13 we observed that nonconservative difference schemes tend to underestimate shock strength and position compared with the corresponding Euler solution. In the VGK (Viscous Garabedian and Korn) method of Reference [47] an empirical factor λ has been introduced to compensate for this deficiency. Lock's partially conservative scheme does indeed produce inviscid potential solutions which are surprisingly close to the correct Euler solution [26]. It is true that the parameter λ is no more than a judicious empirical factor and it is only really helpful if a single value of λ can be found which is valid over a wide range of flow conditions. There is some evidence that a value for the partially conservative factor of about 0.25 or 0.3 does indeed suffice and the calculations shown in Figures 13, 15 and 16 have all been carried out with $\lambda = 0.25$.

It should be noted that the comparisons are with experimental results from the ARA 18" x 8" 2D tunnel. The calculations are for free air conditions and wall constraint effects would be expected to influence the tunnel results. In fact, the tunnel is virtually blockage free owing to the presence of wall slots. Matching at a given C_L compensates for the tunnel incidence effect and the streamwise curvature effect has only a very small influence on the pressures.

Recently Morrison [48] has replaced the SLOR routine in the VGK code by an AF3 algorithm. In every other respect, the program is unchanged and the new VAF (Viscous Approximate Factorisation) method produces identical results. However, the benefit of an approximate factorisation scheme carries over even though the overall convergence for a viscous computation is somewhat slower than an inviscid case. Figure 13 shows the pressure distribution over an aerofoil designed about ten years ago, just prior to the advent of transonic flow field methods. In Figure 14 we present convergence histories for this case obtained from VGK and VAF. Note that the degree of convergence is now measured by the maximum change in ϕ rather than the maximum residual. This has been done since the VGK code does not print out values of the residual. The

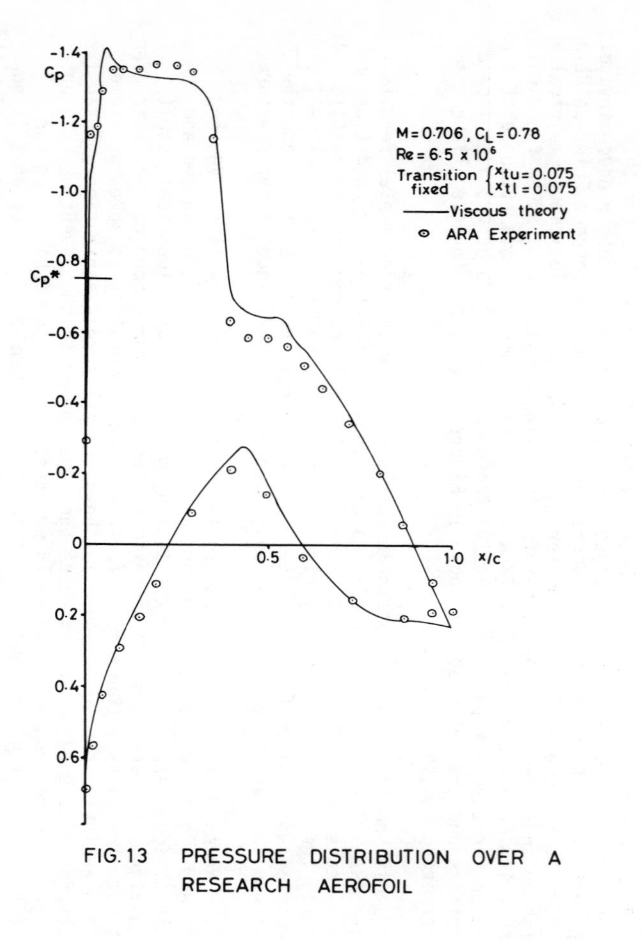

FIG. 13 PRESSURE DISTRIBUTION OVER A RESEARCH AEROFOIL

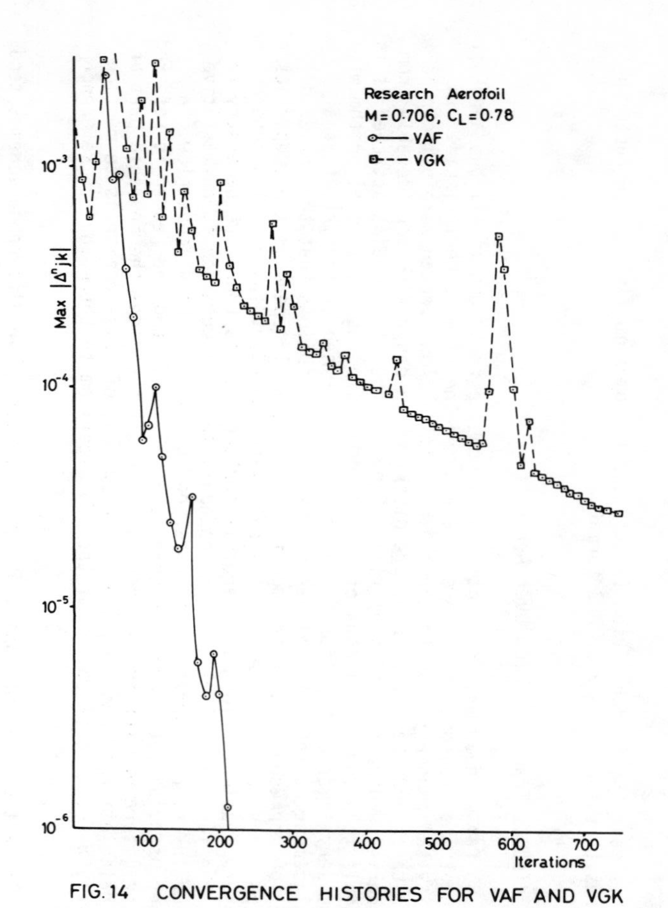

FIG. 14 CONVERGENCE HISTORIES FOR VAF AND VGK

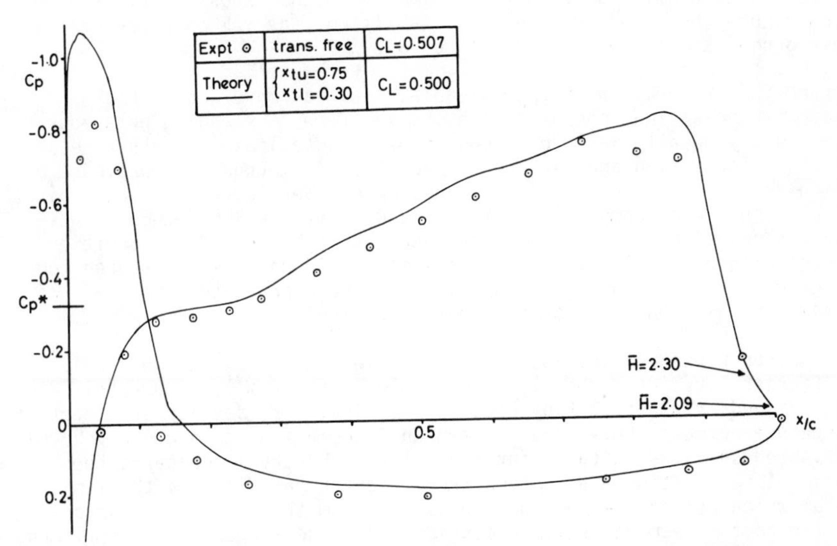

FIG.15 PRESSURE DISTRIBUTION OVER SECTION A(4% t/c) $M = 0.84$, $Re = 3.0 \times 10^6$

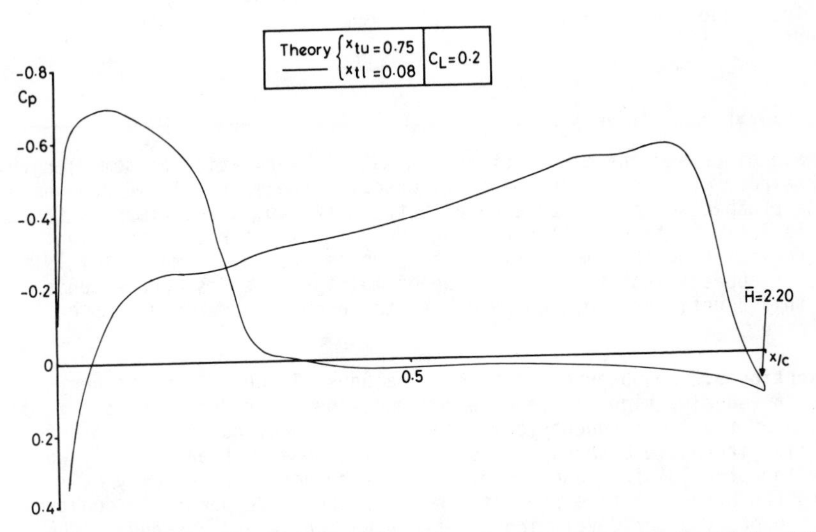

FIG.16 PRESSURE DISTRIBUTION OVER SECTION B (4% t/c) $M = 0.9$, $Re = 3.0 \times 10^6$

two convergence indicators are, however, roughly comparable in numerical value. Over the first 30 to 50 iterations the boundary layer and inviscid solutions are interacting strongly and so initially the VAF method shows little sign of converging. Once the boundary layer has stabilised, the VAF code achieves a fast convergence rate in sharp contrast to VGK.

Our previous remarks about the reliable, assured convergence of approximate factorisation, compared with the sometimes misleading behaviour of relaxation, apply equally well to viscous flow computations. In other respects approximate factorisation and, in particular, the VAF method have proved superior to relaxation. For example, a VAF run can be started from an arbitrary initial solution. When computing a difficult case by the VGK method, however, it is usually necessary to obtain a converged result at an easier condition before proceeding on the final run. Furthermore, VAF has sometimes been found to converge even though convergence difficulties have prevented the corresponding relaxation solution from being obtained.

Two further pressure distributions are presented in Figures 15 and 16 for rather different types of aerofoil. These belong to a family of aerofoil sections designed in about 1975 by Bocci [49]. They were developed as part of a joint venture between the Aircraft Research Association Ltd and Dowty Rotol Ltd as advanced sections suitable for propellers. Figure 15 compares theory and experiment for section A at a freestream Mach number close to the limit $M \simeq 0.85$ at which reliable data can be obtained from the ARA two-dimensional tunnel. The good agreement provided confidence in the adequacy of the viscous theory and recently calculations using VAF have been carried out to expand information on the performance of the ARA-D propeller sections [50]. One example of these computed pressure distributions is shown in Figure 16 for a section with lower camber. It can be seen that large regions of supersonic flow are present with an upper surface shock wave close to the trailing edge. The experience of this exercise [50] suggests that VAF is indeed a robust and reliable technique for providing a blanket investigation of aerofoil performance.

3.8 The Multigrid Strategy

We have discussed the merits of approximate factorisation at some length. In the remainder of this chapter we shall describe the multigrid method which offers the prospect of convergence rates between two and three times as fast as approximate factorisation. After reading some of the literature one might be forgiven for thinking that multigrid is the panacea for all numerical ills. In practice, it appears that the main virtue of multigrid is fast convergence although this benefit can only be gained at the expense of a more complex computer code.

In section 3.2 we observed that the Line Gauss Seidel scheme is very effective in reducing high frequency error modes but provides very poor attenuation of the low frequency components. This behaviour is typical of point or line iterative methods. Pursuing a time dependent analogy, it was suggested that the low frequency modes could be reduced more quickly by increasing the effective time step Δt. We considered the parabolic equation (3.4) as a prototype for developing an iterative scheme and introduced the parameter

$$r = \frac{\Delta t}{h^2}$$

where h is the grid increment which we again consider to be uniform in both coordinate directions. An explicit scheme such as Gauss Seidel or SLOR places a stability bound on r so that, for a fixed grid increment h, the possible size of Δt is restricted. Approximate factorisation methods overcome the limitation because they are constructed as implicit schemes which are unconditionally stable.

The other way out of this difficulty, which also allows one to retain an explicit scheme, is to increase the grid increment h. This is the motivation behind the multigrid technique which relies on a sequence of grids, ranging from fine to very coarse, to eliminate the different error modes. In order to fix ideas, we first consider the method in one dimension and write the error vector e^n as a Discrete Fourier Transform (cf eqn (2.16))

$$e_j^n = \sum_{p=1}^{N} \rho^n(p)e^{ipx} \tag{3.45}$$

where $x = jh = 2\pi j/N$

The spectrum of frequency components $\rho(p)$ which appear in (3.45) are represented schematically in Figure 17a. The principal alias lies between the frequencies $-N/2$ and $N/2$; the other aliases being periodic repetitions of this shape centred on integer multiples of N. After carrying out a small number of fine grid iterations, the high frequency modes (ie those close to $\pm N/2$) will be much reduced although the low frequency end of the spectrum will be largely unaffected. This is shown schematically in Figure 17b. If we now go to a coarser grid, with half the number of grid points, then the principal alias will contain N/2 points extending between the frequencies $-N/4$ and $N/4$. On the coarser grid the iterative scheme will therefore reduce error modes at frequencies close to $\pm N/4$ since these are now the highest frequencies that can be resolved on the coarser grid. The effect of the coarse grid iterations is shown in Figure 17c. The process is continued down to the coarsest grid by which stage the complete error spectrum will have been covered.

We thus see that on each grid the iterative scheme is used as a routine for reducing or smoothing out the error modes within a particular frequency band. For this reason the iterative scheme is usually referred to as a smoothing routine. The sequence of smoothing operations aimed at covering the entire error spectrum can be compared with the cycle of acceleration parameters used in approximate factorisation methods. The goal is the same but multigrid is more efficient because a coarse grid iteration takes less computational effort than a fine grid iteration. If we define a work unit as the computational effort required to carry out one fine grid run, then a cycle of six acceleration parameters means that approximate factorisation takes six work units to make one sweep through the frequency spectrum. For a two dimensional problem the number of grid points is reduced by a quarter with each multigrid level. Thus if we consider the Jameson strategy [4] of one fine grid iteration followed by two iterations on each coarse grid, it will be seen that a multigrid cycle requires

$$1 + 2\left(\frac{1}{4} + \frac{1}{16} + \frac{1}{64} + \ldots\right) \leqslant 1\tfrac{2}{3} \text{ work units}$$

Allowing some extra for overheads we still find that multigrid requires only two work units to fulfil a task for which approximate factorisation takes at least six.

3.9 The Correction Scheme for Linear Problems

We consider a linear difference equation

$$L^h U^h = F^h \tag{3.46}$$

In line with the notation used in the literature [4,35,36] we have adopted the symbol U for the exact solution of the difference equation and u for the approximate solution after a certain number of iterations. The superscript h refers to a computation at a grid level with this grid increment and we write H = 2h for the grid increment on the next coarser grid. The residual is now

$$R^h = L^h u^h - F^h$$

and we may write

$$U^h = u^h + v^h \tag{3.47}$$

where v^h is the correction or error vector (cf eqn (2.15)). After a suitable number of iterations v^h should be dominated by low frequency components. Thus if we insert (3.47) into eqn (3.46) and rearrange the equation we obtain the following equation for the error vector

$$L^h v^h = F^h - L^h u^h \tag{3.48}$$

The error vector v^h and hence $L^h v^h$ should have only a small high frequency content and so it is reasonable to approximate $L^h v^h$ by the corresponding coarse grid equivalent $L^H v^H$. Likewise we can expect the right hand side of (3.48) to be dominated by low frequency components and hence it should be possible to interpolate this quantity onto the coarse grid to obtain the following coarse grid approximation to eqn (3.48)

$$L^H v^H = I_h^H (F^h - L^h u^h) \tag{3.49}$$

The operator I_h^H, sometimes called a restriction operator, denotes an interpolation from the fine grid h to the coarse grid H. Since H = 2h, the coarse grid is found by taking every other point of the fine grid. For simple linear problems we may therefore use the so-called injection mapping

$$I_h^H : G_j^H = G_{2j}^h, \; j = 0 \ldots N \tag{3.50}$$

It should be stressed that the coarse grid equation (3.49) will only be a good approximation to (3.48) if the high frequency components of v^h are small. If this is not the case then the whole multigrid strategy will fail since the high frequencies on the fine grid will be folded or aliased onto the coarse grid spectrum. By the same token it is clearly necessary to work on the coarser grids with the error vector rather than the solution vector u^H.

After a suitable number of iterations on grid H, the error in the approximate solution to v^H can, in a similar way, be represented on the next coarser grid. This can be continued until the coarsest grid is reached. To come back up the cascade of grids it is necessary to interpolate each coarse grid correction onto the next finer grid. Thus when we have obtained a correction term v^H, the old approximation for u^h can be improved in the following way

$$u^h_{NEW} = u^h_{OLD} + I_H^h v^H \tag{3.51}$$

The operator I_H^h is a true interpolation from coarse to fine grid which is sometimes referred to, rather grotesquely, as a prolongation operator.

The application of the multigrid method can be illustrated by the following one dimensional problem

$$u_{xx} = 0, \; 0 \leq x \leq 1$$

with the boundary condition $u(0) = 0$, $u(1) = 1$. A centred difference approximation leads to the set of difference equations

$$\frac{U_{j+1} - 2U_j + U_{j-1}}{h^2} = 0, \; j = 1 \ldots N - 1 \tag{3.52}$$

with the boundary conditions $U_0 = 0$, $U_N = 1$.

This corresponds to eqn (3.46) and the following point iterative scheme may be used to solve (3.52)

$$u_j^{n+1} = \frac{1}{2} (u_{j-1}^{n+1} + u_{j+1}^n)$$

The amplification factor for this scheme is given by

$$|G|^2 = \frac{1}{5 - 4 \cos \xi}$$

where $\xi = ph$. It follows that for low frequency components $\xi = O(h)$,

$$|G| \sim 1 - h^2$$

as one might expect. For the purposes of a multigrid solution we are only interested in the high frequency range, that is $\xi \epsilon [\pi/2, \pi]$. We find that over this range

$$|G| \leq \frac{1}{\sqrt{5}}$$

so that a respectable smoothing rate is obtained. The restriction operator I_h^H used in this example is the injection mapping (3.50) and the coarse to fine grid interpolation is linear, viz

$$I_H^h : \begin{cases} G_{2j}^h = G_j^H \\ G_{2j+1}^h = \frac{1}{2} (G_j^H + G_{j+1}^H) \end{cases}$$

For this problem we adopt the Jameson sequence of applying one iteration for each grid on the way down followed by a further iteration for all but the finest grid on the way up. Convergence histories (plotted against iterations not work units) are presented in Figure 18 for the case N = 32. It can be seen that the application of just one multigrid correction (ie 2 levels) greatly improves the convergence rate. The three level scheme shows very rapid convergence and no further improvement was obtained with four, five or six levels.

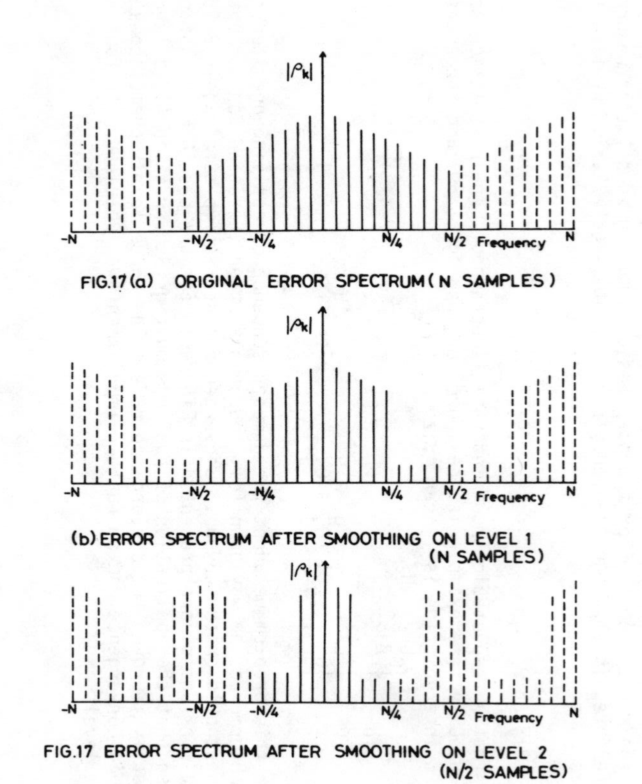

FIG.17 (a) ORIGINAL ERROR SPECTRUM (N SAMPLES)

(b) ERROR SPECTRUM AFTER SMOOTHING ON LEVEL 1 (N SAMPLES)

FIG.17 ERROR SPECTRUM AFTER SMOOTHING ON LEVEL 2 (N/2 SAMPLES)

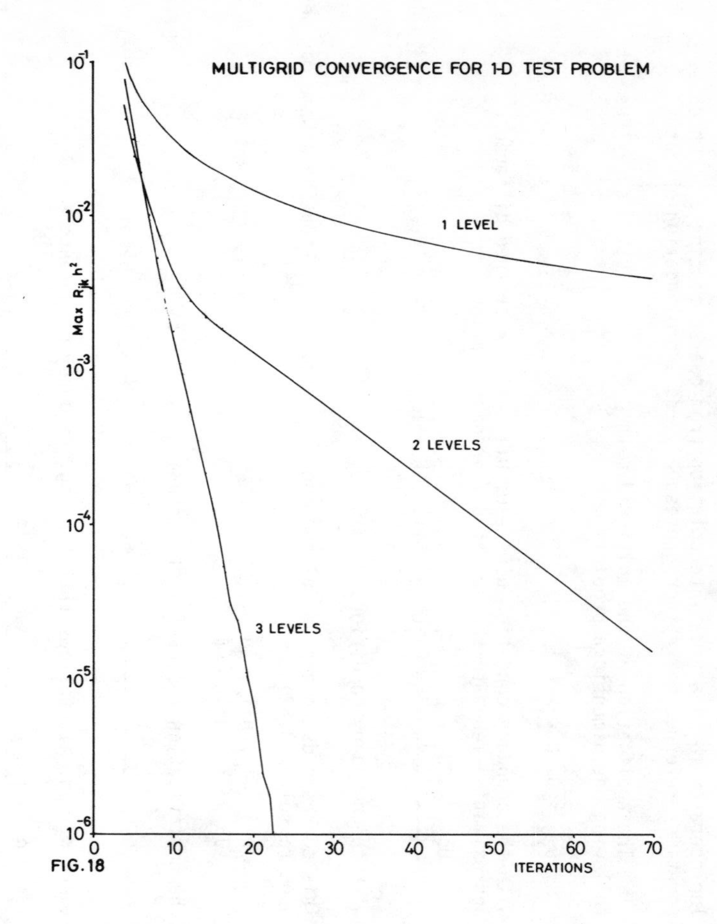

FIG.18

3.10 The Full Approximation Scheme for Nonlinear Problems

In passing from eqn (3.46) to eqn (3.48) we made use of eqn (3.47) and the linear property of L^h viz

$$LU^h = L^h u^h + L^h v^h$$

When L^h is not a linear operator this property is no longer available but it is still necessary to arrange the fine grid equation into a form suitable for approximation on the coarse grid. Thus when L^h is nonlinear we substitute (3.47) into (3.46) and write

$$L^h(u^h + v^h) - L^h u^h = F^h - L^h u^h \tag{3.53}$$

In the linear case this, of course, reduces to eqn (3.48). It is assumed that provided the error vector v^h has a small high frequency content, the same will be true of the two sides of eqn (3.53). Unlike the linear case, this is more an act of faith than a firm conclusion since the nonlinearities in L^h can be expected to introduce some high frequency 'intermodulation products'.

The representation of eqn (3.53) by a coarse grid equivalent then takes the form

$$L^H (I_h^H u^h + v^H) - L^H (I_h^H u^h) = I_h^H (F^h - L^h u^h)$$

which can be arranged as

$$L^H u^H = F^H \tag{3.54}$$

where

$$u^H = I_h^H u^h + v^H$$

and

$$F^H = L^H (I_h^H u^h) + I_h^H (F^h - L^h u^h)$$

The coarse grid equation (3.54) has the same form as (3.46) and hence the same smoothing algorithm can be used to obtain an approximate solution u^H. As with the linear case the process can be continued all the way down to the coarsest grid.

Since the high frequency content of eqn (3.53) is not necessarily small, it is desirable to filter out these components to avoid aliasing errors on the coarse grid representation. Thus the restriction operator I_h^H is chosen to be a weighted average of the adjacent points. In one dimension this could, for example be

$$I_h^H : G_j^H = \frac{1}{4} G_{2j-1}^h + \frac{1}{2} G_{2j}^h + \frac{1}{4} G_{2j+1}^h \tag{3.55}$$

with a similar extension to two and three dimensions. Here one is faced with the same dilemma as the communications engineer; too much filtering will corrupt the required data (ie the low frequency components). On the other hand, too little filtering will also corrupt the data by allowing in unwanted noise. In practice, it appears that a simple weighting, such as (3.55) which involves only the nearest points, is the best compromise.

The variable u^H is a coarse grid approximation to the solution u^h. It follows that the coarse grid approximation to the error vector is

$$u^H_{NEW} - I^H_h u^h_{OLD}$$

and hence for the nonlinear problem eqn (3.51) should be repalced by

$$u^h_{NEW} = u^h_{OLD} + I^h_H (u^H_{NEW} - I^H_h u^h_{OLD})$$

3.11 Jameson's Multigrid Method

For a two-dimensional problem the error spectrum can be represented as the two-dimensional Discrete Fourier Transform (2.16). When we employ a multigrid strategy it is now necessary for the smoothing routine to reduce all frequency components, except for those which belong to the low frequency range in both coordinate directions. If we write (cf eqn (2.19))

$$\xi = p\Delta x \text{ and } \eta = q\Delta y$$

then we can rephrase this requirement by saying that the amplification factor $G(p,q)$ should be acceptably small for all frequency pairs (ξ,η) such that either ξ or η belong to the interval $[\pi/2,\pi]$ (see Figure 19).

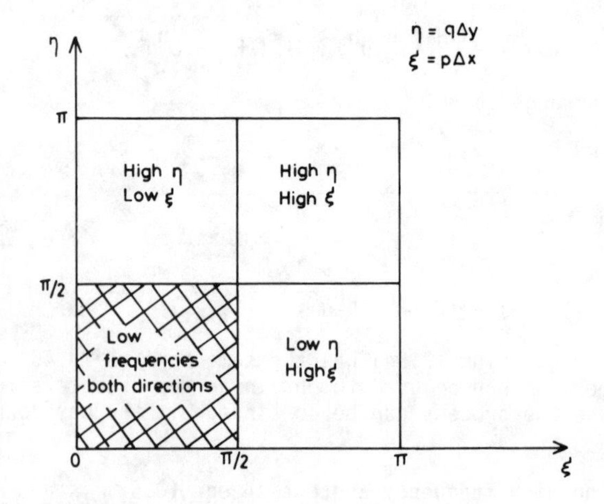

FIG.19 FREQUENCY RANGES IN TWO DIMENSIONS

Consider again the Line Gauss Seidel scheme applied to Laplace's equation. The amplification factor is given by eqn (2.19) with M = 0, viz

$$G(p,q) = \frac{e^{i\xi}}{2 - e^{-i\xi} + 2(\Delta x/\Delta y)^2 (1 - \cos \eta)}$$

and, if $\Delta x = \Delta y = h$, the above smoothing requirement is satisfied. If, however, the grid is nonuniform then this is no longer the case. Consider, for example, a grid such that $\Delta x << \Delta y$. For low frequencies in the x direction (ie $\xi = O(\Delta x)$) we then find that

$$G \sim 1$$

for all frequencies η. In order to overcome this problem, South and Brandt [37] have used multiple line relaxation sweeps in different directions. Arlinger [38] used auxiliary grids in which the grid increment increased in only one coordinate direction. Useful improvements in convergence rate have thus been obtained but it still appears that multigrid methods based on relaxation smoothing routines are not completely satisfactory. To reap the full advantage of the multigrid philosophy, it is necessary to adopt a smoothing routine that reduces high frequency components equally well on all types of grid. This is the essence of Jameson's Multigrid Alternating Direction method [4] which uses an AF1 scheme in place of relaxation. His computed results are astonishingly good and for a typical transonic flow calculation he achieves a residual reduction of three orders of magnitude every 20 work units. This is about $2\frac{1}{2}$ times as fast as the result for AF3 (see section 3.6).

It is perhaps worth noting that the success of a multigrid method is critically dependent on the interpolation operators I_h^H and I_H^h. If either interpolation allows significant errors to creep into the computation then the convergence rate will be seriously impaired. This is particularly important if one is aiming to achieve the very fast convergence rates shown by Jameson, as the author has found when experimenting with an AF3 smoothing routine. Unfortunately, Jameson does not give details in his paper but he has said that the coarse to fine grid interpolation he uses is fourth order accurate [51]. It should also be noted that for transonic flow calculations, Jameson [4] has found it necessary to alter the switching function (cf eqn (2.32)) so that

$$\nu = \max\left\{0,\ 1 - \frac{M_C^2}{M^2}\right\}$$

where the cut off Mach number $M_C < 1$. Thus upwind differencing is introduced when the local Mach number exceeds M_C. This has the effect of smearing the shock which eases the problem of interpolation. Jameson found that the value $M_C^2 = 0.9$ was quite satisfactory. This value for the cut off Mach number appears to have only a small effect on the predicted pressure distribution; the shock is smeared over about four points rather than the three points that is usually the case when the cut off Mach number is $M_C = 1$.

For the nonconservative form of the potential equation the same effect can be achieved by using a modified rotated difference scheme. The second order terms in the potential equation are written (cf eqn (2.23) and (2.43)) in the form

$$(M_C^2 - M^2)\,\phi_{ss} + \phi_{nn}$$

where

$$\phi_{ss} = \frac{u^m u^\ell}{q^2} \quad \text{as before}$$

and

$$\phi_{nn} = g^{m\ell} - M_C^2\,\frac{u^m u^\ell}{q^2} \tag{3.56}$$

For two-dimensional flow the convergence rates that can now be obtained by multigrid are a vast improvement over SLOR. It remains to be seen how well these ideas extend to three dimensions. Writing a three-dimensional multigrid code is a rather daunting task as the grid refinement and the interpolations are much more complicated than in two dimensions. Nevertheless, it is reasonable to assume that a method which clearly works so well in two dimensions should also be very effective for three-dimensional flow calculations. Once this has been achieved we can look forward to the prospect of relatively inexpensive calculations of flows over complete aircraft.

4 GRID GENERATION

The first and perhaps best known transonic code for solving the full potential equation is the aerofoil analysis program of Garabedian, Korn and Bauer [39]. They use Sells' circle plane mapping [52] to transform the space exterior to the aerofoil onto the interior of a unit circle. The introduction of this particular conformal mapping technique extends back a long way [53,54] but its use for generating suitable grids for the numerical solution of a partial differential equation is a recent application.

By taking a rectangular grid in the computing space $\{0 \leq r \leq 1, 0 \leq \theta \leq 2\pi\}$ one obtains a nonuniform distribution of grid points in the space exterior to the aerofoil. This grid has several desirable features. First the complete flowfield is mapped into a finite space. Note that infinity transforms to the centre of the unit circle with the obvious advantage that the farfield boundary condition can be applied easily. Second, the density of grid points varies smoothly with a sparse distribution at large distances from the aerofoil where flow accelerations can be expected to be small. Near the aerofoil surface points are closely spaced with a particularly dense packing at the leading and trailing edges where the flow accelerations are greatest. Thus the distribution of grid points is well suited to the flow variations that are anticipated. Third, since the transformation is conformal it is of necessity orthogonal and the transformed equation therefore has a simpler form with fewer terms than it would have under most nonorthogonal transformations.

Write $z = x + iy$ for the complex variable in physical space and $\sigma = \frac{1}{r} e^{-i\theta}$ for the corresponding point in the space exterior to the unit circle. The mapping derivative $dz/d\sigma$ for a smooth profile can be represented as a Laurent series whose coefficients have to be determined. In order to remove the sharp corner at the trailing edge of an aerofoil, one can either follow Ives [55] and use a Karman-Trefftz transformation to map the aerofoil into a near circle, or one can follow Jameson [17,23] and directly introduce a Schwarz-Christoffel term to give the following form

$$\frac{dz}{d\sigma} = \left(1 - \frac{1}{\sigma}\right)^{1-\varepsilon/\pi} \exp\left\{\sum_{n=0}^{N} \frac{c_n}{\sigma^n}\right\} \tag{4.1}$$

where ε is the included angle at the trailing edge. The c_n are complex coefficients which are determined iteratively. Use of Fast Fourier Transform techniques [55,56] enables this part of the operation to be carried out very efficiently.

The same idea can be applied to other two-dimensional geometries. In other words, one selects auxiliary transformations to map the original profile into a near circle and this is followed by a final mapping like that outlined above. This requires a rather special expertise and a deft handling of various complex variable transformations. The reader is referred to the papers by Arlinger [57] for a nacelle and Ives [55] (see also Arlinger [85]) for multi-element systems to see just how powerful the technique can be when used in the right hands. According to the Riemann mapping theorem it should be possible to find a conformal mapping for any reasonable two-dimensional geometry. In practice, it is not obvious how this can be achieved although the technique of Davis [58] which is based on a generalised Schwarz-Christoffel transformation, looks like a useful method for numerically generating conformal mappings for arbitrary shapes. An excellent review of the conformal mapping method has been presented by Moretti [59].

For mathematical elegance conformal mapping is unsurpassed as a grid generation technique. For algebraic simplicity, however, it would be hard to beat the shearing transformations. Over the region exterior to a profile this mapping from the (x,y) plane to the transformed (ξ,η) plane takes the form

$$x = \xi \qquad -\infty < \xi < \infty$$
$$y = Y(\xi) + \eta \qquad 0 \leq \eta < \infty \tag{4.2}$$

The function $Y(\xi)$ is the equation of the profile shape and the effect of the mapping is to produce an aligned grid such that $\eta = 0$ corresponds to the profile or body surface (see Figure 20).

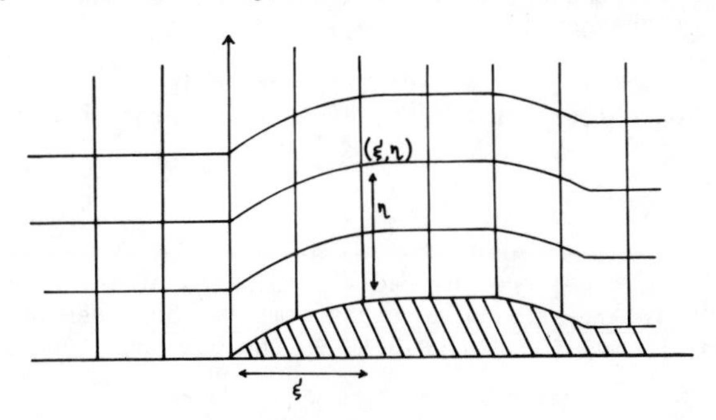

FIG. 20 SHEARED COORDINATES

This can be very useful for cases where the slope $Y'(\xi)$ does not become too large. At the nose of an aerofoil or blunt body the slope becomes infinite. To deal with this situation South and Jameson [60] propose a combination of shearing for the rear end and the following body normal transformation over the forebody

$$x = X(\xi) - \eta \sin \theta$$
$$y = Y(\xi) + \eta \cos \theta \tag{4.3}$$

The computational variable ξ is now the arc distance along the body surface and $\theta(\xi)$ is the angle between the tangent to the body surface, at the point (X,Y) and the x axis (see Figure 21).

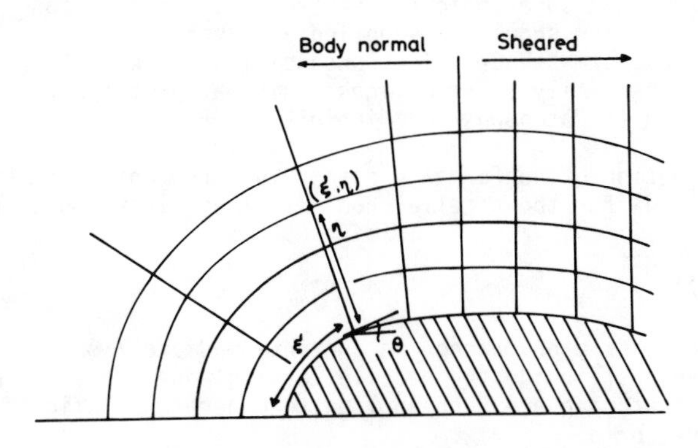

FIG. 21 BODY NORMAL + SHEARED COORDINATES FOR A BLUNT BODY

We note that the transformed coordinate system is still of infinite extent and it is usually desirable to map this into a finite domain. The following stretching transformation [60]

$$\zeta = \frac{\eta}{1+\eta}, \quad \begin{array}{l} 0 \leq \zeta < 1 \\ 0 \leq \eta < \infty \end{array} \tag{4.4}$$

is one possibility that produces a dense distribution of points near the body surface. Clearly, numerous stretchings are available. The important requirement is that the stretching function should be monotonic (ie the derivative should not change sign).

For an internal flow where the radial direction is of finite extent the following modification [61,62] to the shearing transformation can be used

$$\begin{aligned} x &= \xi & -\infty < \xi < \infty \\ y &= (Y_U(\xi) - Y_L(\xi))\eta + Y_L(\xi), & 0 \leq \eta \leq 1 \end{aligned} \tag{4.5}$$

The functions $Y_U(\xi)$ and $Y_L(\xi)$ are the radial ordinates of the upper and lower duct surfaces respectively. A similar transformation has been proposed by Jones and South [63] to map the region between the bow shock and a blunt forebody.

After considering conformal mappings and shearings separately we now turn to the combination of a simple conformal mapping and a shearing transformation. This idea can be very fruitful and has been successfully exploited by both Jameson [17,64] and Caughey [65]. For example, Jameson has treated the aerofoil problem by using the mapping

$$Z = z^{\frac{1}{2}} \tag{4.6}$$

where $Z = X + iY$ is the point in intermediate space corresponding to z in physical space. The origin is a singular point taken just inside the leading edge and the above transformation unwraps the aerofoil to a bump just above the X axis with the flow field mapped onto the upper half plane. A shearing transformation is then employed to generate an aligned coordinate system which is mildly nonorthogonal [17]. The advantage of this transformation for the aerofoil problem is that the coordinate system is better suited to the calculation of flows with a supersonic freestream. For in this case the potential and its derivative should be supplied as Cauchy data at upstream infinity with the downstream boundary left unspecified. Under the parabolic transformation (4.6) one family of grid lines finish at upstream infinity and the other set start and end at downstream infinity.

A similar combination of conformal mapping plus stretching is used by Jameson and Caughey [64] for the nacelle problem. They first employ the conformal mapping

$$z = Z - \exp(-Z) \tag{4.7}$$

to take z space into a strip or channel in the intermediate Z space. The z origin is again taken just inside the nacelle leading edge. A shearing transformation and stretching are then applied to generate an aligned and slightly nonorthogonal grid.

By a judicious choice of conformal mappings, shearings and stretchings it is possible to produce grids for three dimensional configurations such as wings and wing/fuselage combinations [17,66,67,68]. However, a wing/fuselage combination appears to be at the limit where ingenuity alone can derive suitable coordinate transformations. If one goes a stage further in geometric complexity to say, a wing/fuselage/nacelle/pylon combination then one reaches a configuration that appears to be beyond the scope of purely analytic transformations. As we shall argue later this example represents a kind of watershed for flowfield methods in that the problem requires a different approach in both the organisation of the computer code and our understanding of grid effects. It is also an example of considerable aerodynamic interest and importance. The present lack of a satisfactory method for solving the full potential equation for geometries of this complexity is both frustrating and tantalising.

In our preceding discussion of coordinate transformations we have tacitly assumed that the grid is aligned. In other words, any solid boundary or line of symmetry coincides with a coordinate line. If we relax this requirement the problem of grid generation is eased. A nonaligned grid was introduced by Carlson [69] for solving the flow about an aerofoil and later used by Reyhner [70,71] for the nacelle problem. Of course it is still necessary to introduce stretching transformations to pack grid points in areas where large flow accelerations are expected. Nevertheless, the problem of grid generation is less difficult and on a scale comparable to that of finding a grid for solving the TSP equation. This simplification is only achieved at a price, namely the greatly increased difficulty of implementing the flow tangency condition on the body surface. The reader is referred to the literature [69, 70,71] for details of the interpolations that are required and the different possibilities that arise for various positions of the body surface with respect to the grid points. Recently, Catherall [84] has extended this technique to solve the flow over a wing. This is a useful achievement but the further extension to arbitrary three-dimensional shapes involves formidable difficulties in the application of the boundary conditions. There is also the question of the accuracy with which the boundary conditions are approximated on a nonaligned grid. On an aligned grid the application of the flow tangency condition is relatively straightforward. For in this case the contravariant component of velocity which corresponds to a flow velocity normal to the body surface must be zero. In a nonaligned system the points do not usually lie on the body surface and the application of the flow tangency condition often leads to rather complicated difference formulae involving several of the surrounding points. This must surely degrade the accuracy of the computed solution although the results shown by Carlson [69,25] and Reyhner [70,71] do look surprisingly good. Nevertheless, the uncertainty over the accuracy of the computed results as well as the difficulty of implementing the boundary conditions suggests that this particular approach to grid generation is not a serious contender for complicated three-dimensional shapes.

Another strategy for producing grids about arbitrary shapes while retaining alignment with solid boundaries is the numerical generation technique. The best known example of this is Thompson's method [6] which solves a set of elliptic equations to obtain the grid coordinates. In its simplest form this amounts to a numerical determination of the lines of equipotential and equistreamfunction for incompressible flow, an idea that has been suggested by Colehour [72] for ducts and bodies. The importance of Thompson's work lies in the fact that he has turned this idea into a general method for generating grids about arbitrary shapes in both two and three dimensions. Once again, the generality of grid generation is only gained at a price. In this case the penalty is the loss of control over the position and distribution of grid points. By introducing source-like terms into the generation equations, Thompson et al [73,74,75] are able to regain some influence over the positioning of grid points to give, for example, a high point density where large flow accelerations are expected. The work of Thompson has stimulated the development of a number of variants. For example, Steger and Sorenson [76] solve a set of hyperbolic rather than elliptic equations. This approach is useful if the outer boundary is unconstrained which is often the case for external flow problems.

In two dimensions it appears that the numerical approach leads to reasonable grids for almost any configuration. The work of Yu [77] also shows that Thompson's method can produce a respectable grid for a wing plus arbitrary fuselage. However, it is unlikely that the methods of grid control are adequate to generate well behaved grids for the wing/fuselage/nacelle/pylon combination. To cope with a configuration of this complexity, Lee et al [9] and Forsey [43] have adopted the strategy of separating the flowfield into blocks and applying a Thompson mapping for each block. This concept has the advantage of breaking up a complex geometrical problem into a number of simpler units. However, one now has the burden of deciding how the flowfield should be split into the separate blocks. There is also the question of the treatment at the block interfaces. It is reasonable to expect that the distribution of grid points, in effect the coordinate stretching, can change abruptly on passing across such an interface. This can lead to erroneous solutions and ways of dealing with this problem will be required in order to obtain accurate flow predictions. Moreover, a multi-block grid can generate some oddities; for example, a grid node where five rather than four grid lines meet. This type of situation requires careful attention and special difference formulae will probably be required at these rogue points. Lee et al [9] have clearly recognised these problems and appear to have made some headway towards flowfield solutions for complex geometries. It is also worth noting that since the multi-block concept treats different parts of the flowfield as separate entities, it is desirable that the same property should be reflected in the discretised flow equation. This leads naturally to the finite volume method [78,68] in which the transformation derivatives used in the discretised equation at any position depend only on the coordinate mapping for a localised set of points. This removes the requirement for a single global coordinate transformation although some degree of smoothness is still needed. We shall not discuss this further except to say that separation of the grid space into blocks followed by a Thompson mapping in each block and a potential flow solution by the finite volume method, appears to be the most promising combination of techniques for a complex three-dimensinal geometry. It is significant that this is the approach advocated by Lee et al [9].

Underlying any investigation into grid generation is the question of what constitutes a good grid. It would be very useful if we could decide which of two possible grids is better suited to a given geometry. It would be even better if we had the capability to determine what is an optimum grid in some particular sense. Unfortunately, this seems to be an area in which there are many questions but no answers. For example, one feels intuitively that an orthogonal grid is well conditioned and that a highly skewed grid is not. Is it not possible to estimate a priori the corresponding deterioration in the numerical solution for a particular test flow? In other words, can we assign a condition number that would measure the degradation caused by grid skewing? Can we estimate a priori the effect of grid point density on the accuracy of a test flow calculation or the relation, if any, between the solution and grid type? When grid stretching is introduced the first and second derivatives of the stretching function appear in the transformed potential equation. It therefore seems sensible to maintain continuity of the first and second derivatives of the stretching function. What happens when this is not possible? To what extent can we modify the difference formulae that approximate the potential equation so that the computed solution is relatively insensitive to sudden changes in the point distribution?

Answers to these and many other questions are at present unavailable or, at best, only known in a very rudimentary form. What we urgently require is a mathematical theory of grid effects, a theory that would replace broad generalisations by precise statements. At the very least we ought to have a collection of estimates and criteria for measuring the adequacy of different grids. A theory of this kind would certainly help to unravel some of the confusion that exists over the different computed results that are available for two-dimensional geometries. It will be an essential asset if we are not to get hopelessly lost when computations for the more complex three-dimensional geometries are within our compass.

5 APPLICATIONS

In the previous chapters we have described some of the theory on which the codes for calculating transonic potential flow are based. Now we shall present some results obtained from four computer programs which are typical of the many codes available for aerodynamic design purposes. All four codes described here solve the potential equation in nonconservative form and wherever possible we have compared theory and experiment to give an indication of the accuracy that can be attained with a potential flow prediction.

5.1 Axisymmetric Solid Bodies

This computer code [79] calculates potential flow around axisymmetric solid bodies and is based on the method of South and Jameson [60]. It follows that the computing grid is generated by a combination of body normal and sheared cylindrical coordinates for blunt bodies. For pointed bodies, however, a sheared cylindrical system is used throughout the flowfield.

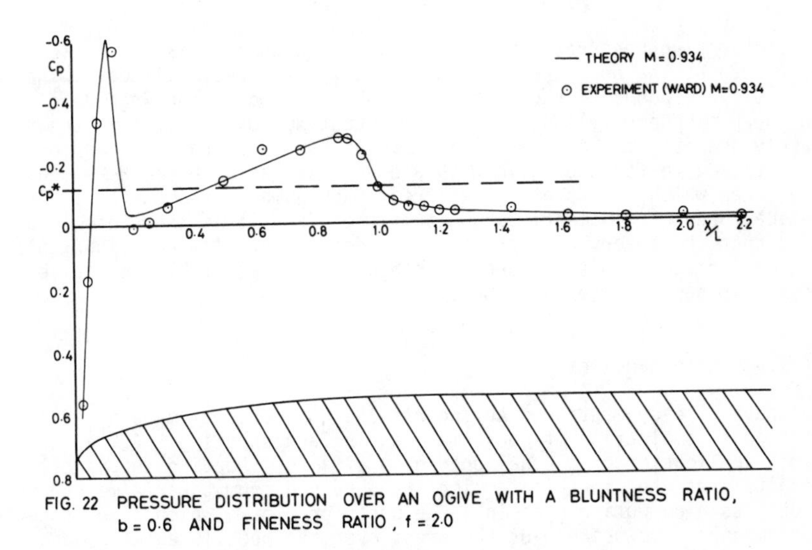

FIG. 22 PRESSURE DISTRIBUTION OVER AN OGIVE WITH A BLUNTNESS RATIO, b = 0.6 AND FINENESS RATIO, f = 2.0

A typical comparison of predicted and experimental pressure distributions is shown in Figure 22 for a spherically blunted ogive. In this example no account has been taken of viscous effects and in particular no boundary layer has been included. The boundary layer over an axisymmetric forebody at zero incidence can be expected to remain thin. It follows that the inviscid flowfield should, in such cases, provide an adequate model of the real flow. This conclusion is supported by the good agreement between the theoretical and experimental pressure distributions. One might therefore expect that an integration of the predicted pressures could provide a reasonable estimate of the wave drag. A comparison between theoretical and experimental wave drag is presented in Figure 23 for three ogives of fineness ratio two and different bluntness ratios.

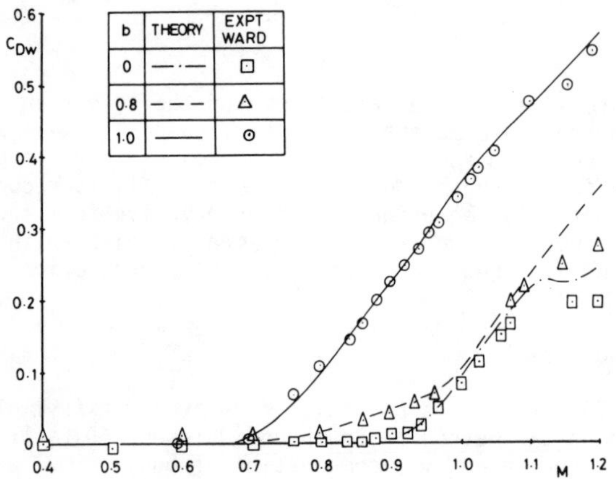

FIG. 23 COMPARISON BETWEEN THEORETICAL AND EXPERIMENTAL WAVE DRAG FOR OGIVE CYLINDERS OF FINENESS RATIO f = 2.0

Note that the bluntness ratio b is defined as r/R where r is the radius of the spherical cap at the nose and R is the radius of the rear cylinder. The special case b = 0 corresponds to the pointed or unblunted ogive and when b = 1 we obtain the hemisphere cylinder. It is interesting to observe that the theory accurately predicts the changes that occur as bluntness ratio is varied. For example, the curve for the ogive with a bluntness ratio, b = 0.8, has a region of drag creep which starts at a freestream Mach number of about 0.7. The theoretical prediction clearly shows this trend and it is therefore to be expected that the theory also predicts the details of flow behaviour that are causing drag creep. This feature and other properties of blunted ogives are discussed further in Reference [80].

5.2 Axisymmetric Nacelles

We have already mentioned Arlinger's use of conformal mapping [57], Caughey and Jameson's combination of conformal mapping and shearing [64], as well as Reyhner's application of a nonaligned grid [70] for treating the nacelle problem. The results presented in this section are obtained from Baker's method [81] which defines a computing grid in three parts or blocks which blend together in a smooth manner. Downstream of the crest over the nacelle exterior and also downstream of the throat inside the nacelle, a sheared coordinate system is used. Over the region between the throat and crest a further simple coordinate system is set up. The resulting grid covers only a finite part of the flowfield and this particular approach to the problem of forming a computing grid is rather ungainly. Nevertheless, the potential method which employs this grid has proved very effective. A later modification, to treat nacelles at small angles of incidence, appears to give good external flow predictions up to about ±10 degrees of incidence.

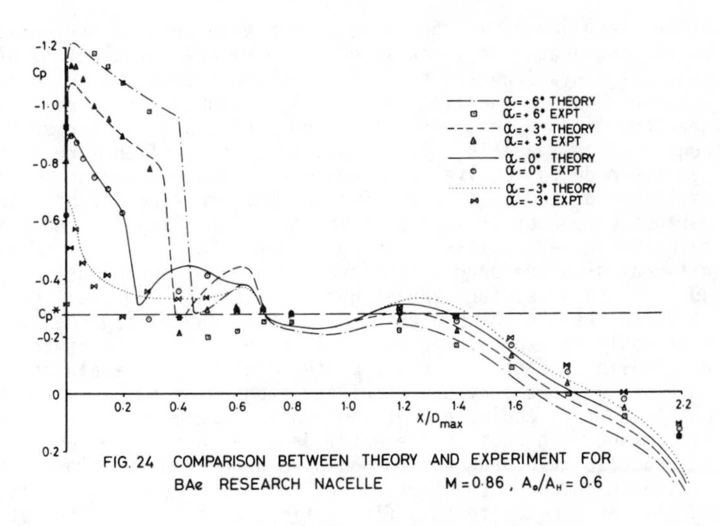

FIG. 24 COMPARISON BETWEEN THEORY AND EXPERIMENT FOR BAe RESEARCH NACELLE $M = 0.86$, $A_o/A_H = 0.6$

A comparison of theoretical and experimental pressure distributions over the exterior of a nacelle, designed with the aid of the nacelle code, is presented in Figure 24. The agreement is generally good apart from an incorrect prediction of the pressures near the trailing edge. This can be explained by the lack of an adequate wake model since the program assumes that the wake extends horizontally downstream of the trailing edge. Figure 25 compares theory and experiment for another nacelle at a static condition (ie zero onset Mach number). The mean throat Mach number for this case is 0.54 and on the nacelle surface it can be seen that a small region of supercritical flow occurs just inside the lip. Again theory and experiment compare well.

Since the method provides a good prediction of the inviscid flowfield throughout the subsonic and transonic regime, it again seems likely that the integrated pressures should give a reasonable estimate of wave drag.

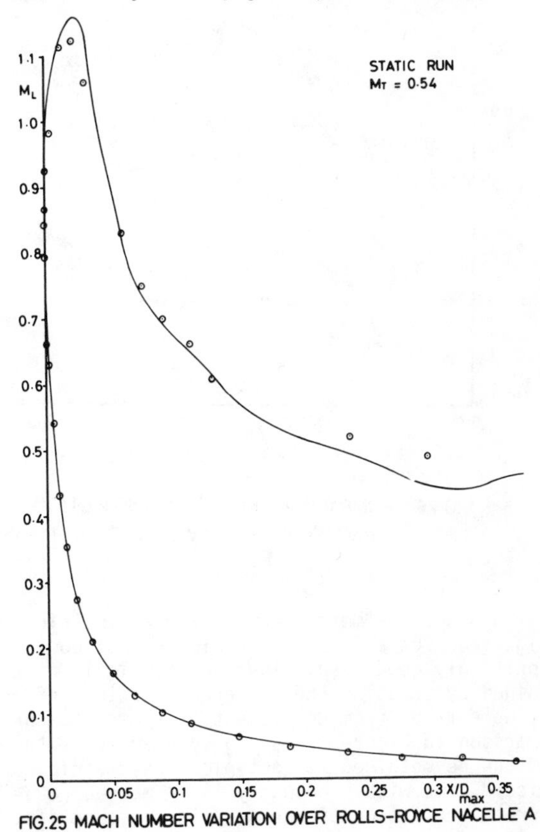

FIG.25 MACH NUMBER VARIATION OVER ROLLS-ROYCE NACELLE A

At the very least, it should be possible to predict the onset of drag rise due to the appearance of shock waves. Theoretical predictions of nacelle drag are generally less accurate than those obtained from the solid body code for two reasons. First, nacelles are usually designed with a relatively high curvature near the lip and this often leads to a high but slender suction peak over this region of the nacelle. It follows that it is often more difficult for a nacelle code to resolve the flow details and produce an accurate integration of the predicted pressures. Second, the breakdown of nacelle drag into its various component is more complicated. The inviscid prediction method described here provides two drag values. The first value represents the external pressure drag which is calculated by an integration of the external surface pressures from the stagnation point to the trailing edge. The second value represents the pre-entry drag which can be regarded as the drag force that would be experienced by the incoming streamtube if this were considered as a solid surface. In the computer code this is evaluated by integrating the internal surface pressures from the stagnation point to the furthest downstream point and then adding a contribution for the momentum loss of the incoming streamtube. The sum of pre-entry drag and external pressure drag represents the nacelle wave drag which should be zero when the flow is entirely subsonic. Figure 26 compares the theoretical and experimental wave drag for a third nacelle at a moderate mass flow ratio (ie at a mass flow ratio for which the flow can be expected to remain attached). The theoretical curve agrees well with experiment and it is apparent that the nacelle code can provide a good indication of the onset of drag rise.

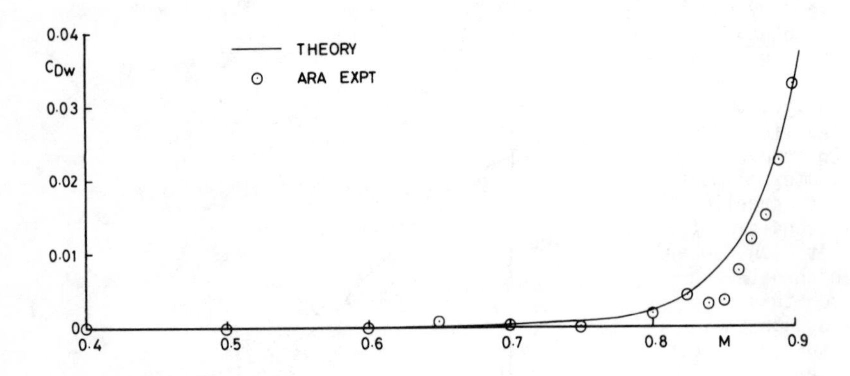

FIG.26 COMPARISON BETWEEN THEORETICAL AND EXPERIMENTAL WAVE DRAG FOR ARA COWL FOUR AT MASS FLOW RATIO $A_0/A_H = 0.6$

In practice, most nacelles are asymmetric and hence the problem of nacelle design requires a fully three-dimensional code. It appears, however, that surprisingly good predictions, particularly of the external flow, can be obtained by treating the different sections of an asymmetric nacelle as though they were taken from equivalent axisymmetric shapes, In particular, the comparison in Figure 27 is a good example of the kind of design information that can be obtained for a typical asymmetric nacelle. By coupling the inviscid code with a boundary layer method [82] it is possible to gain a more accurate prediction of the nacelle pressure distribution. Furthermore, it is

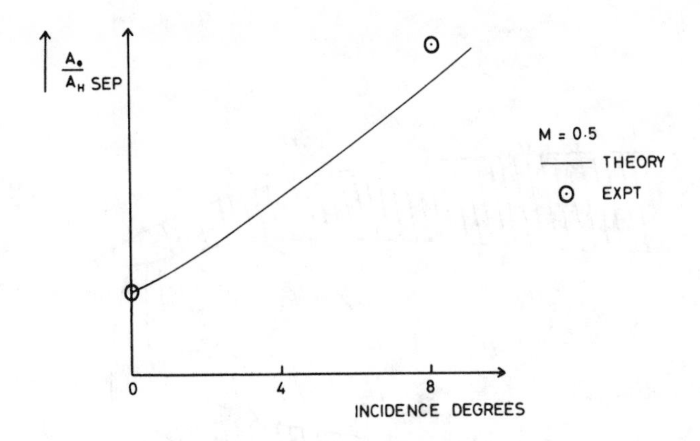

FIG. 27 SEPARATION LOCUS FOR TOP SECTION OF ROLLS-ROYCE NACELLE B

then possible to predict when separations occur, either by increasing the incidence, or keeping the incidence fixed, by decreasing the mass flow ratio. In this manner one can define a separation locus for a given onset Mach number. The separation locus presented in Figure 27 was obtained for the top section of a modern asymmetric nacelle designed using the nacelle code. Only two experimental points are available for this example but it can be seen that both lie remarkably close to the predicted separation line.

5.3 Nozzles and Ducts

We now turn to another aspect of an engine installation, namely design of the fan nozzle. A computer code developed for this purpose by Baker [62] has been extended in a number of ways by Dr E Kitchen of Rolls Royce. He has introduced an improved integration routine to estimate the discharge coefficient and thrust coefficient and has recently coupled the inviscid code with a boundary layer method. It appears that it is now possible to estimate the discharge coefficient of a typical turbofan nozzle to an accuracy of about three decimal places. In one example similar to the geometry shown in Figures 28 and 29, a value C_D = 0.9945 was obtained [82] for the inviscid discharge coefficient. After running the program in viscous mode the reduced value C_D = 0.9890 was predicted. This indicates that a loss of ΔC_D = 0.0055 is caused by flow non-uniformities with a further loss of similar magnitude arising from viscous effects. The corresponding experimental value C_D = 0.9885 compared very well with this theoretical estimate.

A further addition to the nozzle code by Kitchen [82] is the inclusion of a rotational flow component along the lines suggested by Brown et al [61]. This allows one to consider the effect of specifying a non-uniform stagnation pressure at the nozzle entrance. The difference between uniform and non-uniform entrance conditions is clearly demonstrated by the Mach number contours presented in Figures 28 and 29 for a typical nozzle geometry.

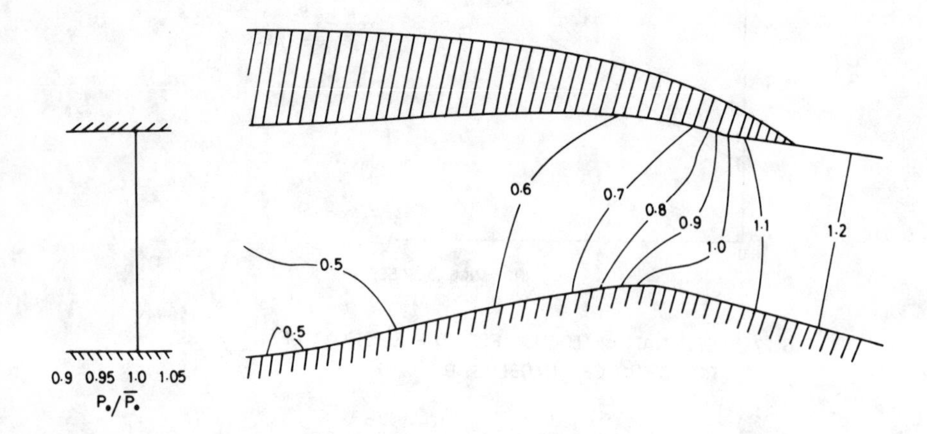

FIG. 28 MACH NUMBER CONTOURS FOR ROLLS-ROYCE FAN NOZZLE WITH UNIFORM ENTRY CONDITIONS

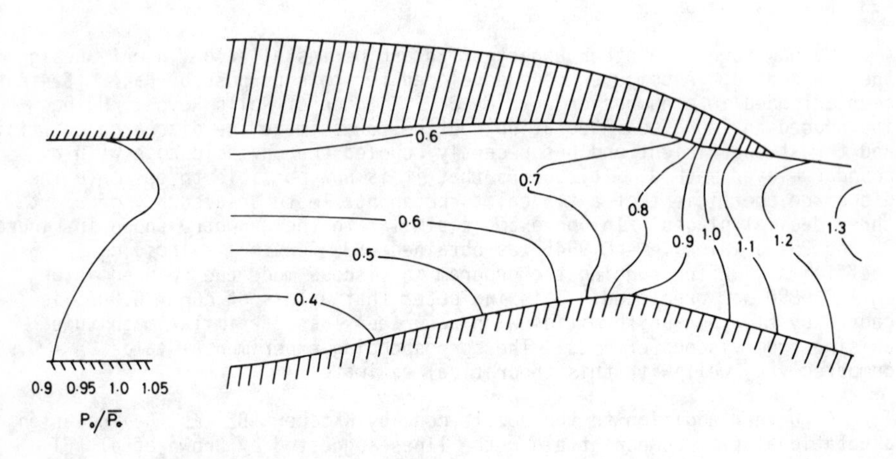

FIG. 29 MACH NUMBER CONTOURS FOR ROLLS-ROYCE FAN NOZZLE WITH NON UNIFORM ENTRY CONDITIONS

5.4 Wing/Body Combinations

Our final example of a transonic potential flow routine is the wing/body method of Forsey [66]. The computer code consists of two separate programs. A grid generation code produces the transformation derivatives and metric coefficients for a given wing or wing/body configuration. These geometric terms are then written onto a disc file for later access by the flow program. Separation of the coordinate generation and flow solution routines provides considerable flexibility since the solution algorithm has been written for a general nonorthogonal grid. It is therefore possible in principle to compute the flow on an entirely different grid without changing the flow program.

The coordinate transformation that has been used is made up of a sequence of mappings which take the flowfield into a finite computational space whilst ensuring an efficient distribution of grid points. The body is first mapped into a slit and this is followed by a spanwise transformation which packs grid planes near the wing root and tip and also brings infinity outboard of the wing tip to a finite position. The aerofoil section corresponding to each computational plane is then mapped conformally onto the interior of a unit circle. The resulting computational space consists of a circular cylinder of unit radius and unit length. A cylindrical polar coordinate system is set up in this cylinder and equally spaced intervals are taken in the three coordinate directions. Certain further modifications to the computing grid are used to deal with wings which have significant amounts of sweep or taper.

Although a fully viscous version of this full potential code is not yet available, viscous effects can be modelled in a crude way. The usual practice is to add a displacement surface, obtained from a two-dimensional boundary layer calculation, to each section of the wing geometry. A full potential calculation is then carried out using the modified geometry. To illustrate the application of this technique we first present some results for a wing whose planform is shown in Figure 30. Predicted and experimental results are shown in Figure 31 and it can be seen that the agreement is generally good. There is some discrepancy at the foot of the shock and near the trailing edge of the wing. However, it is quite possible that an improved model of the boundary layer and wake would rectify this. It is gratifying that the important features of the flow are well predicted, for example, the suction levels and strong shocks inboard of the crank, as well as the double suction peaks over much of the outer wing.

In Figure 33 we compare theory and experiment at one section of the simple panel wing shown in Figure 32. In this example the same case has been computed with and without the fuselage. The dashed line refers to the wing alone calculation and the solid line represents the wing/body result. Apparently the body effect can be quite significant even though the section is at a reasonable distance from the wing/body junction.

This example illustrates very well the flexibility of theoretical methods when compared with wind tunnel testing. No only is it far quicker and less expensive to run a computer code but it is very easy to examine the separate influences of viscosity and body interference. When computer codes for more complex three-dimensional shapes become available this facility will greatly enhance our capability to optimise an aerodynamic design.

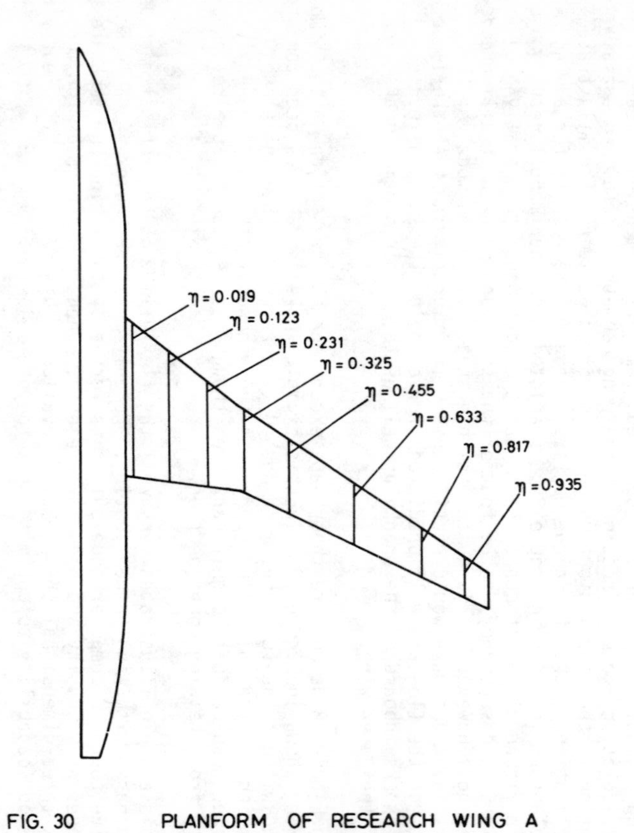

FIG. 30 PLANFORM OF RESEARCH WING A

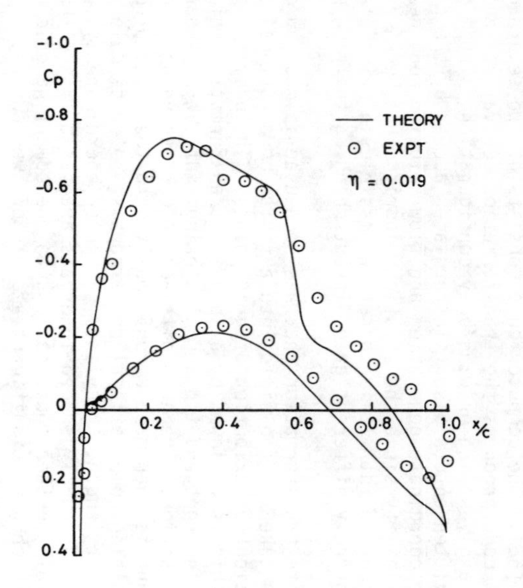

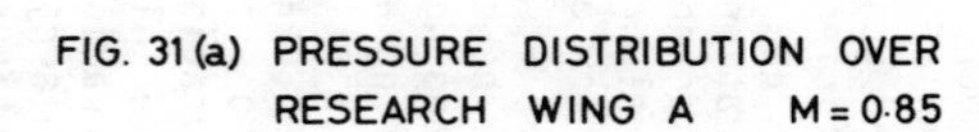

FIG. 31 (a) PRESSURE DISTRIBUTION OVER RESEARCH WING A M = 0·85

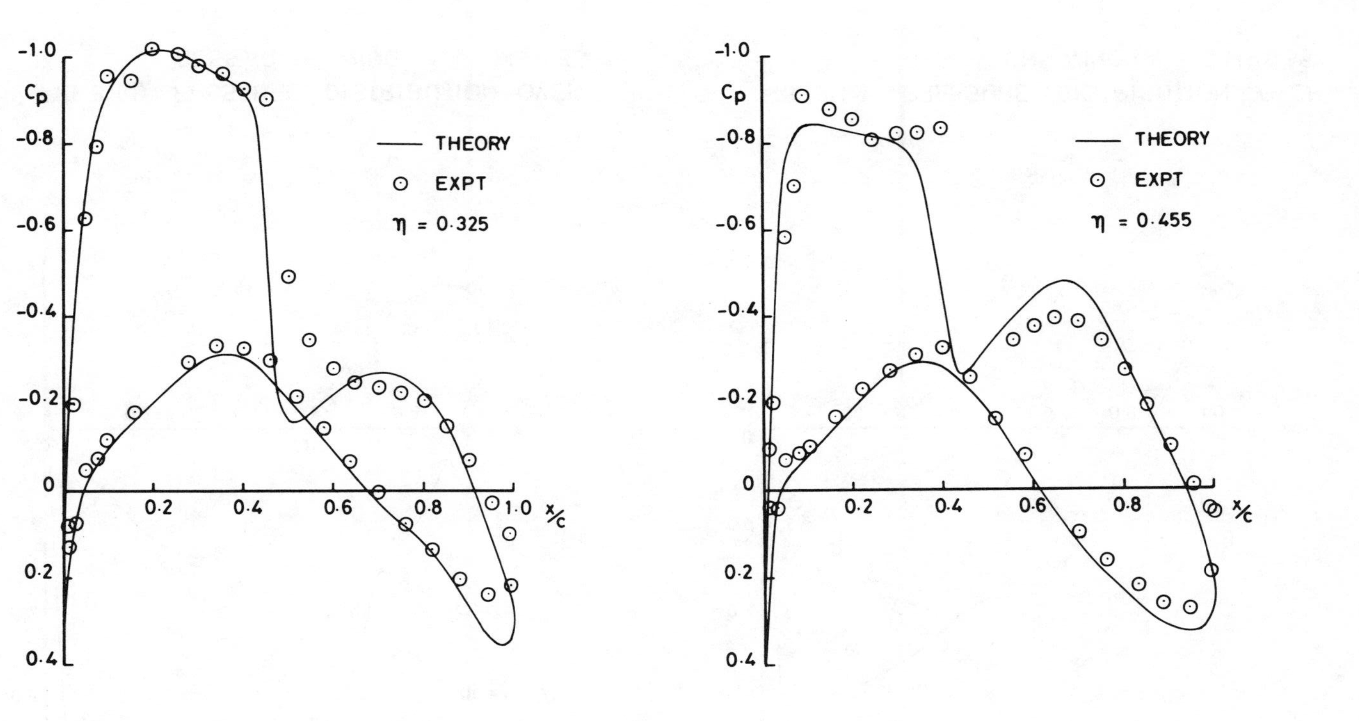

FIG. 31(b) PRESSURE DISTRIBUTION OVER RESEARCH WING A $M = 0.85$

FIG. 31(c) PRESSURE DISTRIBUTION OVER RESEARCH WING A $M = 0.85$

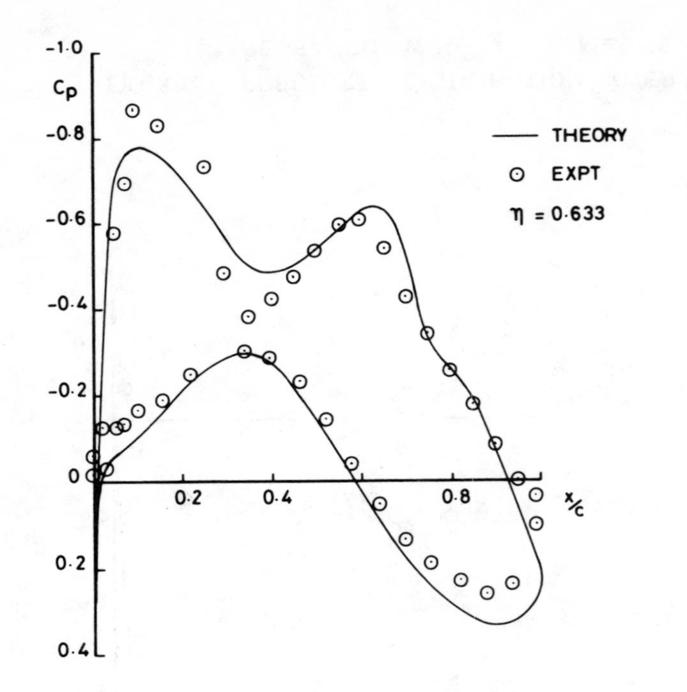

FIG. 31(d) PRESSURE DISTRIBUTION OVER RESEARCH WING A M = 0.85

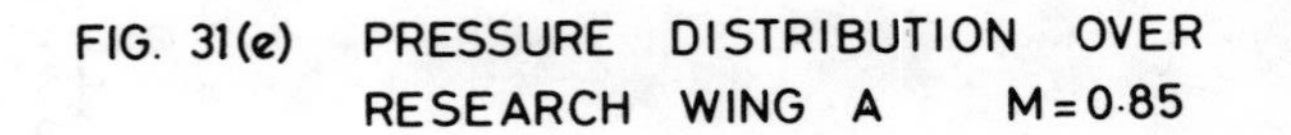

FIG. 31(e) PRESSURE DISTRIBUTION OVER RESEARCH WING A M = 0.85

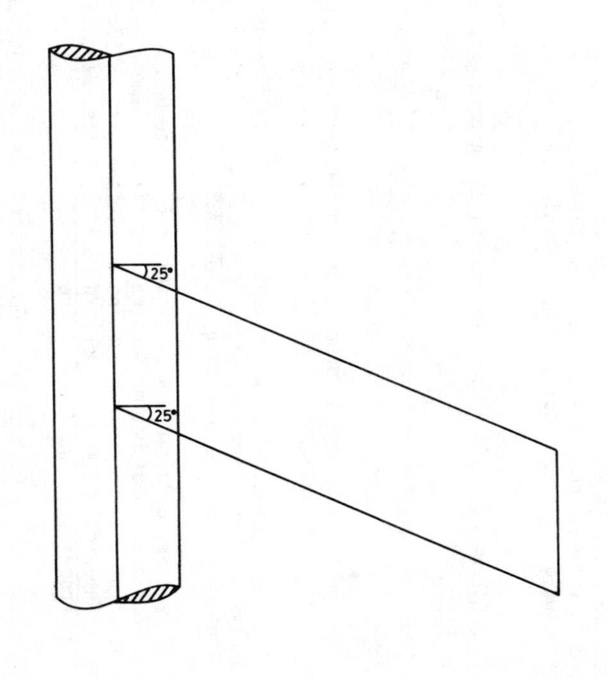

FIG. 32 PLANFORM OF RESEARCH WING B

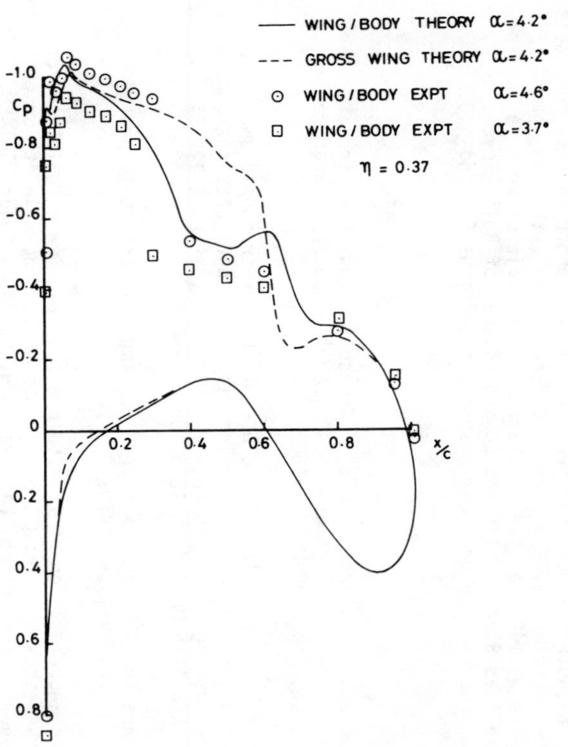

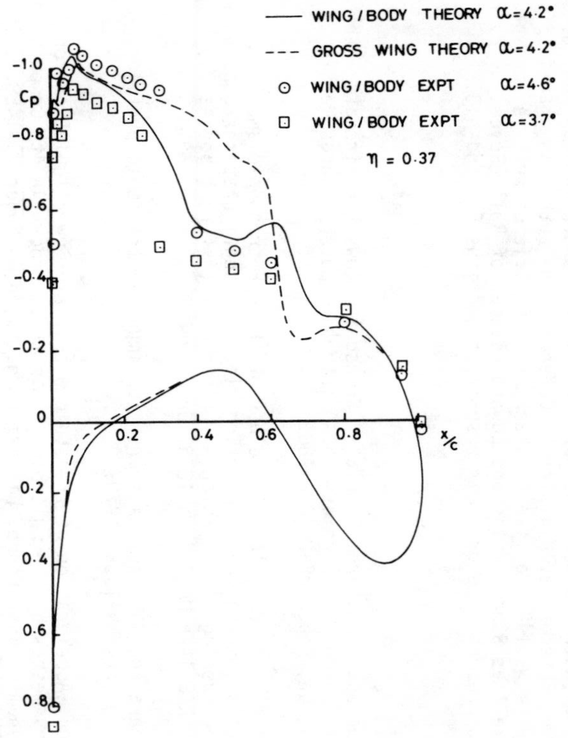

FIG. 33 PRESSURE DISTRIBUTION OVER RESEARCH WING B M = 0·86

ACKNOWLEDGEMENTS

Much of the author's work has been carried out under various contracts from Procurement Executive, Ministry of Defence. The author is grateful for the support. He would also like to thank Frank Ogilvie of British Aerospace, Hatfield as well as Sandra Hiles and Edward Kitchen of Rolls Royce, Derby for permission to present some of their results in sections 5.2 and 5.3.

REFERENCES

1. C W Boppe and M A Stern, Simulated transonic flows for aircraft with nacelles, pylons and winglets, AIAA 18th Aerospace Sciences Meeting, Pasadena, Ca, Paper 80-0130, 1980.

2. C M Albone, Some examples of the use of the RAE transonic small perturbation method for calculating flow past configurations consisting of a wing-fuselage, upon which a variety of interfering bodies may be mounted, RAE Tech Memo Aero 1845, June 1980.

3. H Yoshihara and R Magnus, A search for improved transonic profiles, Convair Report ERR-1536, 1970.

4. A Jameson, Acceleration of transonic potential flow calculations on arbitrary meshes by the multiple grid method, Proc AIAA 4th Computational Fluid Dynamics Conference, Williamsburg, Va, AIAA Paper 79-1458, July 1979.

5. Proc Workshop on Numerical Grid Generation Techniques for Partial Differential Equations, NASA Langley, October 1980.

6. J F Thompson, F C Thomas and C W Martin, Automatic numerical generation of body fitted curvilinear coordinate system for field containing any number of arbitrary two-dimensional bodies, J Comp Physics, Vol 15, pp 299-319, 1974.

7. C R Forsey, M G Edwards and M P Carr, An investigation into grid patching techniques, Proc NASA Langley Workshop, October 1980.

8. E H Atta, Component-adaptive grid embedding, Proc NASA Langley Workshop, October 1980.

9. K D Lee, M Huang, N J Yu and P E Rubbert, Grid generation for general three-dimensional configurations, Proc NASA Langley Workshop, October 1980.

10. A Eberle, Transonic potential flow computations by finite elements: airfoil and wing analysis, airfoil optimisation, Proc DGLR/GARTEur 6th Symposium, Tranosnic Configurations, Bad Harzburg, June 1978.

11. J Periaux, P Perrier, G Poirier, R Glowinski, M O Bristeau and O Pironneau, Application of optimal control and finite element methods to the calculation of transonic flows and incompressible viscous flows, Conference on Numerical Methods in Applied Fluid Dynamics, Reading, January 1978, Proceedings pub Academic Press 1980.

12. G Heckmann, Etude par la méthode des éléments finis des interactions voilure-fuselage-nacelle d'un avion du type Falcon à Mach = 0.79, Proc AGARD Conference on Subsonic/Transonic Configuration Aerodynamics, Munich, May 1980.

13. E M Murman and J D Cole, Calculation of plane steady transonic flows, AIAA Journal, Vol 9, pp 114-121, 1971.

14. A Jameson, Iterative solution of transonic flows over airfoils and wings, including flows at Mach 1, Comm Pure Appl Math, Vol 27, pp 283-309, 1974.

15. P Lax and B Wendroff, Systems of conservation laws, Comm Pure Appl Math, Vol 13, pp 217-237, 1960.

16. A Jameson, Transonic potential flow calculations using conservation form, Proc 2nd AIAA Conference on Computational Fluid Dynamics, Hartford, pp 148-161, June 1975.

17. A Jameson, Transonic flow calculations, VKI Lecture Series 87 on Computational Fluid Dynamics, von Karman Institute for Fluid Dynamics, Rhode-St-Genèse, Belgium, March 1976.

18. T L Holst, A fast conservative algorithm for solving the transonic full potential equation, Proc AIAA 4th Computational Fluid Dynamics Conference, Williamsburg, Va, AIAA Paper 79-1456, July 1979.

19. T L Holst and W F Ballhaus, Fast conservative schemes for the full potential equation applied to transonic flows, AIAA Journal, Vol 17, No 2, pp 145-152, February 1979.

20. H Deconinck and Ch Hirsch, Transonic flow calculations with finite elements, GAMM Workshop on Numerical Methods for the Computation of Inviscid Transonic Flow with Shock Waves, Stockholm, September 1979, Proceedings edited Rizzi/Viviand, pub Vieweg Verlag, 1981.

21. M Hafez, J South and E Murman, Artificial Compressibility Methods for numerical solutions of transonic full potential equation, AIAA Journal, Vol 17, No 8, pp 838-844, August 1979.

22. J L Steger and B S Baldwin, Shock waves and drag in the numerical calculation of isentropic transonic flow, NASA TN D-6997, 1972.

23. F Bauer, P Garabedian, D Korn and A Jameson, Supercritical wing sections II, Lecture notes in Economics and Mathematical Systems, Vol 108, pub Springer 1975.

24. T J Baker and M P Carr, Flow calculations using the nonconservative potential equation, GAMM Workshop, Proceedings edited Rizzi/Viviand, pub Vieweg Verlag 1981.

25. L A Carlson, Test problem for inviscid transonic flow, GAMM Workshop Proceedings edited Rizzi/Viviand, pub Vieweg Verlag, 1981.

26. A Rizzi and H Viviand (eds), GAMM Workshop on Numerical Methods for the Computation of Inviscid Transonic Flow with Shock Waves, Stockholm, September 1979, Proceedings pub Vieweg Verlag, 1981.

27. A Rizzi, Transonic solution of the Euler equations by the finite volume method, IUTAM Symposium Transsonicum II, Göttingen, September 1975, Proceedings pub Springer, 1976.

28. C C L Sells, Solution of the Euler equations for transonic flow past a lifting aerofoil, RAE TR 80065, May 1980.

29. D W Peaceman and H H Rachford, The numerical solution of parabolic and elliptic differential equations, J SIAM, Vol 3, pp 28-41, 1955.

30. N N Yanenko, The method of fractional steps, pub Springer 1971.

31. W F Ballhaus and J L Steger, Implicit approximate factorisation schemes for the low frequency transonic equation, NASA TM X-73082, November 1975.

32. W F Ballhaus, A Jameson and J Albert, Implicit approximate factorisation schemes for the efficient solution of steady transonic flow problems, AIAA 3rd Computational Fluid Dynamics Conference, Albuquerque, N Mex, AIAA Paper 77-634, June 1977.

33. T J Baker, A fast implicit algorithm for the nonconservative potential equation, Open Forum Presentation at the AIAA 4th Computational Fluid Dynamics Conference, Williamsburg, Va, July 1979.

34. R P Federenko, The speed of convergence of one iterative process, USSR Comp Math and Math Phys, Vol 4, pp 227-235, 1964.

35. A Brandt, Multi-level adaptive solutions to boundary-value problems, Math Comp, Vol 31, No 138, pp 333-390, 1977.

36. A Brandt, Multi-level adaptive computations in fluid dynamics, Proc AIAA 4th Computational Fluid Dynamics Conference, Williamsburg, Va, AIAA Paper 79-1455, July 1979.

37. J C South and A Brandt, The multi-grid method: fast relaxation for transonic flows, 13th Annual Meeting of Society of Engineering Science, Hampton, November 1976.

38. B Arlinger, Multigrid technique applied to lifting transonic flow using full potential equation, SAAB Report L-0-1 B439, December 1978.

39. F Bauer, P Garabedian and D Korn, Supercritical wing sections I, Lecture notes in Economics and Mathematical Systems, Vol 66, pub Springer 1972.

40. P R Garabedian, Estimation of the relaxation factor for small mesh size, Math Tables Aid Comp, Vol 10, pp 183-185, 1956.

41. A R Mitchell, Computational methods in partial differential equations, pub Wiley, 1969.

42. T L Holst, An implicit algorithm for the conservative, transonic full potential equation using an arbitrary mesh, AIAA 11th Fluid and Plasma Dynamics Conference, Seattle, Wash, AIAA Paper 78-1113, July 1978.

43. C R Forsey, ARA, Private communication.

44. T J Baker, Potential flow calculation by the approximate factorisation method, J Comp Phys, Vol 42, No 1, pp 1-19, July 1981.

45. D Catherall, Optimum approximate-factorisation schemes for two-dimensional steady potential flows, RAE Tech Memo Aero 1903, May 1981, and AIAA 5th Computational Fluid Dynamics Conference, Palo Alto, Ca, AIAA Paper 81-1018, June 1981.

46. R E Melnik, R Chow and H R Mead, Theory of viscous transonic flow over aerofoils at high Reynolds number, AIAA Paper 70-680, 1977.

47. M R Collyer and R C Lock, Prediction of viscous effects in steady transonic flow past an aerofoil, Aero Qu, Vol 30, pp 485-505, August 1979.

48. J Morrison, A fast algorithm for calculating transonic viscous flow over an aerofoil, ARA Memo to appear.

49. A J Bocci, A new series of aerofoil sections suitable for aircraft propellers, Aero Qu, Vol 28, pp 59-73, February 1977.

50. A J Bocci, Transonic calculations on ARA-D propeller aerofoils, ARA Model Test Note E21/1 to appear.

51. A Jameson, Private communnication.

52. C C L Sells, Plane subcritical flow past a lifting aerofoil, Proc Roy Soc London, Vol 308A, pp 377-401, 1968.

53. T Theodorsen and I E Garrick, General potential theory of arbitrary wing sections, NACA TR 452, 1933.

54. I E Garrick, Conformal mapping in aerodynamics, with emphasis on the method of successive conjugates, Symp on Construct and Appl of Conf Maps, Nat Bureau of Standards, Apply Math Series, Vol 18, 1949.

55. D C Ives, A modern look at conformal mapping including doubly connected regions, AIAA 8th Fluid and Plasma Dynamics Conference, Hartford, AIAA Paper 75-842, June 1975.

56. J W Cooley and J W Tukey, An algorithm for the machine calculation of complex Fourier series, Math Comp, Vol 19, pp 297-301, 1965.

57. B G Arlinger, Calculation of transonic flow around axisymmetric inlets, AIAA 13th Aerospace Sciences Meeting, Pasadena, Ca, AIAA Paper 75-80, January 1975.

58. R T Davis, Numerical methods for coordinate generation based on Schwarz-Christoffel transformations, 4th Computational Fluid Dynamics Conference, Williamsburg, Va, AIAA Paper 79-1463, July 1979.

59. G Moretti, Grid generation using classical techniques, Proc NASA Langley Workshop, October 1980.

60. J C South and A Jameson, Relaxation solutions for inviscid axisymmetric transonic flow over blunt or pointed bodies, AIAA Computational Fluid Dynamics Conference, Palm Springs, pp 8-17, 1973.

61. E F Brown, T J F Brecht and K E Walsh, A relaxation solution of transonic nozzle flows including rotational effects, AIAA/SAE 12th Propulsion Conference, Palo Alto, Ca, AIAA Paper 76-647, July 1976.

62. T J Baker Numerical computation of transonic potential flow through nozzles, Aero Qu, Vol 32, pp 31-42, February 1981.

63. D J Jones and J C South, A numerical determination of the bow shock wave in transonic axisymmetric flow about blunt bodies, NAE Aeronautical Report LR-586, May 1975.

64. D A Caughey and A Jameson, Accelerated iterative calculation of transonic nacelle flowfields, AIAA 14th Aerospace Sciences Meeting, Washington, AIAA Paper 76-100, January 1976.

65. D A Caughey, A systematic procedure for generating useful conformal mappings, Int J Num Meth Eng, Vol 12, pp 1651-1657, 1978.

66. C R Forsey and M P Carr, The calculation of transonic flow over three-dimensional swept wings using the exact potential equation. DGLR Symposium Transonic Configurations, Bad Harzburg, June 1978.

67. J J Chattot, C Coulombeix, F Manie and V Schmitt, Calcul d'écoulements transsoniques autour d'ailes, DGLR Symposium Transonic Configurations, Bad Harzburg, June 1978.

68. D A Caughey and A Jameson, Recent progress in finite volume calculations for wing-fuselage combinations, AIAA 12th Fluid and Plasma Dynamics Conference, Williamsburg, Va, AIAA Paper 79-1513, July 1979.

69. L A Carlson, Transonic airfoil analysis and design using Cartesian coordinates, Proc AIAA 2nd Computational Fluid Dynamics Conference, Hartford, pp 175-183, June 1975.

70. T A Reyhner, Cartesian mesh solution for axisymmetric transonic potential flow around inlets, AIAA 9th Fluid and Plasma Dynamics Conference, San Diego, Ca, AIAA Paper 76-421, July 1976.

71. T A Reyhner, Transonic potential flow computation about three-dimensional inlets, ducts and bodies, AIAA 13th Fluid and Plasma Dynamics Conference, Snowmass, Colorado, AIAA Paper 80-1364, July 1980.

72. J L Colehour, Transonic flow analysis using a streamline coordinate transformation procedure, Palm Springs, Ca, AIAA Paper 73-657, 1973.

73. J F Thompson, F C Thames, C M Mastin and S P Shanks, Use of numerically generated body-fitted coordinate systems for solution of the Navier-Stokes equations, AIAA 2nd Computational Fluid Dynamics Conference, Hartford, pp 63-80, June 1975.

74. J F Middlecoff and P D Thomas, Direct control of the grid point distribution in meshes generated by elliptic equations, AIAA 4th Computational Fluid Dynamics Conference, Williamsburg, Va, AIAA Paper 79-1462, July 1979.

75. J L Steger and R L Sorenson, Automatic mesh-point clustering near a boundary in grid generation with elliptic partial differential equations, J Comp Phys, Vol 33, No 3, December 1979.

76. J L Steger and R L Sorenson, Use of hyperbolic partial differential equations to generate body fitted coordinates, Proc NASA Langley Workshop, October 1980.

77. N J Yu, Grid generation and transonic flow calculations for three-dimensional configurations, AIAA 13th Fluid and Plasma Dynamics Conference, Snowmass, Colorado, AIAA Paper 80-1391, July 1980.

78. A Jameson and D A Caughey, A finite volume method for transonic potential flow calculations, AIAA 3rd Computational Fluid Dynamics Conference, Albuquerque, N M, June 1977.

79. T J Baker and F A Ogle, A computer program to compute transonic flow over an axisymmetric solid body, ARA Memo 197.

80. R Partington and T J Baker, The effect of nose blunting on the wave drag of ogive forebodies, to appear.

81. T J Baker, A numerical method to compute inviscid transonic flow around axisymmetric ducted bodies, IUTAM Symposium Transsonicum II, Göttingen, September 1975, Proceedings pub Springer, 1976.

82. E H Kitchen, Rolls Royce Derby, Private communication.

83. J L Steger, Implicit finite-difference simulation of flow about arbitrary two-dimensional geometries, AIAA Journal, vol 16, No 7, pp 679-686, July 1978.

84. D Catherall, RAE, Private communication.

85. B Arlinger and W Schmidt, Design and analysis of slat systems in transonic flow, 11th ICAS Congress, Lisbon, Portugal, September 1978.

The Calculation of Steady Transonic Flow by Euler Equations with Relaxation Methods

ERIK DICK

ABSTRACT

The well known relaxation method for the transonic potential equation can be extended to Euler equations. This generates methods with first order convergence rate. Several relaxation methods are described.

1. EQUATIONS OF MOTION

1.1. Euler Equations

The Euler equations in conservative form are :

$$\begin{aligned} &\frac{\partial \rho}{\partial t} + \nabla.(\rho\vec{V}) = 0 \\ &\frac{\partial \rho\vec{V}}{\partial t} + \nabla.(\rho\vec{V}\vec{V}) + \nabla p = 0 \\ &\frac{\partial \rho E}{\partial t} + \nabla.(\rho H\vec{V}) = 0 \end{aligned} \tag{1.1}$$

The equations (1.1) describe, together with the equations of state :

$$p = \rho RT \ , \quad e = c_v T \ , \quad h = c_p T \tag{1.2}$$

the unsteady motion of a gas in absence of viscosity, heat conduction and external forces.

For steady motion : $\partial/\partial t = 0$.

When all fluid state variables are continuous, equations (1.1) can be brought in the quasi-linear form :

$$\begin{aligned} &\frac{\partial \rho}{\partial t} + \vec{V}.\nabla\rho + \rho\nabla.\vec{V} = 0 \\ &\frac{\partial \vec{V}}{\partial t} + \vec{V}.\nabla\vec{V} + \frac{1}{\rho}\nabla p = 0 \\ &\frac{\partial p}{\partial t} + \gamma p\nabla.\vec{V} + \vec{V}.\nabla p = 0 \end{aligned} \tag{1.3}$$

The system (1.3) has the general form :

$$\frac{\partial \xi}{\partial t} + A_1 \frac{\partial \xi}{\partial x} + A_2 \frac{\partial \xi}{\partial y} + A_3 \frac{\partial \xi}{\partial z} = 0 \qquad (1.4)$$

with :

$$\xi = \begin{pmatrix} \rho \\ u \\ v \\ w \\ p \end{pmatrix} \qquad A_1 = \begin{pmatrix} u & \rho & 0 & 0 & 0 \\ 0 & u & 0 & 0 & \frac{1}{\rho} \\ 0 & 0 & u & 0 & 0 \\ 0 & 0 & 0 & u & 0 \\ 0 & \gamma p & 0 & 0 & u \end{pmatrix}$$

$$A_2 = \begin{pmatrix} v & 0 & \rho & 0 & 0 \\ 0 & v & 0 & 0 & 0 \\ 0 & 0 & v & 0 & \frac{1}{\rho} \\ 0 & 0 & 0 & v & 0 \\ 0 & 0 & \gamma p & 0 & v \end{pmatrix} \qquad A_3 = \begin{pmatrix} w & 0 & 0 & \rho & 0 \\ 0 & w & 0 & 0 & 0 \\ 0 & 0 & w & 0 & 0 \\ 0 & 0 & 0 & w & \frac{1}{\rho} \\ 0 & 0 & 0 & \gamma p & w \end{pmatrix}$$

From the general theory of partial differential equations, we know that a quasi-linear system of n first order equations in n dependent variables, can be classified with respect to the variable t according to the properties of the characteristic matrix :

$$K = k_1 A_1 + k_2 A_2 + k_3 A_3$$

The system (1.4), with respect to the variable t, is said to be :

hyperbolic : if all eigenvalues of K are real and corresponding eigenvectors can be made lineary independent for all (k_1, k_2, k_3)

parabolic : if all eigenvalues of K are real but corresponding eigenvectors cannot be made lineary independent for all (k_1, k_2, k_3)

elliptic : if all eigenvalues of K are complex

hybrid : when the classification into hyperbolic, parabolic and elliptic is not possible.

It can easily be verified that the Euler equations are hyperbolic with respect to t.

The eigenvalues of K are :

$$\lambda_1 = \lambda_2 = \lambda_3 = k_1 u + k_2 v + k_3 w$$

$$\lambda_4 = \lambda_1 - c$$

$$\lambda_5 = \lambda_1 + c$$

The corresponding lineary independent eigenvectors are the columns of the eigenvector matrix :

$$\tilde{E} = \begin{pmatrix} k_1 & k_2 & k_3 & \rho/c\sqrt{2} & \rho/c\sqrt{2} \\ 0 & -k_3 & k_2 & k_1/\sqrt{2} & -k_1/\sqrt{2} \\ k_3 & 0 & -k_1 & k_2/\sqrt{2} & -k_2/\sqrt{2} \\ -k_2 & k_1 & 0 & k_3/\sqrt{2} & -k_3/\sqrt{2} \\ 0 & 0 & 0 & \rho c/\sqrt{2} & \rho c/\sqrt{2} \end{pmatrix}$$

The meaning of the hyperbolic nature with respect to t of the equations (1.3) can be seen by considering first a particular solution of (1.3) in which the state variables are only functions of time and one spatial coordinate s_o, in the direction (k_1, k_2, k_3).

For such a solution :

$$\frac{\partial}{\partial x} = k_1 \frac{\partial}{\partial s_o}, \qquad \frac{\partial}{\partial y} = k_2 \frac{\partial}{\partial s_o}, \qquad \frac{\partial}{\partial z} = k_3 \frac{\partial}{\partial s_o}$$

(1.4) becomes :

$$\frac{\partial \xi}{\partial t} + K \frac{\partial \xi}{\partial s_o} = 0 \tag{1.5}$$

By introducing new dependent variables $\psi = \tilde{E}^{-1}\xi$, (1.5) becomes :

$$\frac{\partial \psi}{\partial t} + (\tilde{E}^{-1}K\tilde{E}) \frac{\partial \psi}{\partial s_o} = 0 \tag{1.6}$$

with :

$$\tilde{E}^{-1}K\tilde{E} = \Lambda = \begin{pmatrix} \lambda_1 & & & & \\ & \lambda_2 & & & \\ & & \lambda_3 & & \\ & & & \lambda_4 & \\ & & & & \lambda_5 \end{pmatrix}$$

The equation (1.5) resolves into 5 one-dimensional transport equations, which show that the nth component of ψ moves with velocity λ_n along s_o.

The trajectories of these components in the plane (t,s_o) are called characteristics of equation (1.5). There are in each point P three different characteristics with slope :

$$\partial s_o/\partial t = \lambda_1, \lambda_4, \lambda_5$$

as depicted in figure 1.

The state in point P is instantaneously determined by information transfered along the characteristics. As a consequence, the state in point P can only be influenced by the state in the shaded region A (region of dependence) and can only influence the state in the shaded region B (region of influence).

For the multi-dimensional Euler equations, the characteristic surfaces are the loci of the one-dimensional characteristics for variable (k_1, k_2, k_3).

The locus corresponding to λ_1, λ_2 and λ_3 is the trajectory of the fluid particle. The locus corresponding to λ_4 and λ_5 is a cone shaped surface.

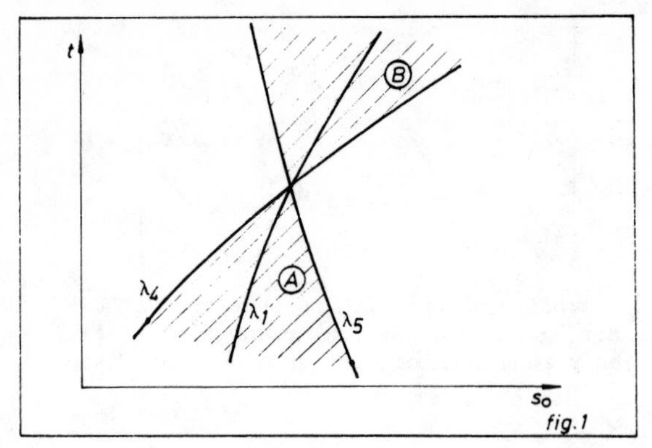

fig.1

The transient Euler equations are not generally hyperbolic with respect to a space variable.

With respect to the x-direction, the characteristic matrix is :

$$K = k_1 A_1^{-1} + k_2 A_1^{-1} A_2 + k_3 A_1^{-1} A_3$$

The eigenvalues of this matrix are real and corresponding eigenvectors can be made orthogonal for $u > c$. For $u < c$, three eigenvalues are real and two eigenvalues are complex. The system is thus hybrid for $u < c$.

For the steady equation :

$$A_1 \frac{\partial \xi}{\partial x} + A_2 \frac{\partial \xi}{\partial y} + A_3 \frac{\partial \xi}{\partial z} = 0 \tag{1.7}$$

the following results hold :

In subsonic flow : $u^2 + v^2 + w^2 < c^2$, the characteristic matrix has three equal real eigenvalues and two complex eigenvalues, with respect to every spatial direction. Hence the system is hybrid for subsonic flow.

In supersonic flow, the equations (1.7) are hyperbolic with respect to all spatial directions within the cone with axis along the velocity and apex angle

$$\alpha = \text{Arccos}\ \frac{1}{M}$$

1.2. Potential Equations

For irrotational steady flow, the equations (1.7) simplify to :

$$\begin{aligned} \nabla.(\rho\vec{V}) &= 0 \\ \vec{\omega} = \nabla \times \vec{V} &= 0 \end{aligned} \tag{1.8}$$

Equations (1.8) are valid if the total enthalpy and the entropy are constant in the whole flow field. Hence for $H = ct$ and $s = ct$:

$$\rho = \rho_o \left(1 - \frac{u^2 + v^2 + w^2}{2H}\right)^{\frac{1}{\gamma-1}} \tag{1.9}$$

The potential equations (1.8) do not contain information transfered along the trajectory of the fluid particle.

This can easily be seen in two dimensions for which equations (1.8) become :

$$\frac{\partial}{\partial x}(\rho u) + \frac{\partial}{\partial y}(\rho v) = 0$$

$$\frac{\partial u}{\partial y} - \frac{\partial v}{\partial x} = 0 \qquad (1.10)$$

By substitution of ρ from (1.9) this yields :

$$\begin{pmatrix} (c^2-u^2) & -uv \\ 0 & -1 \end{pmatrix} \frac{\partial}{\partial x} \begin{pmatrix} u \\ v \end{pmatrix} + \begin{pmatrix} -uv & (c^2-v^2) \\ 1 & 0 \end{pmatrix} \frac{\partial}{\partial y} \begin{pmatrix} u \\ v \end{pmatrix} = 0$$

with respect to x :

$$\frac{\partial}{\partial x} \begin{pmatrix} u \\ v \end{pmatrix} = \begin{pmatrix} -\frac{2uv}{c^2-u^2} & \frac{c^2-v^2}{c^2-u^2} \\ -1 & 0 \end{pmatrix} \frac{\partial}{\partial y} \begin{pmatrix} u \\ v \end{pmatrix}$$

The eigenvalues of the characteristic matrix are :

$$\lambda_{1,2} = \frac{-uv}{c^2-u^2} \pm \frac{c}{c^2-u^2} \sqrt{u^2 + v^2 - c^2}$$

In a subsonic point of the flow field, the eigenvalues are complex. The system (1.10) is then elliptic with respect to x.
In a supersonic point of the flow field, the eigenvalues are real and corresponding eigenvectors can be made orthogonal. The system (1.10) is then hyperbolic with respect to x.

Writing the equations (1.10) with respect to y, gives the same results.

The three-dimensional potential equations (1.8) cannot be analysed by the method for n first order equations in n dependent variables, since in three dimensions the system (1.8) has 4 equations in 3 dependent variables.

The general theory applied to (1.8) shows that there are no characteristic surfaces in a subsonic point and that there exists a characteristic cone in a supersonic point. This can easily be seen by the substitution $\vec{V} = \nabla\Phi$, for which (1.8) reduces to :

$$\nabla.(\rho\nabla\Phi) = 0$$

or in quasi-linear form :

$$(c^2-u^2)\frac{\partial^2\Phi}{\partial x^2} + (c^2-v^2)\frac{\partial^2\Phi}{\partial y^2} + (c^2-w^2)\frac{\partial^2\Phi}{\partial z^2}$$

$$- 2uv\,\frac{\partial^2\Phi}{\partial x\partial y} - 2uw\,\frac{\partial^2\Phi}{\partial x\partial z} - 2vw\,\frac{\partial^2\Phi}{\partial y\partial z} = 0 \qquad (1.11)$$

With a coordinate transformation so that the X-axis is aligned with the velocity, this equation reduces to :

$$(c^2-u^2-v^2-w^2)\frac{\partial^2\Phi}{\partial X^2} + c^2\,\frac{\partial^2\Phi}{\partial Y^2} + c^2\,\frac{\partial^2\Phi}{\partial Z^2} = 0 \qquad (1.12)$$

The characteristic surface of equation (1.12), f = 0, is given by :

$$(c^2-u^2-v^2-w^2)\left(\frac{\partial f}{\partial X}\right)^2 + c^2\left(\frac{\partial f}{\partial Y}\right)^2 + c^2\left(\frac{\partial f}{\partial Z}\right)^2 = 0$$

which is :

$$\frac{X^2}{c^2-u^2-v^2-w^2} + \frac{Y^2}{c^2} + \frac{Z^2}{c^2} = 0$$

In a subsonic point the characteristic surface does not exist, hence the equation is elliptic. In a supersonic point the characteristic surface is a cone with axis along the velocity and apex angle α = Arcsin 1/M, hence the equation is hyperbolic for all directions within the cone with apex angle Arccos 1/M.

1.3. Conclusion

The mathematical character of the steady potential equations is far more simple than the character of the steady Euler equations.
Transonic potential equations are mixed elliptic-hyperbolic, but transonic Euler equations are mixed hybrid-hyperbolic.

2. CLASSIC METHODS OF SOLUTION

2.1. Elliptic Equations

The discritised equations corresponding to second order elliptic equations can efficiently be solved by successive overrelaxation techniques.

Consider for example the potential equation (1.11) in two dimensions :

$$L(\phi) = (c^2-u^2)\frac{\partial^2\phi}{\partial x^2} - 2uv\,\frac{\partial^2\phi}{\partial x\partial y} + (c^2-v^2)\frac{\partial^2\phi}{\partial y^2} = 0 \qquad (2.1)$$

An actual transonic computation should always be carried out on the conservative form of this equation, but for analysis the quasi-linear form is more appropriate.

We discretise by central difference operators :

$$\Phi_{xx} \simeq \frac{\Phi_{i+1,j} - 2\Phi_{i,j} + \Phi_{i-1,j}}{\Delta x^2}$$

$$\Phi_{xy} \simeq \frac{\Phi_{i+1,j+1} - \Phi_{i+1,j-1} - \Phi_{i-1,j+1} + \Phi_{i-1,j-1}}{4\Delta x\Delta y}$$

$$\Phi_{yy} \simeq \frac{\Phi_{i,j+1} - 2\Phi_{i,j} + \Phi_{i,j-1}}{\Delta y^2}$$

The classic first order relaxation scheme then is : starting from an initial state (o), values of Φ at iteration level (n) are substituted by values at iteration level (n+1) according to :

<u>for point relaxation</u> :

$$\frac{c^2-u^2}{\Delta x^2}\left(\Phi^{(n)}_{i+1,j} - 2\tilde{\Phi}_{i,j} + \Phi^{(n+1)}_{i-1,j}\right)$$

$$- \frac{2uv}{4\Delta x\Delta y}\left(\Phi^{(n)}_{i+1,j+1} - \Phi^{(n)}_{i+1,j-1} - \Phi^{(n+1)}_{i-1,j+1} + \Phi^{(n+1)}_{i-1,j-1}\right)$$

$$+ \frac{c^2-v^2}{\Delta y^2}\left(\Phi^{(n)}_{i,j+1} - 2\Phi_{i,j} + \Phi^{(n+1)}_{i,j-1}\right) = 0 \qquad (2.2)$$

<u>for line-relaxation</u> line i is simultaneously calculated.

Φ_{yy} is replaced in (2.2) by :

$$\frac{1}{\Delta y^2}(\tilde{\Phi}_{i,j+1} - 2\tilde{\Phi}_{i,j} + \tilde{\Phi}_{i,j-1})$$

The preliminary value $\tilde{\Phi}_{i,j}$ is then relaxed according to :

$$\Phi^{(n+1)}_{i,j} = (1-r)\,\Phi^{(n)}_{i,j} + r\,\tilde{\Phi}_{i,j} \qquad (2.3)$$

$u_{i,j}$, $v_{i,j}$ and $c_{i,j}$ are calculated as :

$$u_{i,j} = \frac{\Phi^{(n)}_{i+1,j} - \Phi^{(n+1)}_{i-1,j}}{2\Delta x}$$

$$v_{i,j} = \frac{\Phi^{(n)}_{i,j+1} - \Phi^{(n)}_{i,j-1}}{2\Delta y}$$

$$c^2_{i,j} = (\gamma-1)\left(H - \frac{u^2_{i,j} + v^2_{i,j}}{2}\right)$$

The mechanism of the relaxation method can be analysed, as Jameson has shown (1) by considering the iteration from step (n) to (n+1) as advancing in a fictitious time direction.

By Taylor expansion of (2.2)(2.3) into the fictitious time direction and the space directions, the equivalent time dependent equation results, by deleting higher order terms :

$$2c\,\alpha\,\Phi_{xt} + 2c\,\beta\,\Phi_{yt} + \chi\,\Phi_t = L(\Phi) \qquad (2.4)$$

For point relaxation :

$$2c\,\alpha = (c^2-u^2)\frac{\Delta t}{\Delta x}$$

$$2c\,\beta = (c^2-v^2)\frac{\Delta t}{\Delta y} - uv\,\frac{\Delta t}{\Delta x}$$

$$\chi = (\frac{2}{r} - 1)\left((c^2-u^2) + (c^2-v^2)\frac{\Delta x^2}{\Delta y^2}\right)\frac{\Delta t}{\Delta x^2}$$

For line relaxation :

$$2c\,\alpha = (c^2-u^2)\frac{\Delta t}{\Delta x}$$

$$2c\ \beta = -\ uv\ \frac{\Delta t}{\Delta x}$$

$$\chi = (\frac{2}{r} - 1)(c^2-u^2)\frac{\Delta t}{\Delta x^2}$$

In both cases, χ has the general form :

$$\chi = (\frac{2}{r} - 1)K\ \frac{\Delta t}{\Delta x^2}$$

With the following transformation :

$$T = t + \frac{\left(\alpha(c^2-v^2) + \beta uv\right)x + \left(\beta(c^2-u^2) + \alpha uv\right)y}{c(c^2-u^2-v^2)}$$

$$X = \frac{ux + vy}{\sqrt{u^2+v^2}}$$

$$Y = \frac{uy - vx}{\sqrt{u^2 + v^2}}$$

The equation (2.4) becomes :

$$K_1\Phi_{TT} + (c^2-u^2-v^2)\chi\Phi_T = (c^2-u^2-v^2)^2\Phi_{XX} + c^2(c^2-u^2-v^2)\Phi_{YY} \qquad (2.5)$$

with : $$K_1 = \alpha^2(c^2-v^2) + 2\alpha\beta uv + \beta^2(c^2-u^2)$$

K_1 is positive for subsonic flow, hence equation (2.5) is hyperbolic with respect to T for subsonic flow.

The characteristic cone of (2.5) is :

$$\frac{T^2}{K_1} = \frac{X^2}{(c^2-u^2-v^2)} + \frac{Y^2}{c^2(c^2-u^2-v^2)}$$

In the variables (x,y,t) this cone is :

$$c(c^2-u^2-v^2)t^2 + 2\left(\alpha(c^2-v^2) + \beta uv\right)xt + 2\left(\beta(c^2-u^2) + \alpha uv\right)yt - c\left(\alpha y - \beta x\right)^2 = 0 \qquad (2.6)$$

The cone is tangent to the plane $t = 0$ along

$$\alpha y - \beta x = 0$$

with : $t = t \qquad \zeta = \alpha x + \beta y \qquad \eta = \alpha y - \beta x$

(2.6) becomes :

$$2K_1\zeta t + 2K_2\eta t - c(\alpha^2+\beta^2)\eta^2 + c(\alpha^2+\beta^2)(c^2-u^2-v^2)t^2 = 0$$

with : $$K_2 = (\alpha u + \beta b)(\alpha v - \beta u)$$

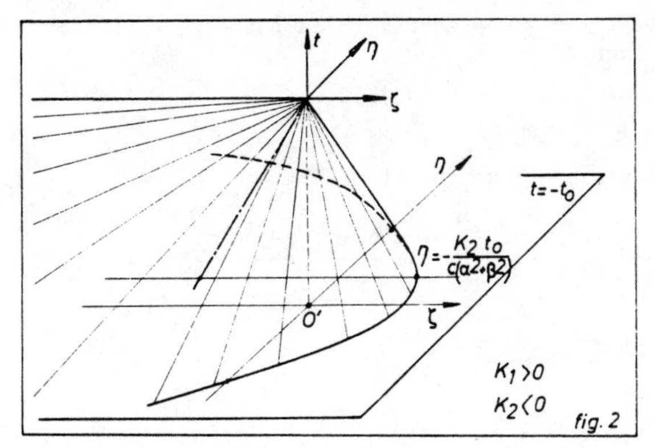

This characteristic cone is depicted in figure 2.

According to the first stability rule of Hirt (2) the numerical domain of dependence should include the domain of dependence of the equivalent time dependent equation. One verifies that this condition is fulfilled.

The second stability rule of Hirt states that the equivalent equation should lead to a stable steady state. One remarks in (2.4) that this will be the case for $\chi > 0$, which implies : $0 \leqslant r \leqslant 2$.

For r not in the vicinity of 2, χ is one order of magnitude larger than α and β. The convergence rate of the scheme is then dominated by :

$$\chi \, \Phi_t = L(\Phi) \tag{2.7}$$

If we linearise equation (2.7) by taking the coefficients as constant, the convergence can be studied by Fourier analysis.

Substitution of a perturbation of the steady solution of (2.7) :

$$\Phi' = e^{\sigma t} \, e^{j\omega_1 x} \, e^{j\omega_2 y}$$

yields :

$$\chi\sigma = -(c^2-u^2)\omega_1^2 + 2uv\omega_1\omega_2 - (c^2-v^2)\omega_2^2 = -\lambda^2(\omega_1,\omega_2) .$$

with $\lambda^2 > 0$

Thus :

$$\sigma(\omega_1,\omega_2) = - \frac{\lambda^2(\omega_1,\omega_2)}{(\frac{2}{r} - 1)K} \frac{\Delta x^2}{\Delta t}$$

For sufficiently long time, the convergence to the steady state is governed by the smallest wave numbers $(\omega_{11}, \omega_{21})$.

Thus for long time :

$$\frac{\Phi'(t + \Delta t)}{\Phi'(t)} = e^{\sigma_1 t}$$

with :

$$\sigma_1 = - \frac{\lambda^2(\omega_{11},\omega_{21})}{(\frac{2}{r} - 1)K} \frac{\Delta x^2}{\Delta t}$$

The number of iterations N, necessary to damp the perturbation a number of magnitudes P, is then :

$$e^{Re(\sigma_1)N\Delta t} = 10^{-P}$$

Hence, the asymptotic convergence rate is :

$$V_c = \frac{1}{N} = -\frac{Re(\sigma_1)\Delta t}{P} = \frac{\lambda^2(\omega_{11},\omega_{21})}{P(\frac{2}{r} - 1)K} \cdot \Delta x^2 = O(\Delta x^2)$$

For $r = 2 - k\,\Delta x$ with $k = O(1)$:

$$\chi = \frac{k\,K\,\Delta t}{2\Delta x}$$

α, β and χ have then the same order of magnitude.

The convergence rate is then influenced by the complete equation (2.4).

The same convergence analysis as before gives :

$$(2c\alpha j\omega_1 + 2c\beta j\omega_2 + \chi) = -\lambda^2(\omega_1,\omega_2)$$

$$Re(\sigma_1) = -\lambda^2(\omega_{11},\omega_{21}) \frac{\chi}{\chi^2 + (2c\alpha\omega_{11} + 2c\beta\omega_{21})^2} = -k' \frac{\Delta x}{\Delta t}$$

Hence :

$$V_c = \frac{-Re(\sigma_1)\Delta t}{P} = O(\Delta x)$$

We conclude that for a critical choice of the relaxation factor, the rate of convergence is $O(\Delta x)$.

We remark that this classic relaxation scheme cannot work in a supersonic point of the flow since $\lambda^2(\omega_1,\omega_2)$ is then no longer positive definite. The classic relaxation method is thus a pure elliptic technique.

2.2. Hyperbolic Equations

The natural technique for solving the discretised equations corresponding to hyperbolic equations is marching in one of the hyperbolic directions.

For the supersonic potential equation, Jameson has constructed a rotated marching scheme.

The equation (2.1) is expressed as :

$$(c^2-u^2-v^2)\Phi_{XX} + c^2\Phi_{YY} = 0$$

X is the direction of the velocity, Y is the perpendicular direction.

$$\Phi_{XX} = \frac{1}{u^2+v^2}(u^2\Phi_{xx} + 2uv\,\Phi_{xy} + v^2\Phi_{yy})$$

$$\Phi_{YY} = \frac{1}{u^2+v^2}(v^2\Phi_{xx} - 2uv\,\Phi_{xy} + u^2\Phi_{yy})$$

Φ_{YY} is discretised in a central way. Φ_{XX} is discretised upwind.
For u > 0 and v > 0 the terms in Φ_{XX} are :

$$\Phi_{xx} \simeq \frac{\Phi_{i,j} - 2\Phi_{i-1,j} + \Phi_{i-2,j}}{\Delta x^2}$$

$$\Phi_{xy} \simeq \frac{\Phi_{i,j} - \Phi_{i-1,j} - \Phi_{i,j-1} + \Phi_{i-1,j-1}}{\Delta x\ \Delta y}$$

$$\Phi_{yy} \simeq \frac{\Phi_{i,j} - 2\Phi_{i,j-1} + \Phi_{i,j-2}}{\Delta y^2}$$

Jameson has shown that this marching scheme can be brought in the form of the relaxation method for subsonic flow, on the conservative equations, for r = 1, when the density is replaced in (1.10) by the density of a suitable chosen upwind point.

These artificial density methods were further developed by Hafez, Murman and South (3), Chattot and Coulombeix (4), Deconinck and Hirsch (5).

Since the Euler equations are hyperbolic with respect to time, the obvious marching direction is the time direction.

In a time marching technique, the unsteady Euler equations are discretised so that the equivalent equation has as lowest order part the Euler equations themselves. The hyperbolic nature of the scheme with respect to time is therefore guaranteed. The steady state is reached after a sufficiently long time.

As a consequence of this mechanism, the internal damping in the time marching technique is only due to higher order terms. Hence the convergence rate is low.

When applied to the single quasi-linear hyperbolic equation :

$$\frac{\partial \xi}{\partial t} + V \frac{\partial \xi}{\partial x} = 0 \tag{2.8}$$

The time marching technique which is best according to convergence rate, has an equivalent equation :

$$\frac{\partial \xi}{\partial t} + V \frac{\partial \xi}{\partial x} - \nu \frac{\partial^2 \xi}{\partial x^2} + \text{HOT} = 0 \tag{2.9}$$

with $\nu = O(\Delta x)$.

In (2.9) terms in $\partial^2\xi/\partial t^2$ and $\partial^2\xi/\partial x\partial t$ are replaced by terms in $\partial^2\xi/\partial x^2$ by substitution of

$$\frac{\partial \xi}{\partial t} = -V \frac{\partial \xi}{\partial x} + \text{HOT}$$

The well known Lax scheme has an equivalent equation of this form :

$$\xi(x,t+\Delta t) = \frac{1}{2}\big(\xi(x+\Delta x,t) + \xi(x-\Delta x,t)\big) - \frac{V\Delta t}{2\Delta x}\big(\xi(x+\Delta x,t) - \xi(x-\Delta x,t)\big)$$

with $\quad \nu = \frac{1}{2}(1 - \lambda^2)\frac{\Delta x^2}{\Delta t} \quad , \quad \lambda = \frac{V\Delta t}{\Delta x}$

Linearising equation (2.8) by taking V as a constant, the convergence rate of the Lax scheme can be found by Fourier analysis.

Substitution in (2.9) of a perturbation of the steady state :

$$\xi' = \Phi(t)\, e^{j\omega x}$$

gives the amplification factor :

$$G = \frac{\Phi(t+\Delta t)}{\Phi(t)} = \cos(\omega\Delta x) - \lambda\, j \sin(\omega\Delta x)$$

$|G| < 1$ for $\lambda < 1$

$|G|$ is the largest for small ω.

For small ω :

$$\ln G = 1 - j\,\lambda\,\omega\,\Delta x - \frac{1}{2}(1 - \lambda^2)\omega^2\Delta x^2 + \text{HOT}$$

Hence :
$$|G| = e^{-O(\Delta x^2)}$$

and
$$V_c = O(\Delta x^2)$$

The Lax scheme applied to the conservative hyperbolic equation :

$$\frac{\partial \xi}{\partial t} + \frac{\partial\, f(\xi)}{\partial x} = 0$$

is :

$$\xi(x,t+\Delta t) = \frac{1}{2}\big(\xi(x-\Delta x,t) + \xi(x+\Delta x,t)\big) - \frac{\Delta t}{2\Delta x}\big(f(x+\Delta x,t) - f(x-\Delta x,t)\big)$$

The corresponding steady state is :

$$f(x+\Delta x) - f(x-\Delta x) = \frac{\Delta x}{\Delta t}\big(\xi(x+\Delta x) - 2\xi(x) + \xi(x-\Delta x)\big)$$

This is an approximation to :

$$\frac{\partial f}{\partial x} = \frac{\Delta x^2}{2\Delta t}\frac{\partial^2 \xi}{\partial x^2}$$

The scheme introduces a non negligible artificial viscosity, since the corresponding Reynoldsnumber is :

$$Re = \frac{V.n\ \Delta x}{\nu} = \frac{V.n\ \Delta x.2\Delta t}{\Delta x^2} = 2\ n\ \lambda$$

in which n is a typical number of node points in one direction of the field.

For $n \simeq 50 \rightarrow Re \simeq 100$.

This artificial viscosity can be corrected according to the scheme introduced by Couston, McDonald and Smolderen (6) :

$$\xi(x,t+\Delta t) = \xi(x,t+\Delta t)\big|_{Lax} - \frac{\beta}{2}\big(\xi(x+\Delta x,t_o) - 2\xi(x,t_o) + \xi(x-\Delta x,t_o)\big)$$

with $\beta = 1 - O(\Delta x)$.

The time level t_o is taken constant during a cycle of time steps. At the end of this cycle the time level t_o is renewed to the current time level.

This scheme has a convergence rate of the same order as the Lax scheme but has an acceptable steady artificial viscosity.

2.3. Conclusion

Time marching schemes have at best a convergence rate $O(\Delta x^2)$. This is one order of magnitude lower than relaxation methods which have a convergence rate $O(\Delta x)$. Classic relaxation methods however cannot be used for hyperbolic equations. Hence steady Euler equations cannot be solved by classic relaxation methods.

The important question arises whether or not relaxation schemes can be extended to hyperbolic equations.

In the sequel we will see that this is indeed possible. It allows the construction of algorithms for Euler equations which have the same convergence rate as the relaxation schemes for the potential equation.

3. NON-FORMAL RELAXATION METHODS FOR EULER EQUATIONS

From the discussion of the relaxation method for the potential equation we have learned that one of the essential features of the relaxation method is that the equivalent equation of the scheme has a convergence rate to steady state of $O(\Delta x)$. From a practical point of view this is the only fundamental requirement.

One can thus define a relaxation method just on the propriety of convergence rate, as Wirz has done (7).:

We denote the steady Euler equations by :

$$E(\xi) = 0 \tag{3.1}$$

These equations are in quasi-linear form :

$$E(\xi) = A_1 \frac{\partial \xi}{\partial x} + A_2 \frac{\partial \xi}{\partial y} + A_3 \frac{\partial \xi}{\partial z} = 0 \tag{3.2}$$

with A_1, A_2, A_3 defined in (1.4).

An arbitrary steady equation of form (3.2) is denoted by :

$$A(\xi) = 0 \tag{3.3}$$

The steady solution of (3.1) and (3.3) is denoted by :

$$\xi = \xi_s$$

H.J. Wirz defines a relaxation scheme of order N_o on (3.3) as any scheme which has the equivalent equation :

$$R(\frac{\partial \xi}{\partial t}) + A(\xi) = 0 \tag{3.4}$$

in which R is a partial differential operator with derivatives of atmost order N_o-1 and for which the solution of (3.4) fulfils :

$$\| \xi(x,y,z,t) - \xi_s(x,y,z)\| \leqslant K(t)e^{-\varepsilon t} \| \xi(x,y,z,o) - \xi_s(x,y,z)\|$$

ε is a real and positive number, K(t) is a polynominal of degree $N_1 \leqslant N_o$.

The artificial time dependent equation (3.4) is then called a relaxation equation and the operator R is called a relaxation operator of order N_o.

For sufficiently long time the norm of a perturbation to the steady state is then damped as :

$$t^{N_1} e^{-\varepsilon t}$$

The number of iterations necessary to damp the perturbation a number of magnitudes P is given by :

$$(t+N\Delta t)^{N_1} e^{-\varepsilon(t+N\Delta t)} = 10^{-P} t^{N_1} e^{-\varepsilon t}$$

For a long time this is :

$$e^{-\varepsilon N\Delta t} = 10^{-P}$$

The asymptotic rate of convergence is :

$$V_c = \frac{1}{N} = \frac{\varepsilon\Delta t}{P} \log e$$

The convergence rate is thus $O(\Delta t)$ which for the discritised equation is the same as $O(\Delta x)$.

In quasi-linear form equation (3.4) is :

$$\sum_{\substack{n=1,N_o \\ m\geq 1 \\ m+n_1+n_2+n_3=n}} B_{n,n_1,n_2,n_3}(\xi,x,y,z,t) \frac{\partial^n \xi}{\partial t^m \, \partial x_1^{n_1} \, \partial y^{n_2} \, \partial z^{n_3}} = A_1 \frac{\partial \xi}{\partial x} + A_2 \frac{\partial \xi}{\partial y} + A_3 \frac{\partial \xi}{\partial z} \tag{3.5}$$

The behaviour of the relaxation equation can easily be studied for the full linear case in which A_1, A_2, A_3, $B_{1,0,0,0}$, $B_{2,1,0,0}$, are constant.

In this case a Fourier-component of a perturbation of the steady state :

$$\xi' = \alpha(t)\, e^{j\omega_1 x}\, e^{j\omega_2 y}\, e^{j\omega_3 z} \tag{3.6}$$

fulfils :

$$\sum_{m=1,N_o} C_m \frac{d^m \alpha(t)}{dt^m} = j\, A_o \alpha(t)$$

with : $A_o = \omega_1 A_1 + \omega_2 A_2 + \omega_3 A_3$

$$C_m = \sum_{\substack{n_1,n_2,n_3 \\ n=m+n_1+n_2+n_3}} B_{n,n_1,n_2,n_3} (j\omega_1)^{n_1} (j\omega_2)^{n_2} (j\omega_3)^{n_3}$$

$\alpha(t)$ can be written as :

$$\alpha(t) = \sum_{i=1,M.N_o} a_i \beta_i (x,y,z) e^{\sigma_i t}$$

in which M is the number of components in α.

The σ_i are the solutions of the eigenvalue problem :

$$\text{Det}\left(\sum_{m=1,N_o} C_m \sigma_i^m - j A_o\right) = 0 \tag{3.7}$$

The β_i are the associated eigenvectors.

A condition for which equation (3.5) is a relaxation equation according to the given definition is :

$$\text{Re}(\sigma_i) < 0 \qquad \text{for} \qquad \omega_1^2 + \omega_2^2 + \omega_3^2 \neq 0 \tag{3.8}$$

The asymptotic rate of convergence is then governed by the σ_i with smallest real part σ_{io}, and is :

$$V_c = \frac{1}{N} = \frac{-\text{Re}(\sigma_{io})\Delta t}{P}$$

The condition (3.6) can only strictly be fulfilled when the operator $A(\xi)$ is elliptic.

We recall that the operator $A(\xi)$ is elliptic with respect to x if all eigenvalues of

$$k_1 A_1^{-1} A_2 + k_2 A_1^{-1} A_3$$

are complex for all real k_1 and k_2.

This is the case if :

$$\text{Det}(k_o I + k_1 A_1^{-1} A_2 + k_2 A_1^{-1} A_3) \neq 0$$

or

$$\text{Det}(k_o A_1 + k_1 A_2 + k_2 A_3) \neq 0$$

for all real k_o, k_1 and k_2.

Thus every direction is elliptic and hence the operator $A(\xi)$ is elliptic if :

$$\text{Det}(k_1 A_1 + k_2 A_2 + k_3 A_3) \neq 0$$

for all real k_1, k_2 and k_3.

For the non-elliptic operator $E(\xi)$

$$\text{Det}(A_o) = 0$$

for some combinations of ω_1, ω_2 and ω_3.

Hence (3.7) cannot be fulfilled.

However we remark that the matrix A_o for Euler equations has only real eigenvalues and corresponding orthogonal eigenvectors.

For the transient Euler equations :

$$\frac{\partial \xi}{\partial t} + E(\xi) = 0$$

A Fourier component of the perturbation of the steady state (3.6) fulfils :

$$I \frac{d}{dt} \alpha(t) + j A_o \alpha(t) = 0$$

The corresponding σ_i according to (3.7) are :

$$\sigma_i + j\,\mu_i = 0$$

in which μ_i are the real eigenvalues of A_o.

This shows that the eigenvector corresponding to $\mu_i = 0$ remains unchanged by the transient Euler equations. This eigenvector is thus part of the steady state. It is thus quite natural that the σ_i corresponding to such eigenvectors in (3.7) vanish.

It is thus sufficient that there are no other eigenvectors associated to $\sigma_i = 0$ in (3.7) but the eigenvectors of A_o corresponding to $\mu_i = 0$.

As a special case of algorithms that fulfil this condition Wirz has constructed methods in which the matrices C_m are polynomials in A_o (7).

He defines a relaxation method of

order two :

$$\frac{\partial\bar{\xi}}{\partial t} + A_1\frac{\partial\xi}{\partial x} + A_2\frac{\partial\xi}{\partial y} + A_3\frac{\partial\xi}{\partial z} = 0$$

$$\frac{\partial\xi}{\partial t} + k(A_1\frac{\partial\xi}{\partial x} + A_2\frac{\partial\xi}{\partial y} + A_3\frac{\partial\xi}{\partial z}) + \frac{1}{\tau}(\xi-\bar{\xi}) = 0 \qquad (3.9)$$

order three :

$$\frac{\partial\bar{\xi}}{\partial t} + A_1\frac{\partial\xi}{\partial x} + A_2\frac{\partial\xi}{\partial y} + A_3\frac{\partial\xi}{\partial z} = 0$$

$$\frac{\partial\tilde{\xi}}{\partial t} + A_1\frac{\partial\bar{\xi}}{\partial x} + A_2\frac{\partial\bar{\xi}}{\partial y} + A_3\frac{\partial\bar{\xi}}{\partial z} = 0$$

$$\frac{\partial\xi}{\partial t} + k(A_1\frac{\partial\bar{\xi}}{\partial x} + A_2\frac{\partial\bar{\xi}}{\partial y} + A_3\frac{\partial\bar{\xi}}{\partial z}) + \frac{1}{\tau}(\xi-\tilde{\xi}) = 0 \qquad (3.10)$$

Elimination of the artificial variables in (3.9) and (3.10) gives :

$$\frac{\partial^2\xi}{\partial t} + k\,A(\frac{\partial\xi}{\partial t}) + \frac{1}{\tau}\frac{\partial\xi}{\partial t} + \frac{1}{\tau}A(\xi) = 0$$

$$\frac{\partial^3\xi}{\partial t^3} - k\,A^2(\frac{\partial\xi}{\partial t}) + \frac{1}{\tau}\frac{\partial^2\xi}{\partial t^2} - \frac{1}{\tau}A^2(\xi) = 0 \qquad (3.11)$$

The second order method is thus a relaxation on $A(\xi)$, the third order method on $A^2(\xi)$.

Since the eigenvalue structure of $A^2(\xi)$ is the same as that of $A(\xi)$, the conditions formulated for $A(\xi)$ are valid for $A^2(\xi)$.

One can easily verify that the equations (3.11) fulfil all conditions formulated by Wirz for $\tau > 0$ and $k > 1$. Hence the convergence rate of these methods is $O(\Delta x)$.

In the sequel we will reserve the name formal relaxation method to methods that fulfil three formal conditions :

1) The order condition.
   In a relaxation method of order N_o, preliminary values $\bar{\xi}_1, \bar{\xi}_2, \ldots, \bar{\xi}_{N_o}$ are calculated for the state at level (n+1), this is on the fictitious time level $t+\Delta t$. These are relaxed according to :

$$\xi(t+\Delta t) = r_1\bar{\xi}_1 + r_2\bar{\xi}_2 + \ldots + r_{N_o}\bar{\xi}_{N_o} + (1-r_1-r_2-\ldots-r_{N_o})\xi(t) \quad (3.12)$$

2) The serial condition.
As soon as the state is calculated in some point (x,y,z) on the level (n+1), the state at level (n) and the preliminary values at level (n) are no longer used.

3) The convergence rate condition.
The asymptotic convergence rate is critically dependent upon the relaxation factors $r_1, r_2, \ldots, r_{N_o}$.
In the vicinity of optimal relaxation, the asymptotic convergence rate is $O(\Delta x)$, elsewhere it is of a higher order.

In order to study the methods of Wirz, we apply the second order scheme to :

$$\frac{\partial \xi}{\partial t} + V\frac{\partial \xi}{\partial x} = 0 \quad (3.13)$$

With the discretisation recommended by Wirz the second order method is :

$$\bar{\xi}(t+\Delta t) - \bar{\xi}(t) + \frac{V\Delta t}{2\Delta x}\big(\xi(x+\Delta x,t) - \xi(x-\Delta x,t)\big) = 0 \quad (3.14)$$

$$\xi(t+\Delta t) - \xi(t) - k'\big(\bar{\xi}(t+\Delta t) - \bar{\xi}(t)\big) + \frac{\Delta t}{\tau'}\big(\xi(t+\Delta t) - \bar{\xi}(t+\Delta t)\big) = 0 \quad (3.15)$$

(3.14) :

$$\xi(t+\Delta t) = \frac{k' + \frac{\Delta t}{\tau'}}{1+\frac{\Delta t}{\tau'}}\bar{\xi}(t+\Delta t) - \frac{k'}{1+\frac{\Delta t}{\tau'}}\bar{\xi}(t) + \frac{1}{1+\frac{\Delta t}{\tau'}}\xi(t)$$

or :

$$\xi(t+\Delta t) = r_1\bar{\xi}(t+\Delta t) + r_2\bar{\xi}(t) + (1-r_1-r_2)\xi(t) \quad (3.16)$$

The relaxation methods of Wirz do not fulfil the serial condition.

(3.14) and (3.16) involve preliminary values at level (n) and (3.14) involves values of ξ at level (n) for which the value at level (n+1) is already known.

Therefore the methods of Wirz are called here non-formal relaxation methods.

(3.14)(3.15) corresponds to (3.9) for :

$$k = \frac{k' + \frac{1}{\tau'}}{1+\frac{1}{\tau'}} \qquad \frac{1}{\tau} = \frac{\frac{1}{\tau'}}{1+\frac{1}{\tau'}}$$

The convergence rate can be studied for small wave numbers on the equivalent equations (3.9).

This is for (3.13) :

$$\frac{\partial}{\partial t}\begin{pmatrix}\xi\\ \bar{\xi}\end{pmatrix} + \begin{pmatrix}\frac{1}{\tau} & -\frac{1}{\tau}\\ 0 & 0\end{pmatrix}\begin{pmatrix}\xi\\ \bar{\xi}\end{pmatrix} + \begin{pmatrix}\lambda V & 0\\ V & 0\end{pmatrix}\frac{\partial}{\partial x}\begin{pmatrix}\xi\\ \bar{\xi}\end{pmatrix} = 0$$

A Fourier component of the perturbation to steady state :

$$\begin{pmatrix} \xi \\ \bar{\xi} \end{pmatrix} = e^{(\sigma)t} \begin{pmatrix} \xi_o \\ \bar{\xi}_o \end{pmatrix} e^{j\omega x}$$

yields :

$$(\sigma) = \begin{pmatrix} -\frac{1}{\tau} - j\omega\lambda V & \frac{1}{\tau} \\ - j\omega V & 0 \end{pmatrix}$$

The eigenvalues of the (σ)-matrix are for small ω :

$$\begin{aligned} \sigma_1 &= -\frac{1}{\tau} - (k-1)j\omega V + HOT \\ \sigma_2 &= -\omega^2 V^2 \tau(k-1) - j\omega V + HOT \end{aligned} \qquad (3.17)$$

The equations (3.17) show that associated to a physical perturbation with velocity V, there are two mathematical perturbations with velocities V and $(k-1)V$. Both are damped $O(1)$ when $\tau = O(1)$. Hence the convergence rate is $O(\Delta x)$.

Since both mathematical perturbations are damped in the same order, the dynamic behaviour of the associated variables $\bar{\xi}$ is not negligible to the dynamic behaviour of the physical variables ξ. This is due to the introduction of dynamic terms in $\bar{\xi}$ in (3.9).

It is absolutely necessary that the velocities of both mathematical perturbations have the same sign as the velocity of the physical perturbation, with respect to the non-linear behaviour of the scheme. If this were not the case a perturbation that disperses physically would cause contraction in the associated variables. Hence expansion shocks would be formed in transonic flow.

This behaviour is very similar to the behaviour of the artificial time dependent systems studied by Essers [8].

Essers studied systems of the general form :

$$\begin{aligned} \frac{\partial\bar{\xi}}{\partial t} + \lambda_1 A(\xi) &= 0 \\ \frac{\partial\xi}{\partial t} + k_1\xi + k_2\bar{\xi} + \mu_1 A(\xi) + \mu_2 B(\bar{\xi}) &= 0 \end{aligned} \qquad (3.18)$$

with the restriction that these equations would not reduce to relaxation equations. It is obvious that (3.18) cannot be brought in the form of a relaxation (3.12) when $\mu_2 \neq 0$.

Essers selected the special form of (3.18) :

$$\begin{aligned} \frac{\partial\bar{\xi}}{\partial t} &= (2-\chi)V\frac{\partial\xi}{\partial x} \\ \frac{\partial\xi}{\partial t} + k\chi(2\xi-\bar{\xi}) &= (2-\chi)V\frac{\partial\xi}{\partial x} + \frac{\chi-1}{2-\chi}V\frac{\partial\bar{\xi}}{\partial x} \end{aligned} \qquad (3.19)$$

with $k > 0$ and $0 \leqslant \chi \leqslant 1$.

In (3.19) the velocities of the mathematical perturbations have both the same sign as the velocity of the physical perturbation and the convergence rate is maximal.

The non-formal relaxation method of Wirz and the artificial time dependent method of Essers have both the special difficulty that the associated variables have dynamics. Boundary conditions have to be imposed on these variables and there is important interaction between physical and associated variables. This non-linear interaction has to be carefully handled in order to avoid expansion shocks. This makes it a non-trivial matter to construct a discretisation scheme that fulfils this entropy condition.

The method of Essers has been successfully applied to transonic potential equations and to one-dimensional Euler equations.

As far as I know, neither the non-formal relaxation methods of Wirz, nor the artificial time dependent method of Essers have until now been successfully applied to equations with hybrid character in steady state as the more-dimensional Euler equations have.

4. FORMAL RELAXATION METHODS FOR EULER EQUATIONS

4.1. Example.

The classic relaxation method for the potential equation has the properties of the in § 3 defined formal relaxation methods.

The first attempt to use relaxation approximatly in this formal sence to Euler equations is due to Désidéri and Tannehill (9). Their scheme is a modification of the well known MacCormack scheme through the introduction of an artificial variable.

For the one-dimensional hyperbolic equation :

$$\frac{\partial \xi}{\partial t} + V \frac{\partial \xi}{\partial x} = 0 \tag{4.1}$$

it is with $\lambda = \frac{V\Delta t}{\Delta x}$:

downwind :

$$\begin{aligned} \xi(x,t+\widetilde{\Delta t}) &= \xi(x,t) - \lambda\,[\xi(x+\Delta x,t) - \xi(x,t)] \\ \bar{\xi}(x,t+\Delta t) &= (1-r_1)\bar{\xi}(x,t) + r_1\xi(x,t+\widetilde{\Delta t}) \end{aligned} \tag{4.2}$$

upwind :

$$\begin{aligned} \bar{\xi}(x,t+\widetilde{\Delta t}) &= \bar{\xi}(x,t+\Delta t) - \lambda\,[\bar{\xi}(x,t+\Delta t) - \bar{\xi}(x-\Delta x,t+\Delta t)] \\ \xi(x,t+\Delta t) &= (1-r_2)\xi(x,t) + r_2\bar{\xi}(x,t+\widetilde{\Delta t}) \end{aligned} \tag{4.3}$$

This scheme reduces to the MacCormack-scheme for $r_1 = 1$ and $r_2 = .5$.
Substitution of (4.3) in (4.2) gives :

$$\begin{pmatrix} \bar{\xi}(x,t+\Delta t) \\ \xi(x,t+\Delta t) \end{pmatrix} = \begin{pmatrix} 1-r_1 & r_1(1+\lambda) \\ r_2(1-r_1)(1-\lambda) & 1-r_2+r_1r_2(1-2\lambda^2) \end{pmatrix} \begin{pmatrix} \bar{\xi}(x,t) \\ \xi(x,t) \end{pmatrix}$$

$$+ \begin{pmatrix} 0 & 0 \\ r_2(1-r_1)\lambda & r_1r_2\lambda(1+\lambda) \end{pmatrix} \begin{pmatrix} \bar{\xi}(x-\Delta x,t) \\ \xi(x-\Delta x,t) \end{pmatrix} + \begin{pmatrix} 0 & -r_1\lambda \\ 0 & -r_1r_2\lambda(1-\lambda) \end{pmatrix} \begin{pmatrix} \bar{\xi}(x+\Delta x,t) \\ \xi(x+\Delta x,t) \end{pmatrix}$$

Substitution of a Fourier component :

$$\begin{pmatrix} \bar{\xi} \\ \xi \end{pmatrix}_\omega = \begin{pmatrix} \bar{\Phi}(t) \\ \Phi(t) \end{pmatrix} e^{j\omega x}$$

yields :

$$\begin{pmatrix} \bar{\Phi}(t+\Delta t) \\ \Phi(t+\Delta t) \end{pmatrix} = \begin{pmatrix} G \end{pmatrix} \begin{pmatrix} \bar{\Phi}(t) \\ \Phi(t) \end{pmatrix}$$

The eigenvalues of the amplification matrix (G) are the solution of :

$$\mu^2 - 2P\mu + Q = 0$$

with :

$$P = 1 - \frac{r_1+r_2}{2} + \frac{r_1 r_2}{2}\left(1 - 2j\lambda\,\sin(\omega\Delta x) - 4\lambda^2 \sin^2(\frac{\omega\Delta x}{2})\right)$$

$$Q = (1-r_1)(1-r_2) \tag{4.4}$$

For $\omega = 0$:

$$\mu_1 = 1$$

$$\mu_2 = (1-r_1)(1-r_2)$$

For small ω :

$$\mathrm{Re}(\ln(\mu_1)) = -r_1 r_2 \frac{(r_1-r_2-r_1r_2)(r_1-r_2+r_1r_2)}{(r_1+r_2-r_1r_2)^3} \lambda^2\omega^2\Delta x^2 \tag{4.5}$$

We conclude from (4.5) that the convergence rate of the scheme is $O(\Delta x^2)$. An amelioration of the convergence rate can be expected from (4.5) when :

$$r_1+r_2-r_1r_2 = 0$$

With this (4.4) yields :

$$P = 1 - r_1r_2 j\lambda\,\sin(\omega\Delta x) - 2r_1r_2\lambda^2\sin^2(\frac{\omega\Delta x}{2})$$

$$Q = 1$$

Since $|\mu_1|\cdot|\mu_2| = 1$, one of the eigenvalues has a modulus larger than 1. Hence the scheme is unstable in the vicinity of the optimal relaxation.

Although the optimal relaxation cannot be reached in this method, we can see that optimal relaxation is closely connected with the change in the equivalent equation caused by an optimal choice of the relaxation factor, as this was also the case for the relaxation method for the potential equation.

The equivalent equations for (4.2) and (4.3) are :

$$\frac{\partial\bar{\xi}}{\partial t} + r_1 V \frac{\partial\xi}{\partial x} + \frac{r_1}{\Delta t}(\bar{\xi}-\xi) = 0 \tag{4.6}$$

$$\frac{\partial \xi}{\partial t} + r_2(1-r_1)V\,\frac{\partial \bar{\xi}}{\partial x} + 2r_1 r_2 V\,\frac{\partial \xi}{\partial x} + \frac{r_2}{\Delta t}(1-r_1)(\xi-\bar{\xi}) = 0 \tag{4.7}$$

These can be combined to :

$$r_1\,\frac{\partial \xi}{\partial t} + r_2(1-r_1)\frac{\partial \bar{\xi}}{\partial t} + r_1 r_2(1-r_1)V\,\frac{\partial \bar{\xi}}{\partial x} + r_1 r_2(1+r_1)V\,\frac{\partial \xi}{\partial x} = 0 \tag{4.8}$$

From (4.2) and (4.3) we conclude that ξ and $\bar{\xi}$ can only deviate from each other by an order Δt since both $\xi(x,t+\Delta t)$ and $\bar{\xi}(x,t+\Delta t)$ are constructed as a relaxation of ξ and $\bar{\xi}$.

Therefore up to lowest order (4.8) is :

$$(r_1+r_2-r_1r_2)\frac{\partial \xi}{\partial t} + 2r_1r_2\,\frac{\partial \xi}{\partial x} + \text{HOT} = 0 \tag{4.9}$$

Out of the vicinity of optimal relaxation, the equivalent equation is similar to the transient equation (4.1).

In the vicinity of optimal relaxation, the order of magnitude of the coefficient of the time derivative term changes, so that in leading order an other equivalent equation is associated to the scheme.

We shall now analyse how this mechanism of change of the coefficients of the time derivative terms can form the basis of a relaxation method for Euler equations.

We shall restrict ourselves to formal relaxation methods. The method of Désidéri and Tannehill is not fully formal since in (4.2) a preliminary value on level (n) is used. This causes again the preliminary values to have dynamics. But as has been shown, the dynamics of these variables and the physical variables are identical. This eliminates non-linear difficulties. In a formal method there is a guarantee that no dynamic behaviour is associated to the preliminary variables.

4.2. Analysis of the Change of the Equivalent Time Dependent Equation.

We consider as an example the discritisation of the equation df/dx = 0, which gives an equivalent equation :

$$a_1\,\frac{\partial f}{\partial t} + a_2\,\frac{\partial f}{\partial x} + b_1\,\frac{\partial^2 f}{\partial t^2} + b_2\,\frac{\partial^2 f}{\partial x \partial t} + b_3\,\frac{\partial^2 f}{\partial x^2} + \text{HOT} = 0 \tag{4.10}$$

Fourier analysis on this equation gives :

$$f = e^{\sigma t}\, e^{j\omega x}$$

$$a_1\sigma + a_2(j\omega) + b_1\sigma^2 + b_2\sigma(j\omega) + b_3(-\omega^2) = 0$$

which yields :

$$\sigma = \frac{-\,a_1 - j\omega b_2 \pm \sqrt{a_1^2 - \omega^2(b_2^2-4b_1b_3) + 2j\omega(a_1b_2-2a_2b_1)}}{2b_1}$$

$$\mathrm{Re}(\sigma_{1,2}) = \frac{-a_1 \pm \frac{1}{2}\sqrt{\sqrt{\left(a_1^2-\omega^2(b_2^2-4b_1b_3)\right)^2 + \omega^2(2a_1b_2-4a_2b_1)^2} + a_1^2-\omega^2(b_2^2-4b_1b_3)}}{2b_1}$$

For a well-posed equivalent equation :

$$\mathrm{Re}(\sigma_{1,2}) < 0$$

The conditions herefore are :

$$a_1b_1 > 0 \tag{4.11}$$

$$b_2^2 - 4b_1b_3 > 0 \tag{4.12}$$

$$a_1a_2b_1b_2 - a_1^2b_1b_3 - a_2^2b_1^2 > 0 \tag{4.13}$$

In general, the order of magnitude of the coefficients is :

$$\begin{gathered} a_1 = O(\Delta t), \qquad a_2 = O(\Delta x) \\ b_1 = O(\Delta t^2), \qquad b_2 = O(\Delta t\Delta x), \qquad b_3 = O(\Delta x^2) \end{gathered} \tag{4.14}$$

With (4.14), σ_1 and σ_2 are in leading order :

$$\sigma_1 = -\,j\omega\frac{a_2}{a_1} - \omega^2\,\frac{a_1a_2b_1b_2 - a_1^2b_1b_3 - a_2^2b_1^2}{a_1^3b_1}$$

$$\sigma_2 = -\,j\omega\left(\frac{b_2}{b_1} - \frac{a_2}{a_1}\right) - \frac{a_1}{b_1}$$

$$\mathrm{Re}(\sigma_1) = O\left(\frac{\Delta x^2}{\Delta t}\right), \qquad \mathrm{Re}(\sigma_2) = O\left(\frac{1}{\Delta t}\right)$$

An initial perturbation resolves into two perturbations. One moves with velocity a_2/a_1 and is damped $O(\Delta x^2/\Delta t)$. The second perturbation is damped $O(1/\Delta t)$.

An analysis which includes higher order terms in the equivalent equation yields the same results. One perturbation is damped $O(\Delta x^2/\Delta t)$ and all other perturbations are damped $O(1/\Delta t)$. The least damped perturbation corresponds to the lowest order part of the equivalent equation.

The damping of this perturbation can be increased by augmenting the order of the coefficients a_1, b_1, c_1, d_1, ... with one order Δx.

The equivalent equation is then :

$$a_1\frac{\partial f}{\partial t} + a_2\frac{\partial f}{\partial x} + b_1\frac{\partial^2 f}{\partial t^2} + b_2\frac{\partial^2 f}{\partial t\partial x} + b_3\frac{\partial^2 f}{\partial x^2} + c_2\frac{\partial^3 f}{\partial t^2\partial x} + c_3\frac{\partial^3 f}{\partial t\partial x^2} + c_4\frac{\partial^3 f}{\partial x^3} + \mathrm{HOT} = 0$$

with :

$$a_1 = O(\Delta t\Delta x), \qquad a_2 = O(\Delta x)$$

$$b_1 = O(\Delta t^2\Delta x), \qquad b_2 = O(\Delta t\Delta x), \qquad b_3 = O(\Delta x^2)$$

$$c_2 = O(\Delta t^2\Delta x), \qquad c_3 = O(\Delta t\Delta x^2), \qquad c_4 = O(\Delta x^3)$$

Fourier analysis gives :

$$\sigma^2(b_1 + j\omega c_2) + \sigma(a_1 + j\omega b_2 - \omega^2 c_3) + (j\omega a_2 - j\omega^3 c_4 - \omega^2 b_3) = 0$$

Up to lowest order of magnitude :

$$\sigma_{1,2} = \frac{-a_1 - j\omega b_2 \pm \sqrt{a_1^2 - \omega^2(b_2^2-4a_2c_2) + 2j\omega(a_1b_2-2b_1a_2)}}{2(b_1 + j\omega c_2)}$$

$Re(\sigma_{1,2}) < 0$ for all ω if :

$$b_1c_2(a_1b_2 - a_2b_1) > 0 \qquad (4.15)$$

and

$$a_1a_2(a_1b_2 - a_2b_1) > 0 \qquad (4.16)$$

Then :

$$Re(\sigma_1) = -\omega^2 \frac{a_2(a_1b_2 - b_1a_2)}{a_1^3}$$

$$Re(\sigma_2) = -\frac{a_1}{b_1}$$

With (4.11), (4.15) and (4.16) are equivalent to

$$a_2c_2 > 0 \qquad (4.17)$$

$$b_1a_2(a_1b_2 - b_1a_2) > 0 \qquad (4.18)$$

(4.18) is a degeneration of (4.13). The only additional condition introduced by the degeneration of the equivalent equation is (4.17).

We conclude :

If the conditions (4.11)(4.12)(4.13)(4.17) are fulfilled, the simultaneous degeneration of the coefficients of $\partial/\partial t$, $\partial^2/\partial t^2$, $\partial^3/\partial t^3$, ... by unaltered order difference, strongly improves the damping.
This concept is illustrated by the discretisation of df/dx = 0 by :

$$-p\overset{(n+1)}{f}(x-\Delta x) + q\tilde{f}(x) + (p-q)\overset{(n)}{f}(x+\Delta x) = 0$$

with $q \neq 0$

$$\overset{(n+1)}{f(x)} = r\tilde{f}(x) + (1-r)\overset{(n)}{f}(x)$$

$$f(x)^{(n+1)} = \frac{p}{q} r\, f(x-\Delta x)^{(n+1)} + (1-r) f(x)^{(n)} + (1 - \frac{p}{q}) r\, f(x+\Delta x)^{(n)}$$

The equivalent equation is :

$$(\frac{q}{r} - p)\Delta t \frac{\partial f}{\partial t} + (2p-q)\Delta x \frac{\partial f}{\partial x} + \frac{1}{2}(\frac{q}{r} - p)\Delta t^2 \frac{\partial^2 f}{\partial t^2} + p\Delta t\Delta x \frac{\partial^2 f}{\partial x \partial t}$$

$$- \frac{q}{2} \Delta x \frac{\partial^2 f}{\partial x^2} + \frac{1}{6}(\frac{q}{r} - p)\Delta t^3 \frac{\partial^3 f}{\partial t^3} + \frac{1}{2} p \Delta t^2 \Delta x \frac{\partial^2 f}{\partial t^2 \partial x}$$

$$- \frac{p}{2} \Delta t \Delta x^2 \frac{\partial^3 f}{\partial t \partial x^2} + \frac{1}{6}(2p-q)\Delta x^3 \frac{\partial^3 f}{\partial x^3} + \text{HOT} = 0$$

With the arbitrary choice 2p-q = 1, the conditions (4.11)(4.12)(4.13)(4.17) become :

$$a_1 b_1 = \frac{1}{2}(\frac{q}{r} - p)^2 \Delta t^3 > 0$$

$$a_2 c_2 = (2p-q)p\, \Delta t \Delta x^2 > 0 \quad \rightarrow \quad p > 0$$

$$b_2^2 - 4b_1 b_3 = \left(p^2 + (\frac{q}{r} - p)q\right)\Delta t^2 \Delta x^2 > 0 \quad \rightarrow \quad 0 < r < 4$$

$$a_1 a_2 b_1 b_2 - a_1 b_1 b_3 - a_2 b_1 > 0 \quad \rightarrow \quad q\left(rp + (1-r)q\right) > 0$$

The optimal relaxation factor is :

$$r_o = \frac{q}{p}$$

First example

$$p = 1.2 \qquad q = 1.4 \qquad r_o = \frac{7}{6}$$

$$f(x)^{(n+1)} = \frac{6}{7} r\, f(x-\Delta x)^{(n+1)} + \frac{1}{7} r\, f(x+\Delta x)^{(n)} + (1-r) f(x)^{(n)}$$

The amplification factor of a harmonic perturbation is :

$$G = \frac{(1-r) + \frac{1}{7} r\, e^{j\phi}}{1 - \frac{6}{7} r\, e^{-j\phi}}$$

with : $\phi = \omega\Delta x$.

Figure 3 shows the corresponding Nyquist-curves for r = 1, 7/6, 1.5.

The Nyquist-curves start for $\omega = 0$ for every relaxation factor r from the point 1, except for the optimal relaxation factor.
All curves end for $\omega\Delta x = \pi$ on the real axis.

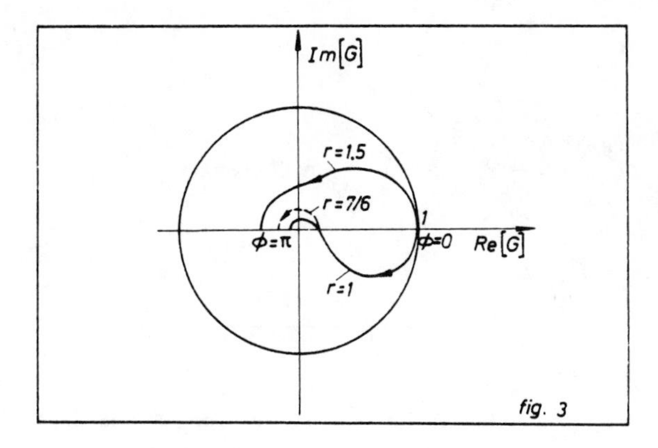

fig. 3

The damping for $r = r_o$ is much stronger than for other r, especially for the small wave numbers. A typical small wave number is :

$$N_\ell \; \omega \; \Delta x = \pi$$

in which N_ℓ is the number of points in the longitudinal direction of the field.

$$N_\ell \approx 50$$

The largest wave number is : $\omega \; \Delta x = \pi$

The moduli of the amplification factors for a small wave number and the largest wave number are depicted in figure 4.

The large wave numbers are strongly damped in a wide range of the relaxation factor. Small wave numbers are only strongly damped in the vicinity of

$$r = r_o$$

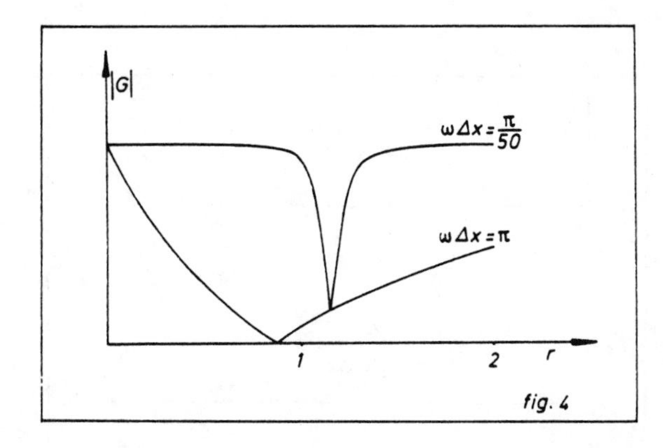

fig. 4

Second example

$$p = .8 \qquad q = .6 \qquad r_o = .75$$

$$\overset{(n+1)}{f(x)} = \frac{4}{3}\, r\, \overset{(n+1)}{f(x-\Delta x)} - \frac{1}{3}\, r\, \overset{(n)}{f(x+\Delta x)} + (1-r)\overset{(n)}{f(x)}$$

The amplification factor is :

$$G = \frac{1 - r - \frac{1}{3}\, r\, e^{j\phi}}{1 - \frac{4}{3}\, r\, e^{-j\phi}}$$

The corresponding Nyquist-curves for r = .5, .75 and 1 are given in figure 5.
The corresponding damping curves are given in figure 6.
This case leads to the same conclusion as the first example.

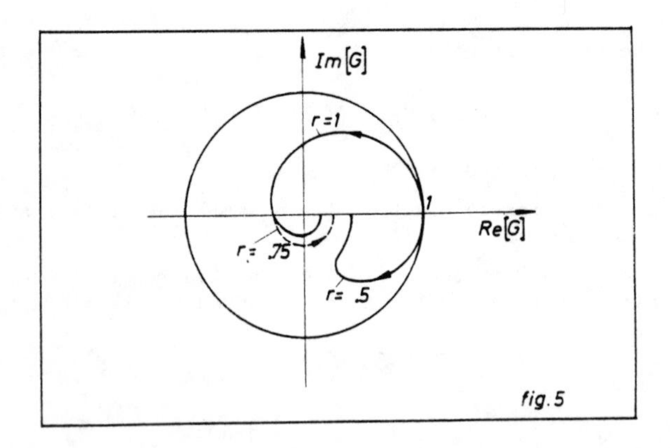

fig.5

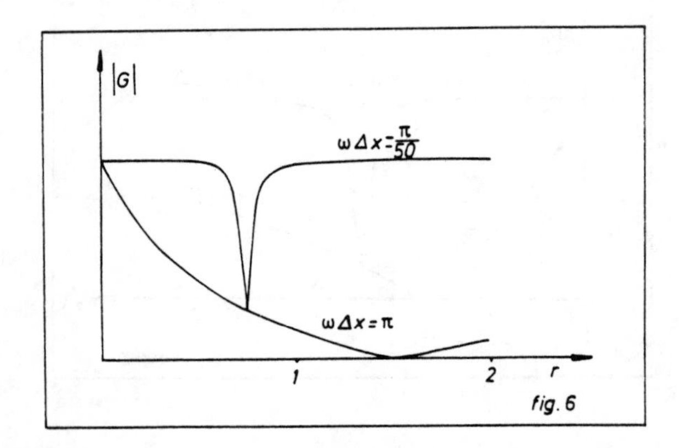

fig.6

Third example

$$p = 1 \qquad q = 1 \qquad r_o = 1$$

$$f(x)^{(n+1)} = r\, f(x-\Delta x)^{(n+1)} + (1-r) f(x)^{(n)}$$

The amplification factor is :

$$G = \frac{1 - r}{1 - r\, e^{-j\phi}}$$

Figure 7 shows the Nyquist curves for r = .5, 1, 1.5.

The curve $r = r_o$ schrinks completely into the origin. So there is complete damping for $r = r_o$ for all wave numbers. As a consequence, the steady state is reached in one iteration. Such a case is called a critical relaxation.

Figure 8 shows the corresponding damping curves.

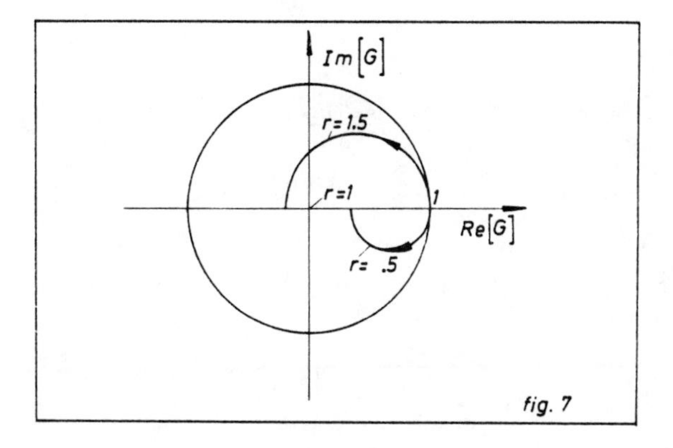

fig. 7

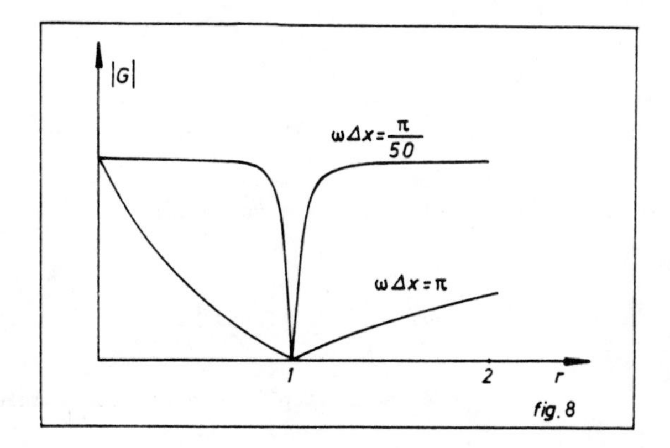

fig. 8

The results of these examples can be generalised for a relaxation scheme of the type :

$$D\,f^{(n+1)}(x) = A\,f^{(n+1)}(x-\Delta x) + B\,f^{(n)}(x) + C\,f^{(n)}(x+\Delta x) \qquad (4.19)$$

with D = A + B + C.

Neumann-analysis gives :

$$G = \frac{B + C\,e^{j\phi}}{D - A\,e^{-j\phi}} \qquad (4.20)$$

$|G| < 1$ if :

$$(A+B)(A+C) \geqslant 0 \qquad (4.21)$$

The equivalent equation is :

$$(D-A)\left(\frac{\partial f}{\partial t}\Delta t + \frac{1}{2}\frac{\partial^2 f}{\partial t^2}\Delta t^2\right) + \frac{\partial f}{\partial x}\Delta x(A-C) + \frac{\partial^2 f}{\partial x\partial t}\Delta t\Delta x\,A - \frac{1}{2}\frac{\partial^2 f}{\partial x^2}\Delta x^2(A+C)$$

$$+ \frac{1}{2}\frac{\partial^3 f}{\partial x\partial t^2}\Delta x\Delta t^2 A + \ldots = 0 \qquad (4.22)$$

The transfert velocity is :

$$V_T = \frac{(A-C)\Delta x}{(D-A)\Delta t}$$

Optimal relaxation is reached for :

$$A = D \qquad (4.23)$$

(4.20) is then :

$$G = \frac{-C(1 - e^{j\phi})}{A(1 - e^{-j\phi})}$$

For stability :

$$|C| < |A| \qquad (4.24)$$

The relaxation is also critical if :

$$B = 0 \quad \text{and} \quad C = 0 \qquad (4.25)$$

One notices that the conditions (4.21)(4.24) are equivalent to (4.11)(4.12) (4.13)(4.17).

4.3. A Relaxation Scheme for a One-dimensional Transport Equation with Variable Transport Velocity

The relaxation principle of 4.2 is not applicable to steady Euler Equations.

In two dimensions these are :

$$A_1 \frac{\partial \xi}{\partial x} + A_2 \frac{\partial \xi}{\partial y} = 0 \qquad (2.26)$$

For line relaxation, the equivalent equations corresponding to the discretisation of $\partial\xi/\partial x$ and $\partial\xi/\partial y$ have the form :

$$\frac{\partial \xi}{\partial x} : \quad a_1 \frac{\partial \xi}{\partial t} + a_2 \frac{\partial \xi}{\partial x} + b_1 \frac{\partial^2 \xi}{\partial t^2} + b_2 \frac{\partial^2 \xi}{\partial t \partial x} + b_3 \frac{\partial^2 \xi}{\partial x^2} + \text{HOT}$$

$$\frac{\partial \xi}{\partial y} : \quad m_1 \frac{\partial \xi}{\partial y} + n_1 \frac{\partial^2 \xi}{\partial t \partial y} + n_2 \frac{\partial^2 \xi}{\partial y^2} + \text{HOT}$$

The equivalent equation corresponding to (4.26) is :

$$a_1 A_1 \frac{\partial \xi}{\partial t} + a_2 A_1 \frac{\partial \xi}{\partial x} + n_1 A_2 \frac{\partial \xi}{\partial y} + \text{HOT} = 0 \qquad (4.27)$$

The characteristic matrix of (4.27) is :

$$- a_1^{-1} A_1^{-1} (k_1 a_2 A_1 + k_2 n_1 A_2) = k_1' I + k_2' A_1^{-1} A_2$$

with eigenvalues :

$$\lambda_1 = \lambda_2 = k_1' + k_2' \frac{v}{u}$$

$$\lambda_{3,4} = k_1' + k_2' \frac{uv}{u^2 - c^2} \pm \frac{k_2' c \sqrt{u^2 + v^2 - c^2}}{u^2 - c^2}$$

Since λ_3 and λ_4 are complex in subsonic flow, the equivalent equation cannot be hyperbolic with respect to t.

When relaxation is applied to the unsteady Euler equations, a term $I \frac{\partial \xi}{\partial t}$ is added in the equivalent equation.

The equivalent equation has the form :

$$(a_1 A_1 + a_1 k\, I) \frac{\partial \xi}{\partial t} + a_2 A_1 \frac{\partial \xi}{\partial x} + m_1 A_2 \frac{\partial \xi}{\partial y} + \text{HOT} = 0$$

for $k > c-u$, this equation is always hyperbolic with respect to time.

Therefore we shall first develop a relaxation scheme for the one dimensional equation

$$\frac{\partial \xi}{\partial t} + U \frac{\partial \xi}{\partial x} = 0 \qquad (4.28)$$

in which U can vary between u+c and u-c, which are both dependent on ξ.

We first prove two theorems.

First non-existence theorem

It is impossible to construct a relaxation scheme for (4.28) as a one step scheme when U can have both positive and negative values.

A general one step scheme is :

$$\xi(x,t+\widetilde{\Delta t}) = \xi(x,t) + Z\left(\ell\ \xi(x-\Delta x,t+\Delta t) + m_1\xi(x,t+\widetilde{\Delta t}) + m_2\xi(x,t) + n\ \xi(x+\Delta x,t)\right) \tag{4.29}$$

with : $Z = U\frac{\Delta t}{\Delta x}$ and $\ell + m_1 + m_2 + n = 0$

$$\xi(x,t+\Delta t) = r\ \xi(x,t+\widetilde{\Delta t}) + (1-r)\xi(x,t)$$

This scheme gives a numerical domain of dependence as depicted in figure 9. This might bring us to think that the scheme can be made stable for :

$$-U_o < U < +\infty$$

in which U_o is a positive value.

We prove now that this scheme cannot be stable.
(4.29) gives :

$$(1-m_1Z)\xi(x,t+\Delta t) = (1-m_1Z)\xi(x,t)$$
$$+ rZ\left(\ell\ \xi(x-\Delta x,t+\Delta t) + (m_1+m_2)\xi(x,t) + n\ \xi(x+\Delta x,t)\right)$$

This is an equation of type (4.19) in which the coefficients are linear polynomials in Z.

The amplification factor is :

$$G = \frac{(1-m_1Z) + (m_1+m_2)rZ + nrZ\ e^{j\phi}}{(1-m_1Z) - \ell rZ\ e^{-j\phi}}$$

$|G| < 1$ cannot be fulfilled for $\phi = \omega\ \Delta x = \pi$

Then :

$$G = \frac{1 - m_1Z + (m_1+m_2)rZ - nrZ}{1 - m_1Z + \ell rZ} = \frac{1 - m_1Z - (\ell+2n)rZ}{1 - m_1Z + \ell rZ}$$

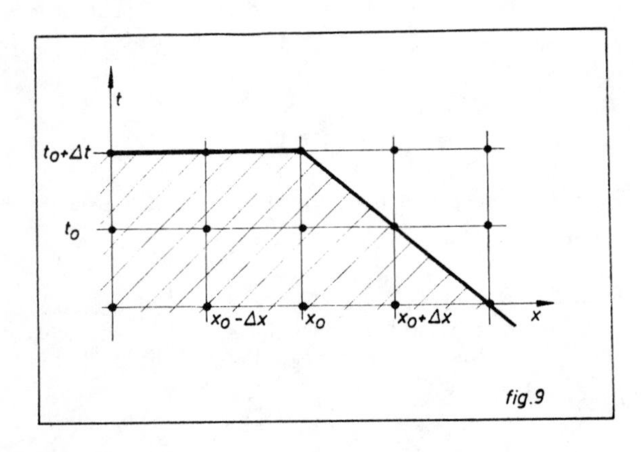

fig.9

$|G| < 1$ if :

$$\left(1 - m_1Z - (\ell+2n)rZ\right)^2 \leqslant \left(1 - m_1Z + \ell rZ\right)^2$$

or

$$\left(2 - 2m_1Z - 2nrZ\right)\left(-(2\ell+2n)rZ\right) \leqslant 0$$

The sign of the left hand side changes in the vicinity of $Z = 0$. Hence the scheme cannot be stable for both positive and negative Z.

Second non-existence theorem

It is impossible to bring a relaxation scheme in critical relaxation for both a positive and a negative velocity.

We first consider a scheme that iterates from left to right :

$$D\ \xi(x,t+\Delta t) = A\ \xi(x-\Delta x,t+\Delta t) + B\ \xi(x,t) + C\ \xi(x+\Delta x,t) \qquad (4.30)$$

with $D = A + B + C$.

This scheme is of the type (4.19). Critical relaxation corresponds to $B = C = 0$ Then the scheme reduces to :

$$\xi(x,t+\Delta t) = \xi(x-\Delta x,t+\Delta t)$$

This form can clearly only correspond to a positive velocity.

For $A = B = 0$, (4.30) reduces to :

$$\xi(x,t+\Delta t) = \xi(x+\Delta x,t)$$

This is not a critical relaxation but a perfect transport in negative direction with velocity $-\Delta x/\Delta t$.

Similarly a scheme that iterates from right to left can only be critically relaxed for a negative velocity.

Such a scheme has the form :

$$D\ \xi(x,t+\Delta t) = A\ \xi(x-\Delta x,t) + B\ \xi(x,t) + C\ \xi(x+\Delta x,t+\Delta t)$$

and can be reduced to :

$$\xi(x,t+\Delta t) = \xi(x+\Delta x,t+\Delta t)$$

Construction of a relaxation scheme

We are limiting ourselves to a predictor-corrector form of the MacCormack-type. The reason for this choice is the form of the method of Désidéri and Tannehill.

Step 1 : predictor :

$$\xi(x,t+\widetilde{\Delta t}) = \xi(x,t) - Z\left(\ell\ \xi(x-\Delta x,t+\Delta t) + \alpha_1 m\ \xi(x,t+\widetilde{\Delta t}) + \beta_1 m\ \xi(x,t) + n\ \xi(x+\Delta x,t)\right) \qquad (4.31)$$

Step 2 : corrector :

$$\xi(x,t+\tilde{\tilde{\Delta t}}) = \xi(x,t) + Z\{n\ \xi(x-\Delta x,t+\Delta t) + \alpha_2 m\ \xi(x,t+\tilde{\tilde{\Delta t}}) + \beta_2 m\ \xi(x,t+\tilde{\Delta t}) + \gamma_2 m\ \xi(x,t) + \ell\ \xi(x+\Delta x,t)\} \qquad (4.32)$$

Step 3 : relaxation :

$$\xi(x,t+\Delta t) = r_1\xi(x,t+\tilde{\Delta t}) + r_2\xi(x,t+\tilde{\tilde{\Delta t}}) + (1-r_1-r_2)\xi(x,t) \qquad (4.33)$$

with $\alpha_1 + \beta_1 = 1$

$\alpha_2 + \beta_2 + \gamma_2 = 1$

In the downwind-upwind version : $\ell = 0$, $m = -1$, $n = 1$.

In the upwind-downwind version : $\ell = -1$, $m = -1$, $n = 0$.

(4.31)(4.32)(4.33) give :

$$\xi(x,t+\Delta t) = \xi(x,t) - \frac{r_1 Z}{(1+\alpha_1 mZ)}\{\ell\ \xi(x-\Delta x,t+\Delta t) + m\ \xi(x,t) + n\ \xi(x+\Delta x,t)\}$$
$$+ \frac{r_2 Z}{(1-\alpha_2 mZ)}\{n\ \xi(x-\Delta x,t+\Delta t) + m\ \xi(x,t) + \ell\ \xi(x+\Delta x,t)\}$$
$$- \frac{\beta_2 m r_2 Z^2}{(1+\alpha_1 mZ)(1-\alpha_2 mZ)}\{\ell\ \xi(x-\Delta x,t+\Delta t) + m\ \xi(x,t) + n\ \xi(x+\Delta x,t)\} \qquad (4.34)$$

We consider two special cases :

<u>Upwind-downwind</u>

$\ell = -1 \qquad m = 1 \qquad n = 0$

$\alpha_1 = \alpha \qquad \alpha_2 = -\alpha \qquad \beta_2 = 1$

$r_1 = r/2 = R \qquad r_2 = r/2 = R$

For these :

$$\xi(x,t+\tilde{\Delta t}) - \xi(x,t) + Z\{\alpha\xi(x,t+\tilde{\Delta t}) + (1-\alpha)\xi(x,t) - \xi(x-\Delta x,t+\Delta t)\} = 0 \qquad (4.35)$$

$$\xi(x,t+\tilde{\tilde{\Delta t}}) - \xi(x,t) + Z\{\xi(x+\Delta x,t) + \alpha\ \xi(x,t+\tilde{\tilde{\Delta t}}) - \xi(x,t+\tilde{\Delta t}) - \alpha\ \xi(x,t)\} = 0 \qquad (4.36)$$

This gives :

$$(1+\alpha Z)\xi(x,t+\tilde{\Delta t}) = (1+\alpha Z)\xi(x,t) - Z\{\xi(x,t) - \xi(x-\Delta x,t+\Delta t)\}$$
$$(1+\alpha Z)\xi(x,t+\tilde{\tilde{\Delta t}}) = (1+\alpha Z)\xi(x,t) - Z\{\xi(x+\Delta x,t) - \xi(x,t+\tilde{\Delta t})\} \qquad (4.37)$$

with $X = \frac{Z}{1+\alpha Z}$ this yields :

$$\xi(x,t+\Delta t) = (RX+RX^2)\xi(x-\Delta x,t+\Delta t) + (1-RX^2)\xi(x,t) - RX\xi(x+\Delta x,t) \qquad (4.38)$$

Downwind-upwind

$$\ell = 0 \qquad m = -1 \qquad n = 1$$

$$\alpha_1 = -\alpha \qquad \alpha_2 = \alpha \qquad \beta_2 = 1$$

$$r_1 = r/2 = R \qquad r_2 = r/2 = R$$

For these :

$$\xi(x,t+\widetilde{\Delta t}) - \xi(x,t) + Z\{\xi(x+\Delta x,t) + \alpha\xi(x,t+\widetilde{\Delta t}) - (1+\alpha)\xi(x,t)\} = 0 \quad (4.39)$$

$$\xi(x,t+\widetilde{\widetilde{\Delta t}}) - \xi(x,t) + Z\{\alpha\xi(x,t+\widetilde{\widetilde{\Delta t}}) + \xi(x,t+\widetilde{\Delta t}) - \alpha\xi(x,t) - \xi(x-\Delta x,t+\Delta t)\} = 0 \quad (4.40)$$

This gives :

$$\begin{aligned}(1+\alpha Z)\xi(x,t+\widetilde{\Delta t}) &= (1+\alpha Z)\xi(x,t) - Z\{\xi(x+\Delta x,t) - \xi(x,t)\}\\ (1+\alpha Z)\xi(x,t+\widetilde{\widetilde{\Delta t}}) &= (1+\alpha Z)\xi(x,t) - Z\{\xi(x,t+\widetilde{\Delta t}) - \xi(x-\Delta x,t+\Delta t)\}\end{aligned} \quad (4.41)$$

and :

$$\xi(x,t+\Delta t) = RX\ \xi(x-\Delta x,t+\Delta t) + (1-RX^2)\xi(x,t) + (-RX+RX^2)\xi(x+\Delta x,t) \quad (4.42)$$

Analysis of the upwind-downwind scheme

The scheme (4.38) has the form (4.19) with :

$$D = 1 \qquad A = RX+RX^2 \qquad B = 1-RX^2 \qquad C = -RX$$

The conditions are :

Stability (4.21) :

$$(1+RX)(RX^2) \geqslant 0 \quad \rightarrow \quad 1+RX \geqslant 0$$

Optimal relaxation (4.23) :

$$1 = RX+RX^2$$

Critical relaxation (4.25) :

$$1-RX^2 = 0 \qquad \text{and} \qquad RX = 0$$

Optimal relaxation is reached for :

$$X = \frac{-1 \pm \sqrt{1 + \frac{4}{R}}}{2}$$

This gives a positive and a negative value :

$$X^+ = \frac{Z^+}{1 + \alpha Z^+} \qquad \text{and} \qquad X^- = \frac{Z^-}{1 + \alpha Z^-}$$

The velocities u+c and u-c can both be optimally relaxed by :

$$Z^+ = (u+c)\frac{\Delta t}{\Delta x} = (M+1)C_o$$

$$Z^- = (u-c)\frac{\Delta t}{\Delta x} = (M-1)C_o$$

with $C_o = c\,\frac{\Delta t}{\Delta x}$.

α, Δt and R are to be chosen by :

$$\frac{(M+1)C_o}{1 + \alpha(M+1)C_o} = \frac{-1 + \sqrt{1 + \frac{4}{R}}}{2}$$

$$\frac{(M-1)C_o}{1 + \alpha(M-1)C_o} = \frac{-1 - \sqrt{1 + \frac{4}{R}}}{2}$$

This yields :

$$\alpha = \frac{R + M\sqrt{R^2 + 4R}}{2}$$

$$C_o = \frac{2}{\sqrt{R^2 + 4R}\,(1-M^2)}$$

Critical relaxation cannot be obtained.

The stability restriction $1+RX \geqslant 0$ gives for

$1+RX^- \geqslant 0$: $\qquad R \leqslant \frac{1}{2} \qquad$ or $\qquad r \leqslant 1$

Only under-relaxation is possible.

The transfert velocity in the equivalent equation (4.22) is :

$$V_T = \frac{A-C}{D-A}\,\frac{\Delta x}{\Delta t} = \frac{2RX + RX^2}{1 - RX - RX^2}\,\frac{\Delta x}{\Delta t}$$

The transfert velocity is given in figure 10 for r = .5.

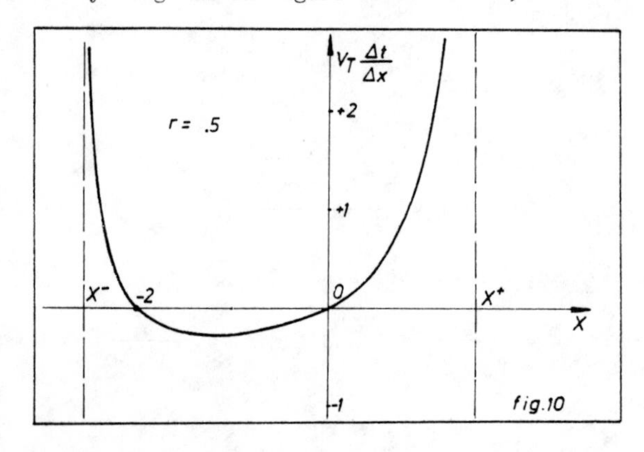

fig.10

A functional relationship between V_T and X as shown in figure 10 is not acceptable. For negative velocities in the vicinity of X^-, there is a sign reverse between the physical and the mathematical velocity. Hence the scheme certainly cannot fulfil the entropy condition. This entropy condition is even more stringent. The transfert velocity has to be a monotonic increasing function of the physical velocity. When this is not fulfilled, a physical dispersing perturbation will not necessarly mathematically disperse.

Since this is a general argument, we can formulate a

third non-existence theorem

It is impossible to construct a relaxation scheme for (4.28) in which a positive and a negative velocity are simultaneously optimally relaxed.

Optimal relaxation implies that the corresponding transfert velocity in the equivalent equation becomes infinite. For a scheme that iterates from left to right, only a positive velocity can be optimally relaxed. A negative velocity cannot be optimally relaxed.

If the negative velocity would be transformed to a transfert velocity $+\infty$, then the entropy condition could not be fulfilled.

If the negative velocity would be transformed to a transfert velocity $-\infty$, the stability conditions would be violated.

The best upwind-downwind scheme is thus a scheme for which the positive velocity u+c is optimally relaxed and for which the transfert velocity corresponding to u-c is maximised. The maximum negative transfert velocity is $-\Delta x/\Delta t$. This is reached for X^- on the stability limit : $X^- = -1/R$.

We choose α, Δt and R by :

$$\frac{(M+1)C_o}{1+\alpha(M+1)C_o} = \frac{-1+\sqrt{1+\frac{4}{R}}}{2}$$

$$\frac{(M-1)C_o}{1+\alpha(M-1)C_o} = -\frac{1}{R}$$

This yields :

$$\alpha = \frac{(R-2-\sqrt{R^2+4R})M - 2 - R + \sqrt{R^2+4R}}{2-2\sqrt{R^2+4R}}$$
$$C_o = \frac{2 - R\sqrt{R^2+4R}}{(2-R+\sqrt{R^2+4R})(M^2-1)} \qquad (4.43)$$

The optimal relaxation for u+c cannot be made critical. The transfert for u-c can be made a perfect transport by choosing :

$$A = RX + RX^2 = 0$$
$$B = 1 - RX^2 = 0$$
$$C = -RX = 1$$

This is reached for R = 1 and $X^- = -1$.

For R = 1, (4.43) becomes :

$$\alpha = \frac{(\sqrt{5}+1)M + 3 - \sqrt{5}}{2(\sqrt{5}-1)} \simeq 1.3090\ M - \ .3090$$

$$C_o = \frac{2(\sqrt{5}-1)}{(\sqrt{5}+1)(1-M^2)} \simeq \frac{.7639}{1-M} \qquad (4.44)$$

When α and C_o are choosen according to (4.44), the upwind-downwind scheme is optimal for $R = 1$, this is for $r = 2$.

Analysis of the downwind-upwind scheme

The scheme (4.42) has the form (4.19) with :

$$D = 1 \qquad A = RX \qquad B = 1-RX^2 \qquad C = -RX+RX^2$$

The conditions are :

Stability (4.21) :

$$(1 + RX - RX^2)(RX^2) \geqslant 0$$

Optimal relaxation (4.23) :

$$RX = 1$$

Critical relaxation (4.25) :

$$1 - RX^2 = 0 \qquad \text{and} \qquad -RX + RX^2 = 0$$

Optimal relaxation can only be reached for a positive velocity. This relaxation is also critical for $R = 1$ and $X^+ = 1$.

The transfert velocity is :

$$V_T = \frac{A - C}{D - A}\frac{\Delta x}{\Delta t} = \frac{2RX - RX^2}{1 - RX}\frac{\Delta x}{\Delta t}$$

This is for $R = 1$:

$$V_T = \frac{2X - X^2}{1 - X}\frac{\Delta x}{\Delta t}$$

The best downwind-upwind scheme is reached when the positive velocity u+c is critically relaxed and the transfert velocity corresponding to u-c is maximised.

$V_T = -\ \Delta x/\Delta t$ for $X^- = \dfrac{1-\sqrt{5}}{2}$

α and Δt have to be choosen according to :

$$\frac{(M+1)C_o}{1 + \alpha(M+1)C_o} = 1$$

$$\frac{(M-1)C_o}{1 + \alpha(M-1)C_o} = \frac{1-\sqrt{5}}{2}$$

This yields :

$$\alpha = \frac{(\sqrt{5}+1)M + 3 - \sqrt{5}}{2(\sqrt{5}-1)} \simeq 1.3090\ M - \ .3090$$

$$C_o = \frac{2(\sqrt{5}-1)}{(\sqrt{5}+1)(1-M^2)} \simeq \frac{.7639}{1 - M^2} \qquad (4.45)$$

The downwind-upwind scheme is better than the upwind-downwind scheme since in the first, the positive velocity u+c can be critically relaxed. The negative velocity u-c has a corresponding transfert velocity that is the same in both cases. For further calculations, we restrict ourselves to the downwind-upwind scheme.

If we want to apply the downwind-upwind scheme to a case in which U can have only positive values, then the parameters cannot be calculated with the formulae (4.45). This case is a model for the supersonic Euler equations. Both velocities u+c and u-c should have a corresponding transfert velocity $+\infty$. When u-c is varying in the field, X^- can for instance be chosen according to figure 11.

For $M < M_o$:

$$X^+ = 1$$

$$X^- = -Q$$

hence :

$$\alpha = (M - 1 + Q_o)/Q_o$$

$$C_o = Q_o/(1 - M^2)$$

with $Q_o = 2Q/(1+Q)$.

For $M > M_o$:

$$X^+ = 1$$

$$X^- = (M-1)/(M - M_o + \frac{1-M_o}{Q})$$

hence :

$$\alpha = (M_o - 1 + Q_o)/Q_o$$

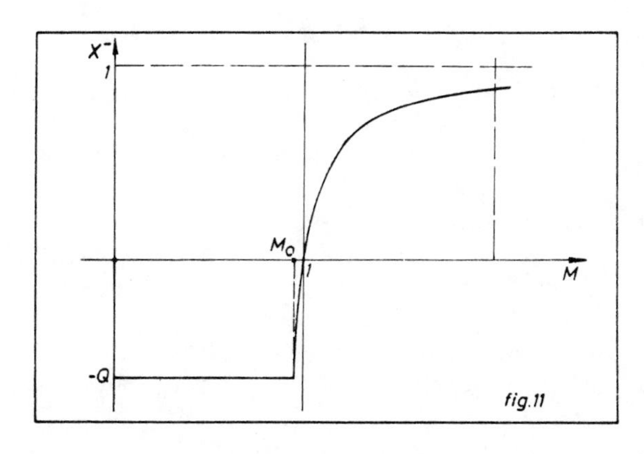

fig.11

$$C_o = Q_o/\left((1-M_o)(1+M)\right)$$

Possible values are : $Q = .615$ and $M_o = .95$.

4.4. Non Linear Analysis.

We apply the scheme (4.39)(4.40) to the one-dimensional equation :

$$\frac{\partial f}{\partial t} + \frac{\partial g}{\partial x} = 0$$

which is the Burgers equation for $g = 1/2\ f^2$.
The first step according to (4.39) is :

$$f(x,t+\widetilde{\Delta t}) - f(x,t) + \frac{\Delta t}{\Delta x}\left(g(x+\Delta x,t) + \alpha\, g(x,t+\widetilde{\Delta t}) - (1+\alpha)g(x,t)\right) = 0 \tag{4.46}$$

In order to solve for $f(x,t+\widetilde{\Delta t})$ we iterate starting from a guess $f(x,t+\overline{\Delta t})$.
With :

$$\Delta f_1 = f(x,t+\widetilde{\Delta t}) - f(x,t+\overline{\Delta t})$$

$$g(x,t+\widetilde{\Delta t}) = g(x,t+\overline{\Delta t}) + \frac{dg}{df}(x,t+\overline{\Delta t})\Delta f_1$$

(4.46) yields :

$$\left(1 + \alpha\frac{\Delta t}{\Delta x}\frac{dg}{df}(x,t+\overline{\Delta t})\right)\Delta f_1 = f(x,t) - f(x,t+\overline{\Delta t}) - \frac{\Delta t}{\Delta x}\left(g(x+\Delta x,t) + \alpha\, g(x,t+\overline{\Delta t}) - (1+\alpha)g(x,t)\right) \tag{4.47}$$

The second step according to (4.40) is :

$$f(x,t+\tilde{\tilde{\Delta t}}) - f(x,t) + \frac{\Delta t}{\Delta x}\left(\alpha\, g(x,t+\tilde{\tilde{\Delta t}}) + g(x,t+\widetilde{\Delta t}) - \alpha\, g(x,t) - g(x-\Delta x,t+\Delta t)\right) = 0 \tag{4.48}$$

Starting from a guess $f(x,t+\overline{\overline{\Delta t}})$:

$$\Delta f_2 = f(x,t+\tilde{\tilde{\Delta t}}) - f(x,t+\overline{\overline{\Delta t}})$$

$$g(x,t+\tilde{\tilde{\Delta t}}) \simeq g(x,t+\overline{\overline{\Delta t}}) + \frac{dg}{df}(x,t+\overline{\overline{\Delta t}})\Delta f_2$$

(4.48) yields :

$$\left(1 + \alpha\frac{\Delta t}{\Delta x}\frac{dg}{df}(x,t+\overline{\overline{\Delta t}})\right)\Delta f_2 = f(x,t) - f(x,t+\overline{\overline{\Delta t}}) - \frac{\Delta t}{\Delta x}\left(\alpha g(x,t+\overline{\overline{\Delta t}}) + g(x,t+\widetilde{\Delta t}) - \alpha g(x,t) - g(x-\Delta x,t+\Delta t)\right) \tag{4.49}$$

In order to iterate the complete scheme we apply the following procedure :

- Let $\overline{f}$ be a guess for $f(x,t+\Delta t)^{(i)}$ on level i. On the first level $\overline{f}$ is $f(x,t)$. Then we pose $f(x,t+\overline{\Delta t}) = \overline{f}$.

- With this value Δf_1 is calculated according to (4.47). An improved value for $f(x,t+\widetilde{\Delta t})$ is then $\bar{\bar{f}} = \bar{f} + \Delta f_1$. We also pose $f(x,t+\overline{\overline{\Delta t}}) = \bar{\bar{f}}$.
- In (4.49) $g(x,t+\widetilde{\Delta t})$ is calculated as $g(x,t+\overline{\overline{\Delta t}}) = g(\bar{\bar{f}})$. With Δf_2 from (4.49) an improved value of $f(x,t+\widetilde{\widetilde{\Delta t}})$ is $\bar{\bar{\bar{f}}} = \bar{\bar{f}} + \Delta f_2$.
- By relaxation an improved value for $f(x,t+\Delta t)$ is :

$$f(x,t+\Delta t)^{(i+1)} = \frac{1}{2}\, r\, \bar{\bar{f}} + \frac{1}{2}\, r\, \bar{\bar{\bar{f}}} + (1-r)f(x,t)$$

This procedure is repeated until the difference between $f(x,t+\Delta t)^{(i+1)}$ and $f(x,t+\Delta t)^{(i)}$ is sufficiently small.

The last value is then the final value for $f(x,t+\Delta t)$.

This procedure is non-lineary well-posed for the Burgers equation.

With $g = \frac{1}{2} f^2$ and $\Delta x/\Delta t = d$ (4.47) is :

$$(d + \alpha\bar{f})\Delta f_1 = d(f-\bar{f}) - \left\{\frac{1}{2} f^2(x+\Delta x,t) + \frac{1}{2}\alpha\bar{f}^2 - \frac{1}{2}(1+\alpha)f^2\right\}$$

hence :

$$\bar{\bar{f}} = \bar{f} + \Delta f_1 = \frac{df + \frac{1}{2}(1+\alpha)f^2 - \frac{1}{2} f^2(x+\Delta x,t) + \frac{1}{2}\alpha\bar{f}}{d + \alpha\bar{f}} \tag{4.50}$$

(4.49) yields :

$$(d+\alpha\bar{\bar{f}})\Delta f_2 = d(f-\bar{\bar{f}}) - \left\{\frac{1}{2}\alpha\bar{\bar{f}}^2 + \frac{1}{2}\bar{\bar{f}}^2 - \frac{1}{2}\alpha f^2 - \frac{1}{2} f^2(x-\Delta x,t+\Delta t)\right\}$$

hence :

$$\bar{\bar{\bar{f}}} = \bar{\bar{f}} + \Delta f_2 = \frac{df + \frac{1}{2}\alpha f^2 + \frac{1}{2} f^2(x-\Delta x,t+\Delta t) - (1-\alpha)\frac{1}{2}\bar{\bar{f}}^2}{d + \alpha\bar{\bar{f}}} \tag{4.51}$$

For final convergence :

$$\bar{f} = \frac{1}{2}\, r\, \bar{\bar{f}} + \frac{1}{2}\, r\, \bar{\bar{\bar{f}}} + (1-r)f$$

In the limiting case : $\Delta x \to 0$ and $\Delta t \to 0$

$$\bar{f} = f + \Delta f$$

$$f(x+\Delta x,t) = f + \varepsilon_1$$

$$f(x-\Delta x,t+\Delta t) = f + \varepsilon_2$$

Δf, ε_1 and ε_2 are small values.

Up to second order in Δf, ε_1 and ε_2 (4.50) yields :

$$\overline{\overline{f}} = \frac{df + \frac{1}{2}(1+\alpha)f^2 - \frac{1}{2}f^2 - f\varepsilon_1 - \frac{1}{2}\varepsilon_1^2 + \frac{1}{2}\alpha f^2 + \alpha f\Delta f + \frac{1}{2}\alpha(\Delta f)^2}{d + \alpha f + \alpha\Delta f}$$

$$= \frac{df + \alpha f^2 - f\varepsilon_1 + \alpha f\Delta f - \frac{1}{2}\varepsilon_1^2 + \frac{1}{2}\alpha(\Delta f)^2}{d + \alpha f}\left(1 - \frac{\alpha\Delta f}{d+\alpha f} + \frac{\alpha^2\Delta f^2}{(d+\alpha f)^2}\right)$$

with :

$$\frac{f}{d+\alpha f} = \frac{f\frac{\Delta t}{\Delta x}}{1 + \alpha f\frac{\Delta t}{\Delta x}} = X$$

hence up to second order, with : $\varepsilon^2 = \alpha\Delta f^2 + 2\alpha X\varepsilon_1\Delta f - \varepsilon_1^2$:

$$\overline{\overline{f}} = f - X\varepsilon_1 + \frac{1}{2}\frac{X}{f}(\alpha\Delta f^2 + 2\alpha X\varepsilon_1\Delta f - \varepsilon_1^2) = f - X\varepsilon_1 + \frac{1}{2}\frac{X}{f}\varepsilon^2$$

Similarly (4.51) gives :

$$\overline{\overline{\overline{f}}} = \frac{df + \frac{1}{2}\alpha f^2 + \frac{1}{2}f^2 + f\varepsilon_2 + \frac{1}{2}\varepsilon_2^2 - \frac{1}{2}(1-\alpha)(f - X\varepsilon_1 + \frac{1}{2}\frac{X}{f}\varepsilon^2)^2}{d + \alpha(f - X\varepsilon_1 + \frac{1}{2}\frac{X}{f}\varepsilon^2)}$$

$$= f + X\varepsilon_2 + X^2\varepsilon_1 + \alpha\frac{X^4}{f}\varepsilon_1^2 - \frac{1}{2}(1-\alpha)\frac{X^3}{f}\varepsilon_1^2 + \frac{1}{2}\frac{X}{f}\varepsilon_2^2 + \alpha\frac{X}{f}\varepsilon_1\varepsilon_2 - \frac{1}{2}\frac{X^2}{f}\varepsilon^2$$

For r = 2 :

$$\overline{f} = \overline{\overline{f}} + \overline{\overline{\overline{f}}} - f = f + \Delta f =$$

$$= f + X\varepsilon_2 + (X^2-X)\varepsilon_1 + \frac{1}{2}\alpha(X-X^2)\frac{\Delta f^2}{f} + \alpha X^2(1-X)\frac{\varepsilon_1\Delta f}{f}$$

$$+ (\alpha X^4 - \frac{1}{2}(1-\alpha)X^3 + \frac{1}{2}X^2 - \frac{1}{2}X)\frac{\varepsilon_1^2}{f} + \frac{1}{2}X\frac{\varepsilon_2^2}{f} + \alpha X^3\frac{\varepsilon_1\varepsilon_2}{f}$$

Up to first order this gives :

$$\Delta f = X\varepsilon_2 + (X^2-X)\varepsilon_1$$

This result is the same as that of the linear analysis.

Up to second order Δf is the solution of :

$$a\Delta f^2 + b\Delta f + c = 0 \tag{4.52}$$

with :

$$a = \frac{1}{2}\frac{\alpha}{f}(X-X^2)$$

$$b = \alpha X^2(1-X)\frac{\varepsilon_1}{f} - 1$$

$$c = (\alpha X^4 - \frac{1}{2}(1-\alpha)X^3 + \frac{1}{2}X^2 - \frac{1}{2}X)\frac{\varepsilon_1^2}{f} + \frac{1}{2}X\frac{\varepsilon_2^2}{f} + \alpha X^3\frac{\varepsilon_1\varepsilon_2}{f} + X\varepsilon_2 + (X^2-X)\varepsilon_1$$

The order of magnitude is : $a = O(1)$, $b = O(1)$, $c = O(\varepsilon)$.

The solutions of (4.52) are :

$$-\frac{b}{2a}\left(1 \pm \sqrt{1 - \frac{4ac}{b^2}}\right)$$

Up to second order these are :

$$(\Delta f)_1 = -\frac{b}{2a}\left(1 + 1 - \frac{1}{2}\frac{4ac}{b^2} - \frac{1}{8}\frac{16\ a^2c^2}{b^4}\right)$$

$$(\Delta f)_2 = -\frac{b}{2a}\left(1 - 1 + \frac{1}{2}\frac{4ac}{b^2} + \frac{1}{8}\frac{16\ a^2c^2}{b^4}\right) = -\frac{c}{b} - \frac{ac^2}{b^3}$$

$(\Delta f)_1$ is a false solution since it has $O(1)$.

$(\Delta f)_2$ has $O(\varepsilon)$.

Up to second order $(\Delta f)_2$ is :

$$\Delta f = X\varepsilon_2 + (X^2-X)\varepsilon_1 + \frac{\varepsilon_1^2}{f}(-\frac{1}{2}X + \frac{1}{2}X^2 - \frac{1}{2}X^3 + \frac{3}{2}\alpha X^4 + \frac{1}{2}\alpha X^5 - \frac{1}{2}\alpha X^6)$$

$$+ \frac{\varepsilon_2^2}{f}(\frac{1}{2}X + \frac{1}{2}\alpha X^3 - \frac{1}{2}\alpha X^4) + \frac{\varepsilon_1\varepsilon_2}{f}(\alpha X^3 + \alpha X^4 - \alpha X^5)$$

$$= X\varepsilon_2 + (X^2-X)\varepsilon_1 + \frac{\varepsilon_1^2}{f}V_1(X) + \frac{\varepsilon_2^2}{f}V_2(X) + \frac{\varepsilon_1\varepsilon_2}{f}V_3(X) \qquad (4.53)$$

By expanding :

$$\Delta f = f(x,t+\Delta t) - f(x,t) = \Delta t\frac{\partial f}{\partial t} + \frac{1}{2}\Delta t^2\frac{\partial^2 f}{\partial t^2}$$

$$\varepsilon_1 = f(x+\Delta x,t) - f(x,t) = \Delta x\frac{\partial f}{\partial x} + \frac{1}{2}\Delta x^2\frac{\partial^2 f}{\partial x^2}$$

$$\varepsilon_2 = f(x-\Delta x,t+\Delta t) - f(x,t) = \Delta t\frac{\partial f}{\partial t} - \Delta x\frac{\partial f}{\partial x} + \frac{1}{2}\Delta t^2\frac{\partial^2 f}{\partial t^2} + \frac{1}{2}\Delta x^2\frac{\partial^2 f}{\partial x^2} - \Delta t\Delta x\frac{\partial^2 f}{\partial t\partial x}$$

$$\varepsilon_1^2 = \Delta x^2(\frac{\partial f}{\partial x})^2$$

$$\varepsilon_2^2 = \Delta t^2(\frac{\partial f}{\partial t})^2 + \Delta x^2(\frac{\partial f}{\partial x})^2 - 2\Delta t\Delta x(\frac{\partial f}{\partial t})(\frac{\partial f}{\partial x})$$

$$\varepsilon_1\varepsilon_2 = \Delta t\Delta x(\frac{\partial f}{\partial t})(\frac{\partial f}{\partial x}) - \Delta x^2(\frac{\partial f}{\partial x})^2$$

(4.53) gives the equivalent equation :

$$(1-X)\Delta t \frac{\partial f}{\partial t} + \frac{1}{2}(1-X)\Delta t^2 \frac{\partial^2 f}{\partial t^2} = (X^2-2X)\Delta x \frac{\partial f}{\partial x} + \frac{1}{2} X^2 \Delta x^2 \frac{\partial^2 f}{\partial x^2}$$

$$- X \Delta t \Delta x \frac{\partial^2 f}{\partial t \partial x} + \left(V_1(X) + V_2(X) - V_3(X)\right)\frac{1}{f} \Delta x^2 \left(\frac{\partial f}{\partial x}\right)^2$$

$$+ V_2(X) \frac{1}{f} \Delta t^2 \left(\frac{\partial f}{\partial t}\right)^2 + \left(V_3(X) - 2V_2(X)\right)\frac{1}{f} \Delta t \Delta x \left(\frac{\partial f}{\partial t}\right)\left(\frac{\partial f}{\partial x}\right) \qquad (4.54)$$

For $1-X \neq O(\Delta x)$:

$$(1-X)\Delta t \frac{\partial f}{\partial t} = (X^2-2X)\Delta x \frac{\partial f}{\partial x} + \text{HOT}$$

$$(1-X)\Delta t \frac{\partial^2 t}{\partial t \partial x} = (X^2-2X)\Delta x \frac{\partial^2 f}{\partial x^2} + \text{HOT}$$

$$(1-X)\Delta t \frac{\partial^2 f}{\partial t^2} = (X^2-2X)\Delta x \frac{\partial^2 f}{\partial t \partial x} + \text{HOT} = \frac{(X^2-2X)^2 \Delta x^2}{(1-X)\Delta t} \frac{\partial^2 f}{\partial x^2} + \text{HOT}$$

(4.54) yields :

$$(1-X)^3 \Delta t \frac{\partial f}{\partial t} + (1-X)^2(2X-X^2)\Delta x \frac{\partial f}{\partial x} - \left[\frac{1}{2}(5X^2-5X^3+X^4) - \frac{\alpha}{2}(X^3-2X^4+2X^5-X^6)\right]\frac{1}{f} \Delta x^2 \left(\frac{\partial f}{\partial x}\right)^2$$

$$= \frac{1}{2} X^2 (1-X)(1+X-X^2)\Delta x^2 \frac{\partial^2 f}{\partial x^2} \qquad (4.55)$$

(4.55) is the same equivalent equation as in the linear case, except for the term in $(\partial f/\partial x)^2$.

The transfert velocity of (4.55) is :

$$V_T = \frac{2X-X^2}{1-X} \frac{\Delta x}{\Delta t} - \frac{\left[\frac{1}{2}(5X^2-5X^3+X^4) - \frac{\alpha}{2}(X^3-2X^4+2X^5-X^6)\right]}{(1-X)^3} \frac{\Delta x^2}{f\Delta t}\left(\frac{\partial f}{\partial x}\right) \qquad (4.56)$$

α can vary from $\simeq -1$ until $+1$. The coefficient of $\Delta f/\Delta x$ varies from :

$$-\frac{1}{2} \frac{5X^2-4X^3-X^4+2X^5-X^6}{(1-X)^3} \frac{\Delta x^2}{f\Delta t}$$

till :

$$-\frac{1}{2} \frac{5X^2-6X^3+3X^4-2X^5+X^6}{(1-X)^3} \frac{\Delta x^2}{f\Delta t}$$

The polynomials in the numerator are both positive for $X < 1$.

For small X, both coefficients are approximations to :

$$-\frac{5}{2} \frac{X^2}{f} \frac{\Delta x^2}{\Delta t} \simeq -\frac{5}{2} f \Delta t$$

the transfert velocity (4.56) has the form :

$$V_T = \frac{2X-X^2}{1-X} \frac{\Delta x}{\Delta t}\left(1 - f^* \frac{\partial f}{\partial x}\right) \qquad (4.57)$$

in which f^* is a positive function of f for $X < 1$.

For a compression wave :

$$\frac{\partial f}{\partial x} < 0$$

All transfert velocities are augmented by the non-linearity.
Large velocities augment more than smaller ones. Hence the forming of a compression shock is favoured.

For an expansion wave :

$$\frac{\partial f}{\partial x} > 0$$

All transfert velocities are decreased. Large velocities decrease more than smaller ones. The dispersion of an expansion wave is counteracted. The non linear behaviour of the scheme has a tendency for forming expansion shocks. This tendency in the form (4.57) can easily be counteracted by the introduction of an explicit artificial viscosity.
The implicit artificial viscosity in (4.55) is not always sufficient. It vanishes for instance in the vicinity of $X = 0$.

An example of an expansion wave with a very small velocity is the expansion wave that moves opposite to the flow in sonic conditions. For such cases an explicit artificial velocity is necessary.

In order not to destroy the accuracy it has to be of the form :

$$D \, \Delta x \, \frac{\partial^2 f}{\partial x^2} \qquad \text{with} \qquad D = 0(1)$$

5. APPLICATION TO ONE-DIMENSIONAL EULER EQUATIONS

5.1. Convergence rate.

The one-dimensional Euler equations are in quasi-linear form :

$$\frac{\partial \xi}{\partial t} + A_1 \frac{\partial \xi}{\partial x} = 0$$

The discretisation according to (4.39) (4.40) is :

<u>downwind step</u>

$$\xi(x,t+\widetilde{\Delta t}) - \xi(x,t) + A_1 \frac{\Delta t}{\Delta x}\Big(\xi(x+\Delta x,t) - (1+\alpha)\xi(x,t) + \alpha\xi(x,t+\widetilde{\Delta t})\Big) = 0$$

This yields :

$$\xi(x,t+\widetilde{\Delta t}) = \xi(x,t) - [X]\Big(\xi(x+\Delta x,t) - \xi(x,t)\Big) \qquad (5.1)$$

with

$$[X] = \Big([I] + \alpha[Z]\Big)^{-1}[Z] \qquad\qquad [Z] = A_1 \frac{\Delta t}{\Delta x}$$

<u>upwind step</u>

$$\xi(x,t+\widetilde{\widetilde{\Delta t}}) - \xi(x,t) + A_1 \frac{\Delta t}{\Delta x}\Big(\xi(x,t+\widetilde{\Delta t}) + \alpha\xi(x,t+\widetilde{\widetilde{\Delta t}})$$

$$- \alpha\xi(x,t) - \xi(x-\Delta x,t+\Delta t)\Big) = 0$$

this yields :

$$\xi(x,t+\tilde{\tilde{\Delta}}t) = \xi(x,t) - [X]\left(\xi(x,t+\tilde{\Delta}t) - \xi(x-\Delta x,t+\Delta t)\right) \qquad (5.2)$$

The equations (5.1) (5.2) are similar to the equations (4.41), when the scalar X is replaced by the matrix [X] . This means that all conditions that were imposed on Z, have to be imposed on the eigenvalues of [Z] .

These are :

$$(u+c)\,\frac{\Delta t}{\Delta x}\,, \quad u\,\frac{\Delta t}{\Delta x}\,, \quad (u-c)\,\frac{\Delta t}{\Delta x}$$

The transfert corresponding to u+c, the largest eigenvalue of A_1 can be critically relaxed. Perturbations transferred along this component are completely damped. The transfert velocity corresponding to u-c is maximised. There is no control on the transfert along u and thus on the convective movement. In the acoustic movement, waves travelling to the right are completely damped, waves travelling to the left are almost undamped but accelerated. The attainment of the steady state is thus not determined by the internal damping but by the expulsion of perturbations along u and u-c. The complete damping along u+c is however absolutely necessary to avoid reflection along u-c at the outflow boundary. Since there is only higher order coupling for small perturbations between the acoustic movement and the convective movement, there is asymptotically no interaction between u and u-c. For sufficiently long time a perturbation along u and u-c is expulsed in a number of iterations which is proportional to the number of elements in the flow field.

The asymptotic rate of convergence is thus $O(\Delta x)$ in the sense that increasing the number of elements in the flow field, only lineary increases the number of iterations necessary to reach a certain level of convergence when starting from the same initial state.

5.2. Discretisation for the one-dimensional equations.

Downwind step

Mass :

$$\frac{\Delta x}{\Delta t}[\rho S(x,t+\tilde{\Delta}t) - \rho S(x,t)] + [\rho u S(x+\Delta x,t) + \alpha\rho u S(x,t+\tilde{\Delta}t)$$
$$- (1+\alpha)\rho u S(x,t)] = 0$$

Momentum :

$$\frac{\Delta x}{\Delta t}[\rho u S(x,t+\tilde{\Delta}t) - \rho u S(x,t)] + [(\rho u^2+p)S(x+\Delta x,t) +$$
$$+ \alpha(\rho u^2+p)S(x,t+\tilde{\Delta}t) - (1+\alpha)(\rho u^2+p)S(x,t)]$$
$$= [S(x+\Delta x) - S(x)]\,\frac{1}{2}\,[(1+\alpha)p(x,t) - \alpha p(x,t+\tilde{\Delta}t) + p(x+\Delta x,t)]$$

Energy :

$$\frac{\Delta x}{\Delta t}[\rho E S(x,t+\tilde{\Delta}t) - \rho E S(x,t)] + [\rho u H S(x+\Delta x,t) + \alpha\rho u H S(x,t+\tilde{\Delta}t)$$

$$- (1+\alpha)\rho uHS(x,t)] = 0$$

with : $\rho E = \frac{p}{\gamma-1} + \frac{\rho u^2}{2} \qquad \rho H = \frac{\gamma p}{\gamma-1} + \frac{\rho u^2}{2}$

linearisation yields :

$$\rho S(x,t+\widetilde{\Delta t}) = \rho S(x,t+\overline{\Delta t}) + S(x)\Delta\rho$$

$$\rho uS(x,t+\widetilde{\Delta t}) + \rho uS(x,t+\overline{\Delta t}) + S(x)[u(x,t+\overline{\Delta t})\Delta\rho + \rho(x,t+\overline{\Delta t})\Delta u]$$

$$(\rho u^2+p)S(x,t+\widetilde{\Delta t}) = (\rho u^2+p)S(x,t+\overline{\Delta t}) + S(x)\ [u^2(x,t+\overline{\Delta t})\Delta\rho + 2\rho u(x,t+\overline{\Delta t})\Delta u + \Delta p]$$

$$\rho ES(x,t+\widetilde{\Delta t}) = \rho ES(x,t+\overline{\Delta t}) + S(x)[\frac{1}{2}\, u^2(x,t+\overline{\Delta t})\Delta\rho + \rho u(x,t+\overline{\Delta t})\Delta u + \frac{1}{\gamma-1}\,\Delta p]$$

$$\rho uHS(x,t+\widetilde{\Delta t}) = \rho uHS(x,t+\overline{\Delta t}) + S(x)[\frac{1}{2}\, u^3(x,t+\overline{\Delta t})\Delta\rho + \frac{3}{2}\,\rho u^2(x,t+\overline{\Delta t})\Delta u + \frac{\gamma}{\gamma-1}\, p(x,t+\overline{\Delta t})\Delta u + \frac{\gamma}{\gamma-1}\, u(x,t+\overline{\Delta t})\Delta p]$$

Thus :

$$\begin{pmatrix} A_1 & A_2 & 0 \\ B_1 & B_2 & B_3 \\ C_1 & C_2 & C_3 \end{pmatrix} \begin{pmatrix} \Delta\rho \\ \Delta u \\ \Delta p \end{pmatrix} = \begin{pmatrix} A \\ B \\ C \end{pmatrix}$$

with

$$A_1 = S(x)\frac{\Delta x}{\Delta t} + \alpha S(x)u(x,t+\overline{\Delta t})$$

$$A_2 = \alpha S(x)\rho(x,t+\overline{\Delta t})$$

$$A = S(x)[\rho(x,t) - \rho(x,t+\overline{\Delta t})]\frac{\Delta x}{\Delta t} + (1+\alpha)S(x)\rho u(x,t) - \alpha S(x)\rho u(x,t+\overline{\Delta t}) - S(x+\Delta x)\rho u(x+\Delta x,t)$$

$$B_1 = S(x)\frac{\Delta x}{\Delta t}\, u(x,t+\overline{\Delta t}) + \alpha S(x)u^2(x,t+\overline{\Delta t}) = u(x,t+\overline{\Delta t})A_1$$

$$B_2 = S(x)\frac{\Delta x}{\Delta t}\, \rho(x,t+\overline{\Delta t}) + \alpha S(x)2\rho u(x,t+\overline{\Delta t})$$

$$B_3 = \frac{1}{2}\,\alpha[S(x) + S(x+\Delta x)]$$

$$B = S(x)[\rho u(x,t) - \rho u(x,t+\overline{\Delta t})]\frac{\Delta x}{\Delta t} + (1+\alpha)S(x)(\rho u^2+p)(x,t)$$

$$- \alpha S(x)(\rho u^2+p)(x,t+\overline{\Delta t}) - S(x+\Delta x)(\rho u^2+p)(x+\Delta x,t)$$

$$+ \frac{S(x+\Delta x)-S(x)}{2}[(1+\alpha)p(x,t) - \alpha p(x,t+\overline{\Delta t}) + p(x+\Delta x,t)]$$

$$C_1 = S(x)\frac{\Delta x}{\Delta t}\frac{1}{2} u^2(x,t+\overline{\Delta t}) + \alpha S(x)\frac{1}{2} u^2(x,t+\overline{\Delta t}) = \frac{1}{2} u^2(x,t+\overline{\Delta t})A_1$$

$$C_2 = S(x)\frac{\Delta x}{\Delta t} \rho u(x,t+\overline{\Delta t}) + \alpha S(x)[\frac{3}{2} \rho u^2(x,t+\overline{\Delta t}) + \frac{\gamma}{\gamma-1} p(x,t+\overline{\Delta t})]$$

$$C_3 = S(x)\frac{\Delta x}{\Delta t}\frac{1}{\gamma-1} + \alpha S(x)\frac{\gamma}{\gamma-1} u(x,t+\overline{\Delta t})$$

$$C = S(x)[\rho ES(x,t) - \rho ES(x,t+\overline{\Delta t})]\frac{\Delta x}{\Delta t}$$

$$+ (1+\alpha)S(x)\rho uH(x,t) - \alpha S(x)\rho uH(x,t+\overline{\Delta t}) - \rho uH(x+\Delta x,t)$$

Artificial viscosity is added to the momentum equation. In this downwind step, the viscosity term is discretised upwind in order to avoid a term in $x + 2\Delta x$.

The term $\frac{\partial}{\partial x}(\rho u^2 + p)S$ is replaced by :

$$\frac{\partial}{\partial x}\left((\rho u^2+p)S\right) - \frac{\partial}{\partial x}(D_x \Delta x\ S\ \frac{\partial u}{\partial x})$$

p is thus replaced by :

$$p - D_x \Delta x \frac{\partial u}{\partial x}$$

and after discretisation by :

$$p(x) - D_x u(x) + D_x u(x-\Delta x)$$

B is replaced by :

$$B - D_x S(x+\Delta x)u(x) + D_x S(x+\Delta x)u(x+\Delta x) + D_x S(x)u(x-\Delta x) - D_x S(x)u(x)$$

Upwind step

Mass :

$$\frac{\Delta x}{\Delta t}[\rho S(x,t+\widetilde{\widetilde{\Delta t}}) - \rho S(x,t)] + [\rho uS(x,t+\widetilde{\Delta t}) + \alpha\rho uS(x,t+\widetilde{\widetilde{\Delta t}})$$

$$- \alpha\rho uS(x,t) - \rho uS(x-\Delta x,t+\Delta t)] = 0$$

Momentum :

$$\frac{\Delta x}{\Delta t}[\rho uS(x,t+\widetilde{\widetilde{\Delta t}}) - \rho uS(x,t)] + [(\rho u^2+p)S(x,t+\widetilde{\Delta t})$$

$$+ \alpha(\rho u^2+p)S(x,t+\widetilde{\widetilde{\Delta t}}) - \alpha(\rho u^2+p)S(x,t) - (\rho u^2+p)S(x-\Delta x,t+\Delta t)]$$

$$= [S(x) - S(x-\Delta x)]\frac{1}{2}[p(x,t+\tilde{\Delta}t) + \alpha p(x,t+\tilde{\tilde{\Delta}}t) - \alpha p(x,t) + p(x-\Delta x,t+\Delta t)]$$

Energy :

$$\frac{\Delta x}{\Delta t}[\rho ES(x,t+\tilde{\tilde{\Delta}}t) - \rho ES(x,t)] + [\rho uHS(x,t+\tilde{\Delta}t) + \alpha\rho uHS(x,t+\tilde{\tilde{\Delta}}t) - \alpha\rho uHS(x,t) - \rho uHS(x-\Delta x,t+\Delta t)] = 0$$

The linearisation in the upwind step is similar to the linearisation in the downwind step. The coefficients A_1, A_2, B_1, B_2, C_1, C_2 and C_3 have the same expressions when the superscript $-$ is replaced by $=$.

$$B_3 = \frac{1}{2}\,\alpha[S(x) + S(x-\Delta x)]$$

$$A = S(x)[\rho(x,t) - \rho(x,t+\overline{\overline{\Delta t}})]\frac{\Delta x}{\Delta t} + S(x-\Delta x)\rho u(x-\Delta x,t+\Delta t) + \alpha S(x)\rho u(x,t) - (1+\alpha)S(x)\;u(x,t+\overline{\overline{\Delta t}})$$

$$B = S(x)[\rho u(x,t) - \rho u(x,t+\overline{\overline{\Delta t}})]\frac{\Delta x}{\Delta t} + S(x-\Delta x)(\rho u^2+p)(x-\Delta x,t+\Delta t) + \alpha S(x)(\rho u^2+p)(x,t) - (1+\alpha)S(x)(\rho u^2+p)(x,t+\overline{\overline{\Delta t}}) + \frac{S(x)-S(x-\;x)}{2}[p(x-\Delta x,t+\Delta t) - \alpha p(x,t) + (1+\alpha)p(x,t+\overline{\overline{\Delta t}})]$$

$$C = S(x)[\rho E(x,t) - \rho E(x,t+\overline{\overline{\Delta t}})]\frac{\Delta x}{\Delta t} + S(x-\Delta x)\rho uH(x-\Delta x,t+\Delta t) + \alpha S(x)\rho uH(x,t) - (1+\alpha)S(x)\rho uH(x,t+\overline{\overline{\Delta t}})$$

In order to introduce artificial viscosity $p(x)$ is replaced by :

$$p(x) - D_x u(x+\Delta x) + D_x u(x)$$

and B by :

$$B - D_x S(x)u(x) + D_x S(x)u(x+\Delta x) + D_x S(x-\Delta x)u(x-\Delta x) - D_x S(x-\Delta x)u(x)$$

5.3. Example.

The flow is calculated for a nozzle divided in 28 constant Δx segments. The section is S_o between nodes 1 and 3 and between nodes 21 and 29. Between nodes 3 and 21 the section is :

$$S(i) = S_o\left\{ .9 + \;.1 \times \left[2\left(\frac{i-12}{9}\right)^2 - \left(\frac{i-12}{9}\right)^4\right]\right\}$$

The outlet pressure is fixed on .718025 x inlet total pressure.
The exact solution has then a shock on node 16.

The calculation is done with a relaxation factor r = 2, a damping term in the artificial viscosity D_x = .01 and a transition machnumber M_o = .95.

The value of X^- is : $X^- = - Q = - .615$.

Since the formulae (4.45) are very sensitive to machnumber, M has to be multiplicated by a safety factor S = .995 before it is used in (4.45). Overestimation of M causes X^- to exceed the stability limit $X^- = - .618$.
For each nodal calculation the iterations are stopped when the increments per iteration fulfil :

$$\Delta\rho^2 + \Delta p^2 + \Delta u^2 < C = .0001$$

A convergence norm is calculated as the mean absolute deviation in the nodal points between the calculated machnumber and the exact machnumber. The initial state is a uniform flow with M = .6.

The convergence history for the norm is depicted in figure 12.

The steady state is reached very abruptly. This is due to the expulsion mechanism on which the convergence is based.

The calculations are repeated for the same geometry but subdivided in 56,84 and 112 elements. The norm is still calculated in the nodal points of the first case. The accuracy is defined as the value of the norm.

D_x = .01 S = .995 Q = .615 C = .0001 M_o = .95

number of elements	28	56	84	112
optimal relaxationfactor	2.000	2.000	1.996	1.992
fielditerations for an accuracy .01	55	104	153	201
idem for .001	-	119	168	218
nodal iterations for an accuracy .01	1828	6428	13382	23173
idem for .001	-	7262	14637	25073
final accuracy	.00201	.00053	.00044	.00027

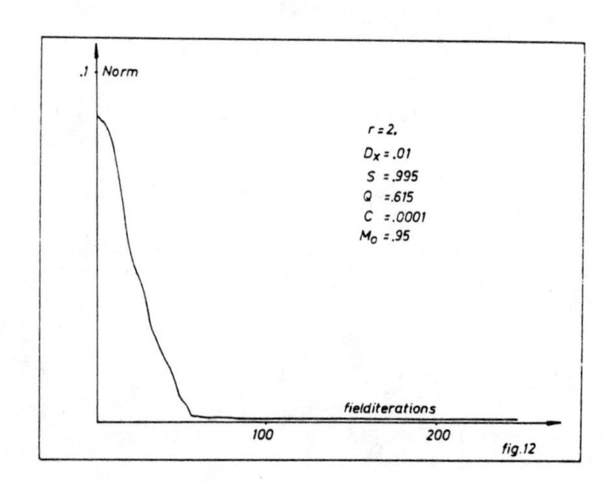

fig.12

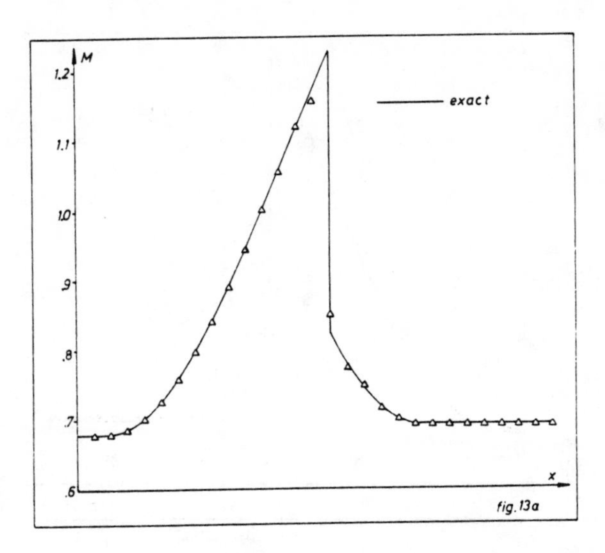

fig.13a

These results show clearly the linear convergence rate. The global rate is linear although there is only proved by theory that the asymptotic rate is linear.

This illustrates, as is well known, that rough perturbations disappear almost immediatly from the flow field, so that the global convergence is determined by the asymptotic convergence.

Figure 13 shows the machnumber for the different calculations. The gain in accuracy with the increase in number of elements is clear.

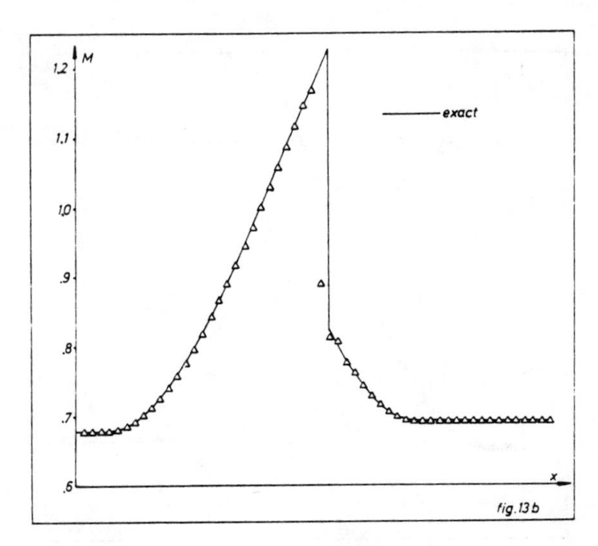

fig.13b

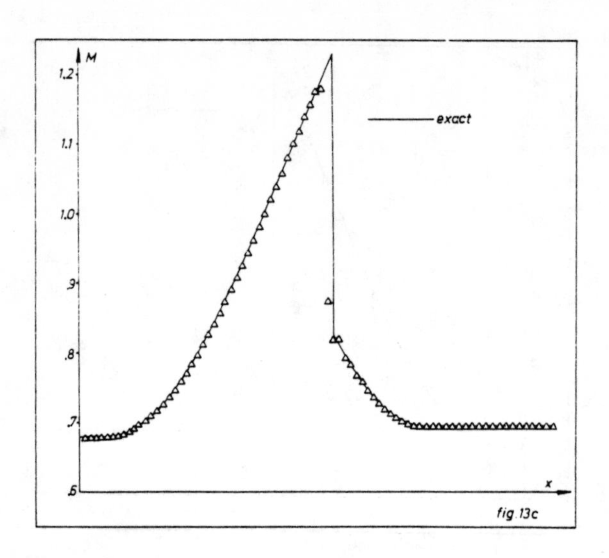

fig. 13c

5.4. Choice of the parameters.

The algorithm has 6 parameters. Four of them : relaxation factor r, safety factor S, largest negative X and transition machnumber M_o have a universal value.

The convergence rate increases when these parameters approach their theoretical values : $r = 2$, $S = 1$, $X^- = -Q = -.61803$, $M_o = 1$.

These values cannot be reached for stability reasons. Practical values are : $r = 1.99$, $S = .995$, $Q = .615$, $M_o = .95$.

The convergence factor C and the damping factor D_x have a problem dependent value.

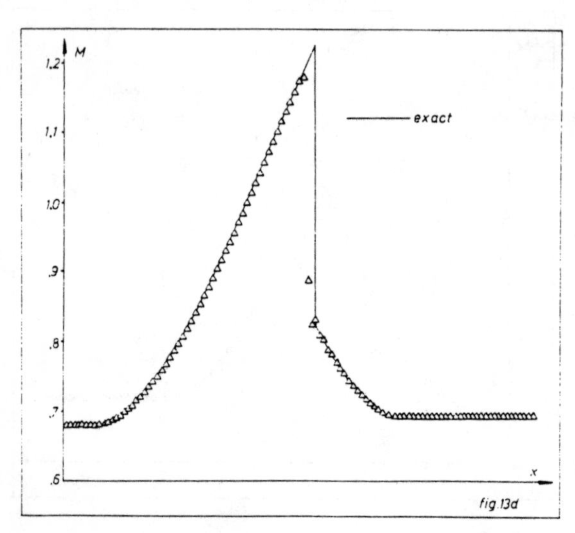

fig. 13d

The convergence factor should be low enough to insure sufficient convergence in a nodal calculation. Insufficient convergence in a nodal calculation causes instability. However a convergence factor that is too safe augments the computational effort without influencing the convergence rate. A rule of thumb is that after a number of fielditerations equal to the number of elements in the field, only one iteration is done per nodal calculation.

The dampingfactor D_x has no influence on the convergence rate, but has a detremental effect on the accuracy. It should be kept as low as possible but high enough to avoid expansion shocks. The necessary value is strongly dependent on the shock strenght.

6. TWO-DIMENSIONAL APPLICATIONS

6.1. Discretisation.

The one-dimensional scheme can immediatly be extended to two-dimensional applications for line relaxation on a mesh in which the transversal lines are straight and parallel as depicted in figure 14.

In the formulae (4.45) the machnumber is to be replaced by the component of the machnumber in the direction perpendicular to the transversal lines.

The discretisation can easily be done by the finite volume technique. The upwind and the downwind volume corresponding to the mesh point (i,j) are depicted in figure 15.
In order to have formal similarity with the one-dimensional algorithm, the time derivative term has to be distributed on the points (i+1,j) (i,j) (i-1,j) proportional to the coefficients of these points for the flux trough AB.

The stability analysis on this two-dimensional dicretisation reveals that there is a slight instability in the transversal direction. This instability can easily be compensated by a small artificial viscosity in this direction.

It can for instance be done by introducing in the momentum equations a corrected viscosity of the form : for the X-momentum :

$$D_{y1}\,\Delta y\,\frac{\partial^2 u}{\partial y^2}\,(t^*) - D_{y2}\,\Delta y\,\frac{\partial^2 u}{\partial y^2}\,(t_o)$$

t^* is taken to be $t+\widetilde{\Delta t}$ in the downwind step and $t+\widetilde{\widetilde{\Delta t}}$ in the upwind step.

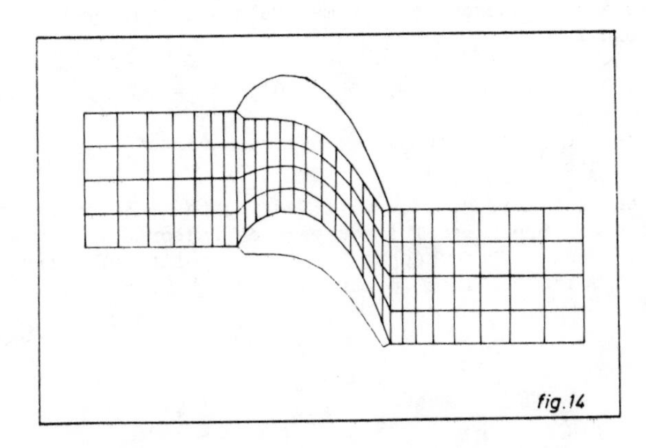

fig.14

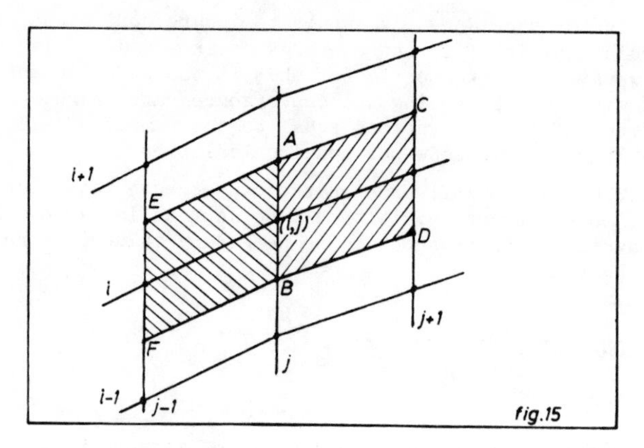

fig.15

t_o is a time level that is fixed for a cycle of N_v iterations and renewed after that cycle.
$D_{y1} - D_{y2}$ is sufficiently small.

Typical values are :

$$D_{y1} = 1 \qquad D_{y2} = .99$$

With these values the cycle should have approximatly as many iterations as there are nodes in the transversal direction of the flow field, in order to insure stability.
This viscosity mechanism has no influence at all on the convergence rate since the convergence is driven by the expulsion mechanism in the longitudinal direction and not by the damping mechanism in the transversal direction.

6.2. Transonic flow in a choked nozzle. Verification of the convergence rate.

The same nozzle as was used in the one-dimensional example is subdivided into 4x28 and 8x56 elements with constant hight on a transversal line. The calculation is also done for a discritisation with 8x28 and 4x56 elements. In the first case the width of the nozzle is doubled, in the second case the lenght is doubled. This guarantees that the form of the elements is similar in all calculations. As a consequence the optimal values of the problem dependent parameters are equal.
These parameters are :
$r = 2$, $S = .995$, $Q = .615$, $M_o = .95$, $D_x = .01$, $D_{y1} = 1$, $D_{y2} = .99$, $C = .013$.

$N_v = 5$ when there are 5 nodes in the transversal direction.
$N_v = 9$ when there are 9 nodes.

The initial state is a uniform flow with $M = .6$, the outflow pressure is .718025 x inlet total pressure. The convergence factor C is now defined as the mean incremental change of the machnumber in the points of a transversal line. Figure 16 shows the isomachlines for the 4x28 and 8x56 cases.

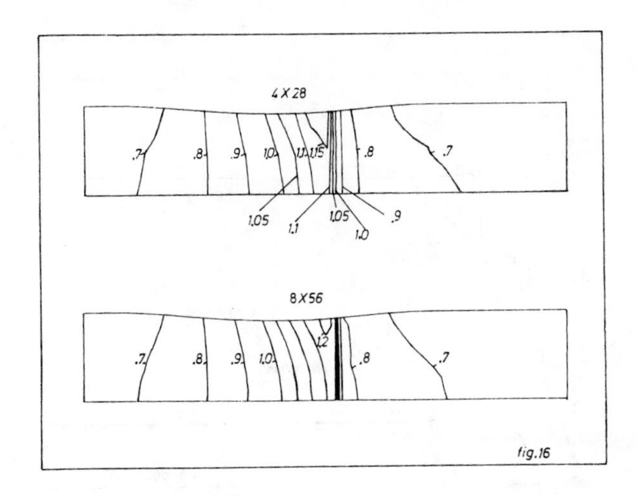

fig.16

The results of these calculations are :

number of elements	4x28	8x28	4x56	8x56
final value of the mean machnumber	.78444	.78375	.78860	.78868
number of fielditerations for a mean machnumber = .999 x final value	57	58	132	135
corresponding line iterations	1742	1764	7483	7207
corresponding equivalent fielditerations	64.5	65.3	136.0	131.0
final value T_{o2}/T_{o1}	.99938	.99987	.99954	.99958
final value $\dot{m}_2/\dot{m}_1$	.99950	.99945	.99953	.99993

An equivalent fielditeration is defined as 27 or 55 line iterations.
These results show clearly the full conservativity of the used finite volume method. The results reveal clearly the linear convergende rate in Δx and proove that the number of nodes in the transversal direction has no influence at all on the convergence rate.

6.3. Gain in efficiency in comparison to a method with convergence rate $O(\Delta x^2)$.

The nozzle with 4x28 elements is calculated with the corrected viscosity time marching method. The optimal parameters for this case are determined experimentally. These are : viscosity correction coefficient β = .96 and number of iterations in one cycle N_v = 12.

The convergence history is shown in figure 17 for the relaxation method and for the time marching method for three typical points.

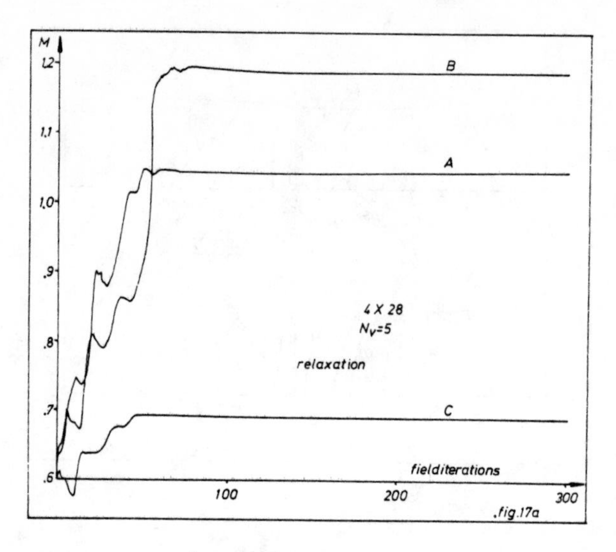

A = sonic area

B = before the shock

C = behind the shock

A comparable degree of convergence is reached in 80 iterations for the relaxation method and 1200 time steps for the time marching method. The relaxation program needs herefore about 265 s on Siemens 7755 (540 KOPS) the time marching program 1760 s. The utilised time marching program is however not optimised. Therefore this time consumption has to be scaled taking into account cited time consumptions in the literature. The program used needs .0101 Cpu-s per node and per iteration for a 4 equation system with elements with 8 surrounding nodes. In [10] Van Hove needs .00061 Cpu-s/n.i on CDC 6500 for a 3 equation system (total enthalpy is constant) with elements with 4 surrounding

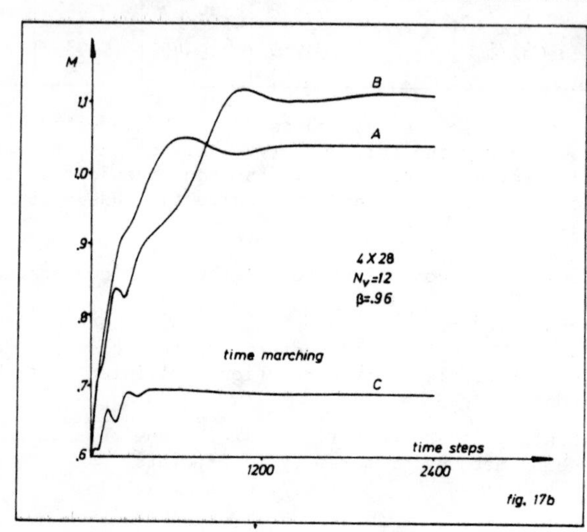

nodes. Couston needs .00140 Cpu-s/n.i for a 4 equation system and elements with 6 surrounding nodes.

Converting these to 4 equations and 8 surrounding nodes on Siemens 7755 which is about two times slower than CDC 6500 this gives :

$$\text{for Van Hove : } .00122 \times \frac{4\times8}{3\times4} = .0033 \text{ Cpu-s/n.i}$$

$$\text{for Couston : } .00280 \times \frac{4\times8}{4\times6} = .0037 \text{ Cpu-s/n.i}$$

The program of Couston has a slightly greater consumption since it takes into account the height variation of the cascade.
This means that when an optimised time marching program would be used for the same equations and the same elements as is used for the relaxation method, 1200 time steps could be reached after about 575 s on Siemens 7755.

This means that the relaxation method is about two times faster on a 4x28 grid. On a 12x84 grid this is a factor 6.

6.4. Verification of the accuracy.

Figure 18 shows the grid and the isomachlines for the well known Gamm-test case [11]. The outlet pressure is constant in the whole outlet section and corresponds to an isentropic machnumber .85.

The isomachlines coincide completely with the results of Lerat & Sides and Veulliot & Viviand obtained by time marching methods [11].

6.5. Cascade calculations.

Figure 19 shows the grid for a gasturbine cascade and the resulting isomachlines.

The inlet angle $\beta_1 = 30°$, the outlet pressure corresponds to an isentropic machnumber $M_{s2} = 1.21$.
Figure 20 shows the grid for a steamturbine cascade and the resulting isomachlines.
The inlet angle $\beta_1 = 66°$, the outlet pressure corresponds to $M_{s2} = 1.40$.

Figure 21 shows the isentropic machnumbers corresponding to the surface pressure in comparison with measurements of Sieverding [12]. There is some difference between the calculated and measured pressure due to viscous effect.
The shock structure in the turbine calculations is not physically realistic due to the absence in the inviscid calculation of trailing edge separation.

7. CONCLUSIONS

By the preceding theory and computational examples it is proved that the relaxation method can be extended to Euler equations.
This gives a method with convergence rate $O(\Delta x)$. The gain in computational efficiency in comparison with time marching techniques is clear. The relaxation idea thus gives the possibility to construct fast algorithms for inviscid transonic rotational flows.

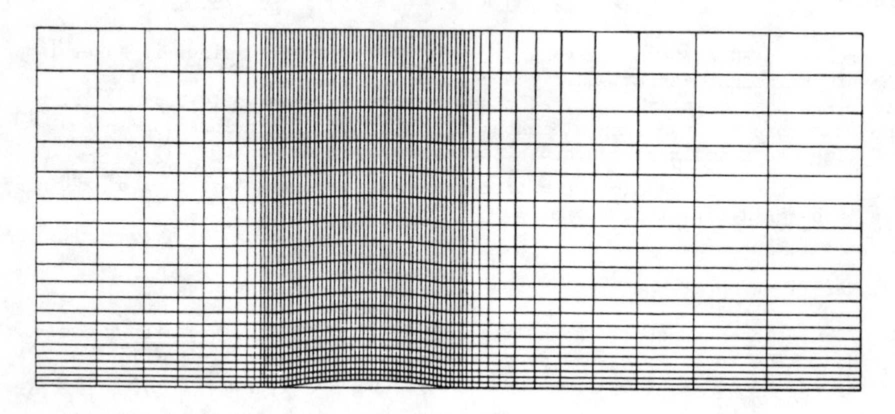

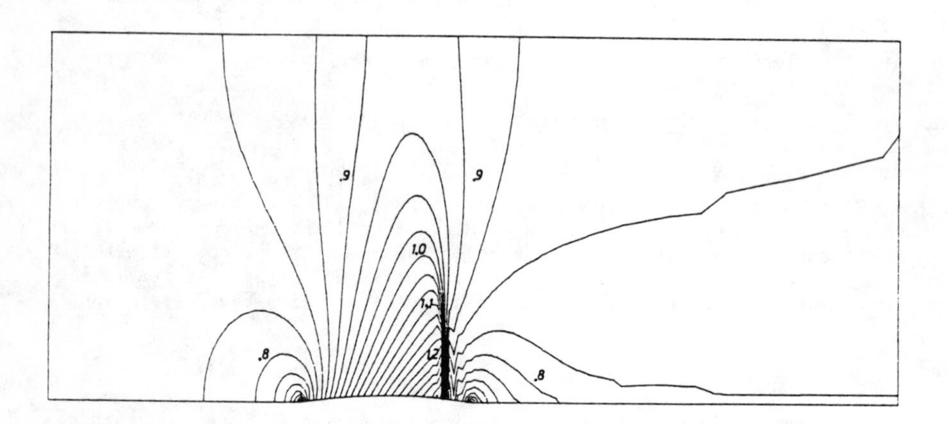

fig.18

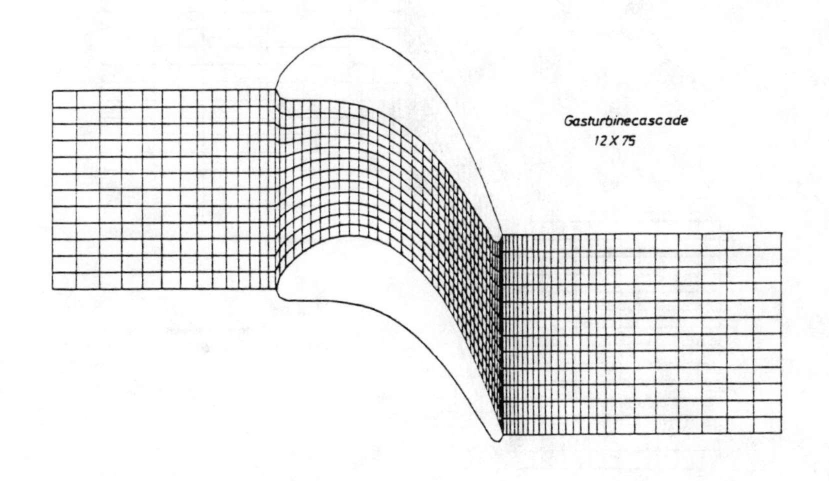
Gasturbinecascade
12 X 75

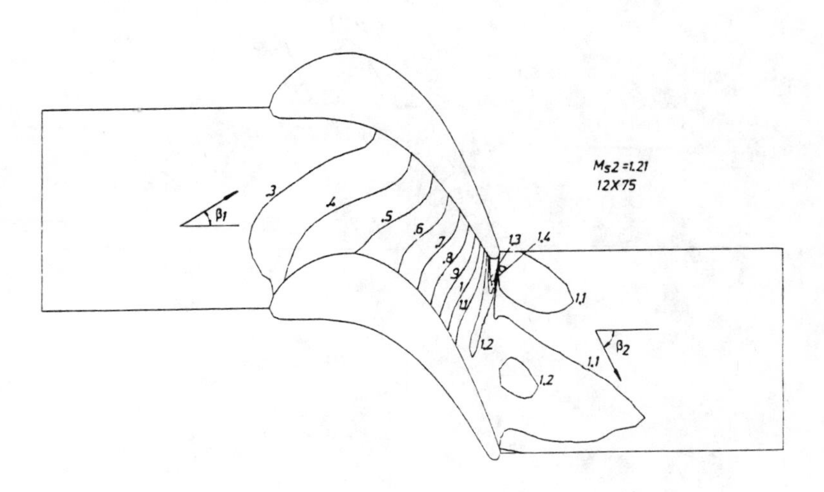
M_{S2} =1.21
12X75
.3
.4
.5
.6
.7
.8
.9
1
1.1
1.2
1.3
1.4
1.1
1.1
1.2
β_1
β_2

fig.19

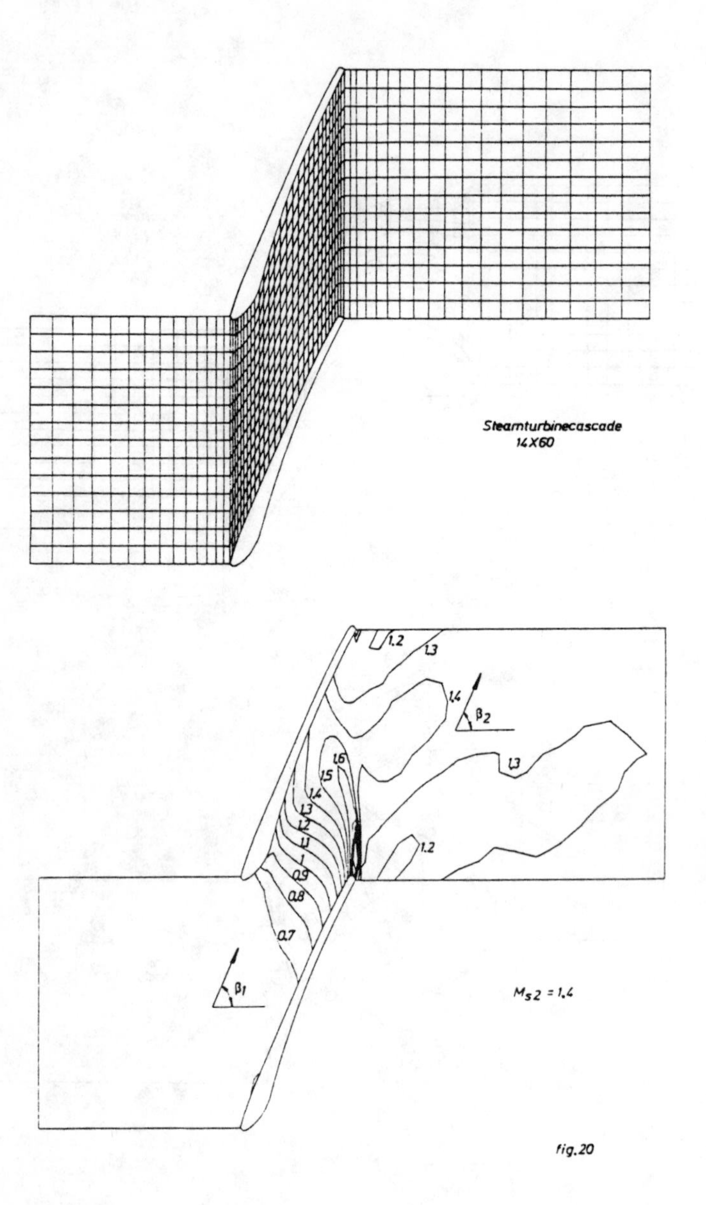
Steamturbinecascade
14X60
1.2
1.3
1.4
β_2
1.3
1.6
1.5
1.4
1.3
1.2
1.1
1
0.9
0.8
0.7
1.2
β_1
$M_{s2} = 1.4$

fig.20

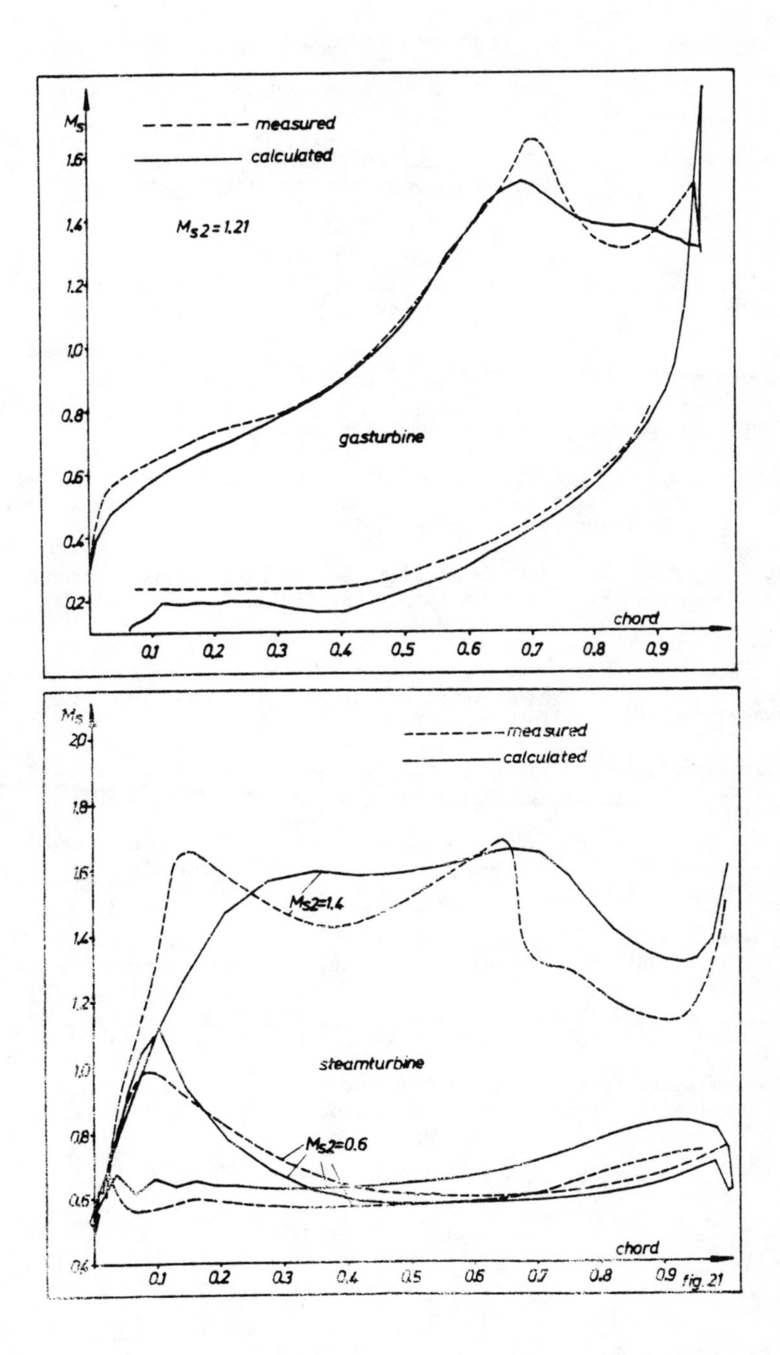
M_s
measured
calculated
$M_{s2} = 1.21$
gasturbine
chord
1.6
1.4
1.2
1.0
0.8
0.6
0.4
0.2
0.1 0.2 0.3 0.4 0.5 0.6 0.7 0.8 0.9
M_s
measured
calculated
M_{s2}=1.4
steamturbine
M_{s2}=0.6
chord
2.0
1.8
1.6
1.4
1.2
1.0
0.8
0.6
0.4
0.1 0.2 0.3 0.4 0.5 0.6 0.7 0.8 0.9
fig. 21

REFERENCES

1. A. Jameson, Numerical Computation of Transonic Flows with Shock Waves, Proc. Symp. Transsonicum II, Göttingen, pp. 385-414, 1975.

2. C.W. Hirt, Heuristic Stability Theory for Finite-Difference Equations, J. Comp. Phys., vol. 2, pp. 339-355, 1968.

3. M. Hafez, E.M. Murman, J.C. South, Artificial Compressibility Methods for Numerical Solution of Transonic Full Potential Equation, AIAA paper 78-1148, 1978.

4. J.J. Chattot, C. Coulombeix, Relaxation Method for the Full Potential Equation, in A Rizzi, H. Viviand (eds.), Numerical Methods for the Computation of Inviscid Transonic Flow with Shock Waves, pp. 37-44, Vieweg, Braunschweig, 1981.

5. H. Deconinck, Ch. Hirsch, Finite Element Methods for Transonic Flow Calculations, in E.H. Hirschel (ed.), Proc. Third GAMM Conf. Numerical Methods in Fluid Mechanics, pp. 66-77, Vieweg, Braunschweig, 1980.

6. M. Couston, P.W. Mc Donald, J.J. Smolderen, The Damping Surface Technique for Time-Dependent Solutions to Fluid Dynamic Problems, VKI-TN-109, Von Karman Institute, Sint Genesius Rode, Belgium, 1975.

7. H.J. Wirz, Relaxation Methods for Time Dependent Conservation Equations in Fluid Mechanics, VKI-LS-97, Von Karman Institute, Sint Genesius Rode, Belgium, 1977.

8. J.A. Essers, New Fast Super-Dashpot Time-Dependent Techniques for the Numerical Simulation of Steady Flows, Computers and Fluids, vol. 8, pp. 351-368, 1980.

9. J.A. Desidéri, J.C. Tannehill, Over-relaxation Applied to the Mac-Cormack Finite-Difference Scheme, J. Comp. Phys. vol. 23, pp. 313-326, 1977.

10. W. Van Hove, Time Marching Methods for Turbomachinery Flow Calculations. Methods of Improving Convergence, VKI-LS-1979-7, Von Karman Institute, Sint Genesius Rode, Belgium, 1979.

11. A. Lerat, J. Sides, Finite Volume Methods for the Solution of Euler Equations, J.P. Veuillot, H. Viviand, Computation of Steady Inviscid Transonic Flows using Pseudo-Unsteady Methods, in A. Rizzi, H. Viviand (eds.), Numerical Methods for the Computation of Inviscid Transonic Flow with Shock Waves, pp. 45-57, pp. 142-152, Vieweg, Braunschweig, 1981.

12. C. Sieverding, Experimental Data on two Transonic Turbine Blade Sections and Comparison with Various Theoretical Methods, VKI-LS-59, Von Karman Institute, Sint Genesius Rode, Belgium, 1973.

Index